Work related musculoskeletal disorders (WMSDs): a reference book for prevention

D0862905

Work related musculoskeletal disorders (WMSDs): a reference book for prevention

SCIENTIFIC EDITORS

Ilkka Kuorinka
Lina Forcier

AUTHORS

Mats Hagberg
Barbara Silverstein
Richard Wells
Michael J. Smith
Hal W. Hendrick
Pascale Carayon
Michel Pérusse

Taylor & Francis
Publishers since 1798

UK Taylor & Francis Ltd, 4 John St, London WC1N 2ET
USA Taylor & Francis Inc., 1900 Frost Road, Suite 101, Bristol, PA 19007

British Library Cataloguing in Publication Data

A catalogue record for this book is available from the British Library

ISBN 0 7484 0131 8 (Hard)
   0 7484 0132 6 (Soft)

Library of Congress Cataloging in Publication Data are available

Cover design by Amanda Barragry
Illustrations in chapter 3 by Chantal Lichaa

Typeset by EXCEPT*detail* Ltd, Southport

Printed in Great Britain by Burgess Science Press, Basingstoke, on paper which has a specified pH value on final paper manufacture of not less than 7.5 and is therefore '*acid free*'.

Contents

1

Introduction

Work related musculoskeletal disorders, or WMSDs, have become a major problem in many industrialized countries. In 1981, at the time of the first conference on machine pacing and occupational stress (Salvendy and Smith, 1981), it was thought that the number of repetitive jobs would decline in future; this should also have led to diminishing negative consequences, including WMSDs. This has not been the case, and some government agencies expect WMSDs to be one of the major work related problems in the future. The available data show that the incidence of recorded WMSDs is increasing in most of the industrialized world, although some countries seem to have been more effective in controlling them.

WMSDs have the face of Janus. On one hand they are diseases like any other, having their natural history, diagnostic criteria and therapy. On the other hand, having been designated 'work related', they have become the subject of special legislation and compensation. This has an influence on how they are understood and how, for example, companies and labour react to them.

The extent of the problem is not precisely known. This is partly because in various countries the criteria for compensability of WMSDs, and therefore the definition of a 'WMSD', are different. Also, the situation in the workplace may differ from the official picture: the number of musculoskeletal symptoms and disorders that individuals consider to be significantly influencing their work and private lives is usually higher than the figures in compensation statistics. The third reason is that the definitions and diagnostic criteria are not uniform. The same disorder may in one context be recorded as work related, while in another it is not.

The high cost of WMSDs has alarmed government agencies and companies. What is not generally recognized, however, is that the direct costs of compensated cases are only a fraction of the total costs, which also include replacement and retraining of personnel, disruptions in production, etc. Although few data are available, it has been assumed that total costs may be two to three times direct compensation costs. There are hidden costs even when cases do not meet compensation criteria, for workers still experience significant musculoskeletal symptoms and discomfort. Poor quality, lower work performance and decreased motivation are hypothetically part of these hidden costs.

The nature of WMSDs and their work relatedness has initiated a sometimes heated discussion among health and safety professionals and other interested parties. Compensation costs have largely dominated the debate, to the detriment of the issue of preventing WMSDs.

This book contains scientific information that will help prevent WMSDs. It was produced at the request of the Institut de recherche en santé et en sécurité du travail du Québec (Québec Research Institute on Occupational Health and Safety), which decided in 1991 to launch various activities to prevent WMSDs in the province of Québec. The project, of which this book is the result, contained among other parts, the creation of an important database on WMSD research, and was financed and supported by the IRSST. Part of the strategy was to invite an international expert group to prepare a scientific evaluation of WMSDs to back up other activities (including the possibility of preparing guidelines for industry). This initiative, originated by the ergonomics group of the National Research Council of Canada gained support from other parts of Canada*.

The authors want to express special thanks to the management of the IRSST, especially to Jean Yves Savoie, Chief Executive Officer and F.-Pierre Dussault, Director, External Research, for their support and trust throughout the 3-year writing process.

The scientific evaluation prepared by the international expert group forms the essence of this book. The expert group's goal was twofold. The first objective was to examine the work relatedness of WMSDs in the light of the existing literature. The second objective was to explore and synthesize information, avenues and approaches that could help in the prevention of WMSDs. While the literature is not solid on all aspects of the problem, analogies from other areas of safety and health have been used to illustrate other approaches.

The content of the book is the result of a collective effort by everyone, experts and scientific editors alike. Numerous face-to-face meetings have shaped the organization and the content of individual chapters. However, each chapter has had one or more main authors who have specifically given it flavour. The scientific editors were responsible for compiling the material and have also contributed to the writing of various texts, tables and figures throughout the book.

In essence the opinions expressed in the book have been formulated during a process by which the expert group agrees on the general conclusions, viewpoints and the organization of the material, although individual opinions may differ with regard to certain details.

---

*Notably the Ontario Workplace Health and Safety Agency.

Acknowledgements

The expert group consisted of: Professor Mats Hagberg, National Institute of Occupational Health, Sweden; Dr Barbara Silverstein, Safety and Health Assessment & Research for Prevention (SHARP) Division, WA Department of Labor and Industries, USA; Associate Professor Richard Wells, University of Waterloo, Canada; Professor Michael J. Smith, University of Wisconsin, USA; Professor Hal W. Hendrick, University of Southern California, USA; Associate Professor Pascale Carayon, University of Wisconsin, USA; Professor Michel Pérusse, Université Laval, Canada: *The scientific editors were:* Dr Ilkka Kuorinka, IRSST, Canada; Dr Lina Forcier, IRSST, Canada. *Several persons also contributed to the book in various ways:* Dr Anneli Leppänen, Institute of Occupational Health, Finland; Dr Monique Lortie, Université du Québec à Montréal, Canada; Dr Michel Rossignol, Jewish General Hospital of Montréal and Équipe de santé publique de Montréal, Canada. *A team of dedicated persons has produced excellent work throughout the project in different tasks:* Odette Falardeau, Project Secretary; Louise Gagnon, Information Officer. *Others at the IRSST who have been instrumental in bringing this project and this book to its successful completion:* Danièle Gastonguay, Project Secretary from May 1991 to December 1991; Bernard Jobin, Librarian; Jacques Blain, Consulting Librarian; Lynda Cloutier, Library Technician; Denise Mallette, Library Secretary; Lise Brière, Administrative Secretary; Christine Lecours, Administrative Secretary. *Members of the advisory committee have also provided valuable comments throughout this project:* Dr Marcel Asselin, Joel Carr, Gary Cwitco, Robert Demers, Jocelyne Everell, Roger Langlois, Normand Nault, Bob Webb. *Others whose help and criticisms were very much appreciated:* Thomas J. Armstrong, Harvey Checkoway, Lawrence J. Fine, Carol Glegg, Pamela Ireland, Terry Knowles, John Last, Anne Moore, Robert Norman, Joseph N. Nearing, Gisela Sjøgaard, Larry Stoffman. *For their work on the figures, illustrations and tables, special thanks* to Dominique Desjardins for supervising the graphic work of the book, to Chantal Lichaa for the illustrations in chapter 3, and to Martine Gamache for her graphic work on the figures in various chapters.

The authors/editors deliver this book to its readers confident that the prevention of WMSDs is feasible, useful and profitable. It takes knowledge and understanding of the basic principles and tenacity in realizing the goals of prevention in an organization. Collaboration by concerned parties, including management, employees and the competent authorities, is an important tool to achieve prevention.

2

WMSDs: conceptual framework

2.1 Introduction: what are WMSDs?

WMSDs are by definition a work related phenomenon. The literature shows unarguably that certain jobs and certain work related factors are associated with the manifold risk of contracting a WMSD compared with other population groups or groups not exposed to these risk factors (e.g. Silverstein, 1985; Ayoub and Wittels, 1989; Stock, 1991; Hagberg, 1992). The definitions used for WMSDs (diagnostic criteria or case definitions) are not consistent and require, in each individual case and in each population study, careful identification of the symptoms, signs and findings used.

In this book, we use the term 'work related musculoskeletal disorder' (WMSD) as a descriptor for disorders and diseases of the musculoskeletal system having a proven or hypothetical work related causal component, and we deal mainly with neck and upper-limb disorders. The chosen term corresponds roughly to other concepts like 'cumulative trauma disorder' (CTD), 'repetitive strain injury' (RSI), 'occupational cervicobrachial disease' (OCD), 'occupational overuse syndrome' (OOS) and others. In this respect, 'WMSD' stands as an umbrella term for grouping together specific work related musculoskeletal disorders. To provide some context for the reader, a list of symptoms and disorders classified in the literature under CTDs, RSIs, etc. is presented in appendix I. It is felt that the concept chosen for this book, 'WMSD', better corresponds to the World Health Organization's definition and concept of work related disease (see section 3.1); it avoids the confusion of including both the postulated cause (e.g. 'cumulative' in CTD or 'repetitive' in RSI) and the effect ('disorder' in CTD or 'injury' in RSI) in the same term.

Our definition of WMSDs excludes musculoskeletal accidents. The work related low back pain (LBP) problem should, strictly speaking, be discussed within the framework of WMSDs. However, it is not in this document. LBP research is now a field of its own, its results and concepts are used here mainly as examples and analogies when appropriate or useful, and/or when knowledge on WMSDs is lacking. For our purposes, WMSD will usually imply a work related musculoskeletal disorder of the upper limbs, since there is very little research available on lower limbs. However, whenever information on WMSDs of the lower limbs is available it has been included here.

2.2 Natural history of WMSDs

2.2.1 Introduction

WMSDs are a heterogeneous group of disorders; the natural history of most of the individual disorders is only vaguely known. It is assumed, however, that repeated efforts (movements, postures, etc.), static work, continuous loading of the tissue structures or lack of recovery time trigger or cause a pathological process that then manifests itself as a WMSD. WMSDs were first observed in jobs characterized by repeated or continuous loading, but later were also observed in other occupational groups. Not much is known of the biochemical or immunological mechanisms, although hypotheses on the role of immuno-logical reactions against substances released by injured tissues have been advanced. The hypothesized pathophysiology of some of the disorders called WMSDs is mentioned in chapters 3 and 4. The multifactorial nature of WMSDs has prompted several authors to attempt to produce a model to unify the often incongruous observations on these disorders. In the following text, one general pathological model presents the conceptual description of the disease process itself. The generic models of WMSDs presented later make an attempt to describe the mechanisms and external factors that trigger or influence the pathological process.

2.2.2 A pathological model for natural history

The natural history of WMSDs can be viewed in the context of a classical scheme of pathology (Leavell and Clark, 1965), containing the following elements:

- the disorder is understood as a process, which has causal factors and may end up a more or less defined pathological entity;
- usually a complex set of various stimuli influence the causal factors;
- the organism reacts to the stimuli. The outcome of this reaction may in a certain phase become an observable disorder (disease), or it may lead to recovery and possibly to improved resistance to renewed stimuli.

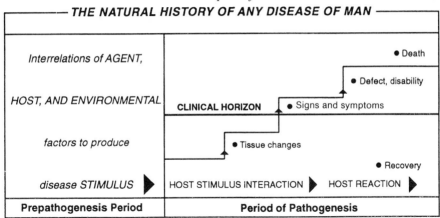

Figure 2.1   The natural history of human disease (part of a diagram from Leavell and Clark, 1965).

2.2.2.1 Period of pre-pathogenesis

In the pre-pathogenesis period, environmental factors and other pathophysiological factors are assumed to trigger a pathological condition. This introduces the pathogenesis phase.

2.2.2.2 Period of pathogenesis

During the period of pathogenesis two theoretical alternatives arise. One, the process may continue, signs and symptoms of a disorder will be detected and, if the process is not halted, more serious consequences may appear. Two, the process may be brought to a halt if the host reaction (mobilization of hormonal or immunological resources, etc., or adaptation) is able to stop the process. Further, the host reaction may improve the adaptive capacity of the organism so that the organism is stronger to meet any subsequent attack.

The scheme presented above is purely theoretical. First, WMSDs may not behave as a process wherein the next phase can be predicted from the earlier one (e.g. tendinitis may not be predictable from previous local pain). This assumption is, however, dominant in the models currently applied to WMSDs. Second, very little is known of the interaction of the different factors that trigger the pathological process. Third, we do not know in detail how the reparative processes function.

2.2.3 Generic models for WMSDs

Numerous attempts have been made to model WMSDs. The models usually comprise a varied number of elements and generally aim at describing some specific aspect of the pathological process. Most of the existing models assume a dose-response relationship between the amount of strain on the organism and the outcome (WMSD). The following models are presented as examples of different approaches to modelling WMSDs.

A model proposed by Armstrong *et al.* (1993) takes the external factors (i.e. work requirements) as a starting point. These factors produce an internal dose (e.g. tissue loads, metabolic demands, etc.), which produces a physiological response (reaction by the organism). The capacity of the individual to withstand and react to external factors is considered an effect-modifying factor. A conceptually new addition to this model is the assumption of cascading adaptation: the exposure-dose-response experience may form the starting point for a new cycle, which repeats itself. Armstrong's model essentially depicts the adaptation process in WMSDs.

Moore *et al* (1991) and more recently, Tanaka and McGlothlin (1993) presented a model that explains the etiology of a type of WMSD, carpal tunnel syndrome (CTS) by frictional load inside the carpal tunnel and the tendon sheaths. This friction is assumed to be a product of three biomechanical factors: internal force, repetitiveness and wrist angles. The model is an example of a mechanical basis for explaining WMSDs and aims at developing quantitative guidelines, which would then be used to prevent WMSDs.

The causal mechanisms possibly involved in the development of WMSDs have also been explained from a job design and stress theories point of view (Smith and Sainfort, 1989). This model integrates the psychological and biological aspects of work within an ergonomic framework. The theory states that poorly designed working conditions (and other workplace features such as technology) can produce a 'stress load' on the person and an 'imbalance' in the job design. This stress load can have both physiological and psychological consequences, such as biomechanical loading of muscles or joints, increased levels of catecholamine releases or adverse psychological mood states. The extent of the stress load is influenced by an individual's psychological 'perception' of the load, which are a product of the physical characteristics of the load and the individual's personality, past experiences and social situation. However, the load is also greatly influenced by its objective physical properties independent of the perception of those properties.

In this model, psychological perceptions are affected by the organizational environment and psychosocial factors of the workplace. Negative psychological perceptions can lead to adverse psychological and physiological strain reactions. These in turn can lead to physical problems such as muscle tension or elevated catecholamine and cortisol production. In addition, they can lead to inappropriate behaviour at work, for instance poor work methods, using excessive force to accomplish a task or failing to rest when fatigued. Such influences may lead to health problems, including WMSDs.

2.2.3.1 A generic model of prevention

The following is a synthesis of different models and incorporates the deliberations of the authors of this book. There are two essential features in the model

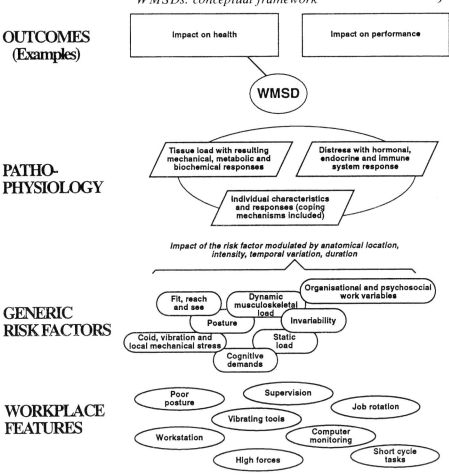

Figure 2.2   A generic model of prevention.

depicted in Figure 2.2: the workplace–individual dichotomy and the cascading nature of the different elements (when an element is changed, the new state becomes a basis for the next cycle). Organizational and psychosocial factors are understood partly as factors controlling other elements (e.g. work organization influences the mechanical load), but that also have specific influences (e.g. through stress mechanisms). The model consists of three layers of elements, namely pathophysiological phenomena, generic risk factors and workplace features. Although not included in this model, external factors are also important and are discussed in chapter 6.

The *pathophysiological layer* consists of phenomena related to the injury and disease process. While almost all risk factors that may trigger a pathological process are part of everyday life, this part of the model implicitly contains the idea of adaptation: the pathophysiological process is triggered if the risk factor in question exceeds the organism's capacity to tolerate that risk factor. The organism may also adapt to the situation and be more resistant during the next cycle.

Generic risk factors are a group of factors traditionally considered as being associated with higher risk of developing WMSDs. They include postures, static load, psychosocial work variables, etc. These risk factors are assumed to interact, eventually cumulate and form cascading cycles. Generic risk factors are assumed to be directly responsible for pathophysiological phenomena which depend on the location, intensity, duration, repetitiveness, etc. of risk factors. It is implicitly proposed that risk factors in a number of cases have a U-shaped function with respect to pathophysiological phenomena, meaning that a certain optimum level of the influence of the risk factor is necessary for health (in this sense, it is not quite correct to speak of risk factors here); both insufficient and excessive musculo-skeletal loads, for example, have deleterious effects on the organism.

Workplace features comprise the factors dealt with in workstation design, work organization and other components that are traditionally the focus of ergonomics, production engineering, etc. These are the factors that preventive efforts are mainly intended to correct.

2.2.4 Course of WMSDs

Our current knowledge of WMSDs is insufficient to allow a general description of the course of WMSDs. This is true for most of the individual disorders that can be grouped under WMSDs (see appendix I). According to a conventional assumption, the beginning of the process is fatigue and discomfort after strenuous exercise. Under certain conditions, the symptoms persist, resulting in a disease state and possibly incapacity and invalidity. There is no scientific proof, however, that this model of a course of WMSDs can be generalized. Also, the time periods for each phase of the development of WMSDs are obscure. A disease state may develop rapidly in the course of a few hours or days, or it may last weeks or months. Pathological phenomena similar to WMSDs are found in populations not exposed to specific work related factors. Some of those cases can be traced back to the same type of mechanical stress as that found in work, e.g. in sports (Walter *et al.*, 1989). For example, lateral epicondylitis is synonymous with tennis elbow; the elbow structures may be stressed as much by work motions as by playing tennis.

2.3 Some WMSD-related concepts and characteristics

2.3.1 Work related disease – a World Health Organization concept

The World Health Organization (WHO, 1985) divides diseases that have a relation to work into two categories. The first category comprises actual occupational diseases, and includes more or less well-defined disease states attributable to a specific causal agent, for example asbestos fibres causing asbestosis. The second category is work related diseases, including diseases for which work related factors play a partial causal role. WMSDs are work related diseases, but many of them are also considered occupational diseases depending on local customs and legislation.

2.3.2 Some commonly used terms relevant to WMSDs

The term 'disorders' (musculoskeletal) is used as a descriptor for pathological entities in which the functions of the musculoskeletal system are disturbed or abnormal. Disorders are contrasted with diseases, which are considered defined pathological entities having observable impairments in body configuration and function. Illness is used as a synonym of sickness: an unhealthy condition with a connotation of someone feeling sick.

Discomfort, fatigue and pain are the most common first symptoms associated with WMSDs. Signs, like loss of function, limited movement or loss of muscle power, are less common. Discomfort (discomfort is defined as physical or mental distress, and is a synonym of inconvenience) indicates a perceptual, subjective phenomenon that is more diffuse than pain. In practice, discomfort is often defined by the methodology used (e.g. Corlett and Bishop, 1976). Fatigue is a complex and diffuse term. In practice it has been defined as an incapacity to continue strenuous physical or mental work at the same rate as previously, i.e. an unwanted, transient sensation that influences the individual's motivation and capacity to continue the work as before. In this context, pain is defined as an unpleasant sensory and emotional experience associated with actual or potential tissue damage, or described in terms of such damage.

Terminology on work organizational, psychosocial and behavioural factors is not well established. In this text we use the terms as follows:

- Work organization is the objective nature of the work process. It deals with the way in which work is structured, supervised and processed.
- Psychosocial factors at work describe the subjective aspects of work organisation and how they are perceived by workers and managers.
- Behavioural factors at work define the way in which workers carry out their work, without any value judgement as to 'goodness' or 'badness'.

In this text, an ergonomics programme is defined as a programme to identify, control and prevent both WMSDs and their risk factors. It therefore consists of, among other things, a surveillance programme (see chapter 5), a training programme (see chapter 8), and a medical management programme (see chapter 9).

Definitions for many other terms used in this text are provided in appendix II.

2.3.3 Problems with WMSD research and interpretation of results

2.3.3.1 Various types of biases possible in research on WMSDs

In a study, several biases are possible if the methodology is not well controlled.

This should be taken into account when scientific and other reports are analyzed and interpreted.

SELECTION BIAS

Primary selection bias (only healthy workers selected into the workplace) and secondary selection bias (workers no longer healthy selected out of jobs) are difficult problems to overcome in all occupational health studies, particularly in cross-sectional studies. This selection bias tends to result in underestimation of the effect of physically strenuous work on musculoskeletal health.

RESPONSE BIAS

Occupational studies with low participation rates may be quite unrepresentative of the target worker population.

INFORMATION/MEASUREMENT/DETECTION BIAS

Studies that use surrogate measures such as job titles for risk factors or extrapolate risk factor measures from a few 'representative' workers to an entire group may under- or overestimate exposure to risk factors. Misclassification of the outcome (WMSDs in our case) may also occur and may be related to poor validity or reliability of clinical tests.

CONFOUNDING AND EFFECT MODIFICATION BIAS

Confounding distorts the interpretation of the results of a study on the association between a risk factor and an outcome. Gender, for example, a confounding variable in certain studies, could be related both to the job and the outcome, CTS. In most studies, on WMSDs or other outcomes, some amount of confounding is inevitable and must be carefully considered. The most important confounder for WMSDs appears to be gender, although some later studies have not confirmed this (see subsection 9.8.3). Other confounders to be considered in a research study on WMSDs could be age and various other factors discussed in chapters 3 and 4 and section 9.8.

Effect-modifying variables, on the other hand, modify the effect of the risk factor under study (e.g. years on the job) on the outcome (e.g. WMSDs). For example, ageing could be an effect modifier in certain studies: if old age diminishes physical activity and increases joint degeneration, it is difficult to draw conclusions on the effect on WMSDs of years of physical exertion on the job (since this would also be related to age).

2.3.3.2 Contribution of individual factors

Individuals are different with respect to the triggering and development of WMSDs. Susceptibility to disease and reaction to work-risk factors vary from

one person to another. Some individuals are .
musculoskeletal disease than others. This susceptibility m.
appear earlier or in an unusual location. Whether thi.
permanent characteristic of an individual (hypothetically h
and structure of his/her connective tissue or muscle, etc.)
modifiable (other illness that the individual is currently suffe.
cally a lack of motor skill, risk-taking behaviour, higher thresh
or work strain, etc.) is not known at present. Individual charact   ..ke age,
gender, anthropometric and anatomical characteristics, tissue type, alcohol and
smoking habits, etc. are discussed in detail in chapter 9, section 9.8. Finally, it
should be briefly mentioned that evaluating the risk of contracting a WMSD
associated with individual risk factors is problematic. There is no known
effective and reliable method to test an individual for WMSD 'potential', i.e. to
select more resistant workers or eliminate those who are less resistant through
pre-employment or preplacement screening (see section 9.7).

2.3.3.3 Surrogate outcomes

In the course of their work, people are exposed to various physical and
psychological factors causing different reactions/outcomes. Although we deal
here with WMSDs as outcomes, it is worth noting that there are also second-
order outcomes, which can be defined as surrogate outcomes. The same
conditions at work that produce WMSDs may also cause other effects. These
include diseases other than WMSDs, absenteeism (sickness or other), work
accidents and high compensation costs, high turnover, poor working climate,
dissatisfaction, poor quality, etc. Surrogate outcomes may serve as general
indicators of the situation at work, and/or may be associated with WMSDs and
their risk factors. However, a great deal of caution is necessary when attempting
to interpret the meaning of surrogate outcomes. The association between
WMSDs and surrogate outcomes is in most cases hypothetical and liable to be
influenced by several factors.

2.4 Prevention of WMSDs: important concepts

The basic objective of this book is to analyze WMSDs with a view to preventing
these disorders. The various aspects of prevention discussed in chapters 5 to 9
show that there is no one single action that alone would be sufficient to eliminate
WMSDs. The complex and multi-factorial nature of these disorders requires a
multi-factorial preventive strategy. Several examples of successful prevention of
WMSDs and related ailments have been published (Luopajärvi *et al.*, 1982;
Westgaard and Aarås, 1985; Oxenburgh, 1991). In the example of Spilling et al.
(1986), economic calculations show that the return on investment in prevention

WMSDs is not only positive, but forms a basis for future development of production economics.

THE ROLE OF STANDARDIZATION

In some countries the prevention of WMSDs is linked with normative regulations. The European Union is about to prepare standards aimed at influencing industrial practices related to WMSDs (e.g. guidelines for the furniture, 89/391/CE, 90/207/CE, VDU: EN 29241). In the United States, OSHA standards have the same purpose. The ANSI Z-365 committee, also in the US, is about to prepare a consensus on upper limb CTDs. An international standard, ISO 9000 (1987), aims at regulating the principles of quality assurance and management. Contrary to many earlier standards, the ISO 9000 series is a procedural standard, which describes principles, goals and procedures to assure and demonstrate quality in industry. Although the ISO 9000 standard does not state it explicitly, it can be inferred that a production system associated with high prevalence of WMSDs can hardly comply with this standard. What role the ISO 9000 standard will play in the prevention of WMSDs remains to be seen.

THE ROLE OF PARTICIPATORY MANAGEMENT IN PREVENTION

Participatory management practices have shown their usefulness in the prevention of accidents (Simard *et al.*, 1993) and of WMSDs. There are theoretical foundations (e.g. Smith and Sainfort, 1989) that explain why participatory approaches are efficient, and practical reasons for their usefulness. The prevention of WMSDs requires knowledge of many production details, sometimes known only to experienced personnel. This requires collaboration between management, WMSDs specialists and the workers concerned. More on this topic can be found in chapters 6 and 7.

2.5 Conclusion

Work related musculoskeletal disorders are, from both the etiological and pathological points of view, a heterogenous group of disorders (diseases) necessitating a careful definition of diagnostic criteria, the nature of risk and other factors. The assumed conceptual framework presented in the book focuses on the question of work relatedness. The authors feel that a straightforward mechanical model is too limited for a proper understanding of WMSDs. Although such mechanical factors as repetition, force and posture are important first-line risk factors, there are also other plausible factors that may provoke a disorder or indirectly influence other risk factors (e.g. organizational and psychosocial factors). It is also assumed that the various risk factors interact, forming cascading states of the organism that either promote or impair musculoskeletal health.

Risk factors are presented in an operational context: what are the risk factors and how are they identified? The discussion extends to analysing how prevention can be achieved and how it operates in an organizational context. The scope is systemic and macroergonomic, for prevention is most effective if all aspects, from organizational to ergonomic and biomechanical, are taken into account simultaneously. In most cases this is the only way to achieve positive results.

3

Evidence of work relatedness for selected musculoskeletal disorders of the neck and limbs

3.1 Introduction

The aim of this chapter is to examine and interpret the scientific evidence available on the association between work and the development of musculo-skeletal disorders of the neck and limbs. First, this introduction will briefly define the context and approach used to look at this relationship. Then the following sections (3.2 to 3.7) will examine, by tissue type, various examples of musculoskeletal disorders and their association with work. One section (3.8) will also look at unspecified and multiple-tissue disorders. It is in this section that studies on RSI (repetitive strain injuries), CTD (cumulative trauma disorders), OCD (occupational cervico-brachial disorders) etc. will be found. Finally, a summary of chapter 3 will be provided (section 3.9).

This chapter will focus on work relatedness. Chapter 9 contains a section that briefly examines the contribution of personal susceptibilities to the development of musculoskeletal disorders; this is an issue quite often raised, but as shown in chapter 9, does not diminish the overall impact of work on the development of musculoskeletal disorders. The focus is on an epidemiologic review in human working populations rather than animal or human laboratory studies.

3.1.1 Definition of work relatedness

In contrast to 'occupational' diseases, where there is a direct cause and effect relationship between hazard and disease (e.g. asbestos and asbestosis, lead and lead poisoning), the World Health Organization expert committee described 'work related' diseases as multifactorial, where the work environment and the performance of work contribute significantly, but as two of a number of factors, to the causation of disease:

> ... they may be partially caused by adverse working conditions; they may be aggravated, accelerated or exacerbated by work place exposures; and they may impair working capacity. It is important to remember that personal characteristics, other environmental and socio-cultural factors usually play a role as risk factors for these diseases.
>
> (*Identification and Control of Work related Diseases*, WHO Technical Report Series 714, 1985, Geneva)

There has been increasingly widespread recognition of the multifactorial nature of work related musculoskeletal disorders, with varying levels of attention to individual, psychosocial and physical factors that may contribute to the development or prevention (salutary buffering effect) of these disorders. These factors have not been studied simultaneously with equal rigour in any scientific investigations. There are a variety of types of studies that can be used to evaluate the 'work relatedness' of different disorders of the musculoskeletal system: clinical case series, epidemiologic studies and laboratory studies. These will be discussed later in this introduction under 'Types of evidence' (3.1.4).

3.1.2 Studies of work populations: how valid are the results of any study for providing evidence on the work relatedness of musculoskeletal disorders?

There is no 'perfect' population study of occupational exposures and health outcomes. The strengths of population studies are evaluated by looking at four major components:

1. Selection bias. How representative are the study participants of the study population (participation rate) and the target population (those believed to be at risk)? In occupational studies, there are at least two kinds of selection biases that occur: (a) selection of 'healthy workers' into the work population (healthy enough to work when hired); and (b) selection of 'sick' workers out of the active workforce (often lost to follow-up). These biases tend to lead to underestimation of the magnitude of the association between work and disease.
2. Information or misclassification bias. How well do the methods used to characterize exposure and health effect actually measure what they are supposed to be measuring? Are all participants and exposures measured in the same way? Are some of the referents actually cases? Are some of the 'exposed' subjects really 'unexposed'?
3. Confounding or effect modification bias. Are there other factors that may actually explain any supposed relationship between work and disease? A confounding factor is one that is related to both the disease and the exposure of interest (in occupational morbidity studies, gender is often a confounder). The association between exposure and disease is increased as a third factor or effect modifier (e.g. age) is increased.

Table 3.1: Evaluation basis: studies selected to provide evidence on the work relatedness of musculoskeletal disorders of the neck and limbs

---

Studies selected:

- were evaluated using the four components described above under 'Studies of work populations',
- provided original data,
- dealt with a work population, were related to work or had analogies with work (e.g. sport studies),
- were published in English,
- were published as peer-reviewed articles.

Cases studies and literature reviews were excluded.

---

4. The power of the study design to detect significant differences in disease if they are truly present. If the method used to measure disease is very precise and the association with exposure is dramatic (fivefold difference), then a smaller sample size may be adequate to identify these differences. However, if there is much misclassification of exposure or disease, and a modest association between the two (twofold or less difference), then a much larger sample size will be required.

The four components described above were used, among other criteria, to evaluate individual studies and determine whether they would provide evidence for the discussion on the work relatedness of musculoskeletal disorders in this chapter (Table 3.1).

3.1.3 Criteria for causality: how do the results of studies contribute evidence to the relationship between musculoskeletal disorders and work related activities?

As our understanding of the complex nature of any disease process increases, we are continually debating the 'evidence' for causality (Weed, 1986). This debate will probably continue for a long time. For the purposes of this chapter, the framework proposed by Susser (1991) was modified to examine the relationship between work and musculoskeletal disorders.

1. Do the results of the studies show an association between disease and work exposure? If the answer is yes:
   (a) How strong is the association with work?
   (b) How specific is the association with work?
2. Do they show a temporal relationship? (Did the work exposure occur before the effect?) The cause must at least have occurred by the time the effect occurred.
3. Is there consistency in the association?
   (a) Is the association with work replicated in more than one study and in other circumstances (i.e. the 'replicability' of the association)?

(b) Does the association with work survive rigorous tests of alternative hypotheses (i.e. the 'survivability' of the association)?

4. Can a change in disease be predicted by a change in work exposure (predictive performance)? If there is an increase in musculoskeletal disorders, is it related to an increase in work related stressors? If there is a decrease in work related stressors, is there a consequent reduction in disease?

For the purposes of this discussion, it is important to evaluate work relatedness in terms of exposure (the presence of a factor in the environment external to the worker, measured in terms of duration and intensity), burden (the amount of the factor that exists in the worker's body or specific target organ, affected by retention and recovery time) and dose (the amount of the factor that remains in the worker's target organ for some time interval) (Checkoway *et al.*, 1989). The time integrated measure of exposure is referred to as cumulative exposure. These concepts have most often been used in discussions of chemical substances, but are useful in looking at the physical and psychosocial characteristics of work as well. Most often, we measure exposure as a surrogate for burden and dose. This does not take into account the individual worker's capacity to modify the burden or reduce or increase the dose through a variety of feedback mechanisms.

Since exposure is likely to be used in studies to evaluate the impact of work, expressions like exposure-effect and exposure-response will be found throughout this examination of causality. Effect and response are often used as synonyms but a distinction is motivated in occupational epidemiology as suggested by Hernberg (1992). Effect is defined as the biological change caused by an exposure and response is defined as the proportion of a population having values showing an abnormal effect. The exposure-effect relationship refers to the association between the numerical values for both the exposure and the effect. The exposure-effect relationship describes the average effect of each exposure level. The exposure-response relationship denotes the association between each exposure level and the porportion of the population having abnormal values.

5. Coherence of evidence. To what extent does the association with work fit with pre-existing theory and knowledge on how the musculoskeletal disorder could develop (theoretical, factual, biological, statistical)?

3.1.4 Types of evidence: what types of studies are available and can contribute evidence to the relationship between musculoskeletal disorders and work?

Clinical case series in which clinicians describe common characteristics of their patients have often been used to generate hypotheses about risk factors for diseases. For example, there are numerous reports about disorders among performing artists, such as carpal tunnel syndrome among pianists or left-sided cubital tunnel syndrome among right-handed violinists (Hochberg *et al.*, 1983; Lederman, 1986). However, there have been no studies of pianists with a comparison group to determine whether pianists are at higher risk than others. Nonetheless, these case series are extremely important. A classic example of this arose when a rare disease (angiosarcoma of the liver) was identified in vinyl chloride workers. With more common diseases it is important to evaluate how well or poorly the clinicians' population reflects the general population or the 'at risk' population. Cultural and gender differences in the reporting of pain and in seeking treatment have been well documented (Bates, 1987; Buckelew *et al.*, 1990). For the most part, clinical case series will not be addressed further in this chapter.

Laboratory studies are useful for isolating specific exposures of interest and testing their individual or combined effects on a relevant outcome. These studies are valuable for hypothesizing mechanisms that contribute to the development of the disease of interest. Rarely is the outcome the production of the disease in humans. The applicability of laboratory findings to working populations depends on a variety of study design issues.

Epidemiologic studies are used to evaluate patterns of morbidity in populations by occupation, industry or risk factor. Usually a reference population is also studied for comparison purposes. Case control studies identify 'cases' and 'referents' and then go back in time to try to estimate exposures of interest and see whether there are any differences between groups. The greatest challenges in this kind of study are finding an appropriate referent group and reducing exposure misclassification. Prospective cohort studies identify a new or healthy group of workers and follow them over time, measuring exposures and seeing who develops the disease of interest. While cohort studies tend to provide the most convincing evidence of causality, they are extremely difficult and expensive to conduct in the workplace. Cross-sectional studies are 'snap shots' in which exposure and outcome are measured across a population at one point in time. Intervention studies (sometimes including natural experiments in the workplace, in which exposure is increased or decreased and a change in disease can be measured) are a challenge in that investigators rarely have control over the intervention process, but are merely the recorders of events in a dynamic environment.

Epidemiologic studies of work related musculoskeletal disorders tend to suffer from three major problems:

1. Cross-sectional study designs are the most common type of studies available and yet do not identify a temporal pattern and are most subject to a 'healthy worker effect', particularly in high-exposure jobs (Punnett *et al.*, 1985; Silverstein *et al.*, 1987; Ostlin, 1989). Many cross-sectional studies have a retrospective occupational history component, ranging from previous job title to self-estimates of percentage time in various postures or other exposure-related activities.
2. Lack of consistency in case definition among studies (Kuorinka and Viikari-Juntura, 1982; Gerr *et al.*, 1991). Those studies that tend to combine all areas of the upper limb into one case definition may mask important differences in exposure.

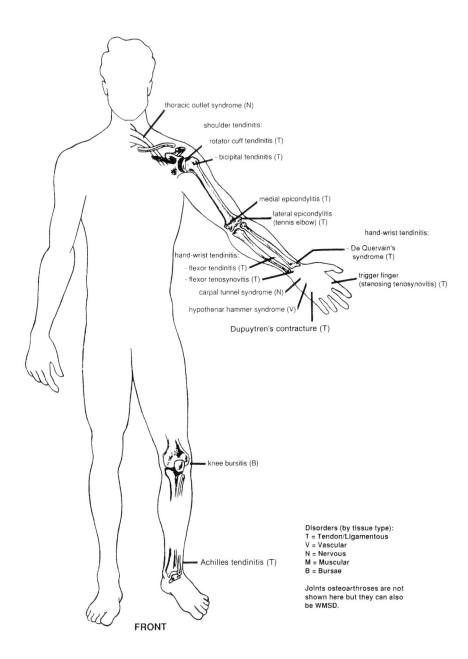

Figure 3.1 Examples of musculoskeletal disorders which may be work related (Front).

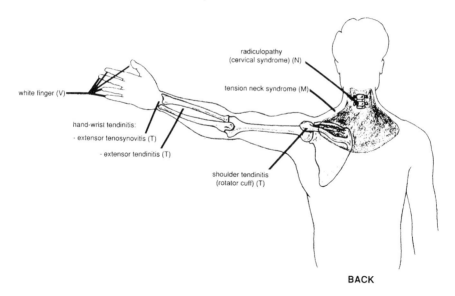

Figure 3.1 Examples of musculoskeletal disorders which may be work related (Back).

3. Imprecise estimates of work exposures due to high resource demands for detailed job analysis. Job titles are usually used as surrogate variables for occupational exposures in studies. A job title rarely describes homogeneous tasks performed in the same manner by all workers with that job title. Also, the job of origin may not be the current job, resulting in misclassification of exposure, most likely resulting in an underestimate of the association between work and the musculoskeletal disorder under study.

3.1.5 Illustrative review of the literature for evidence of work relatedness

Not all studies available on musculoskeletal disorders were included. Instead, a selection basis was established to help identify those studies that would contribute evidence to answer the question at hand: whether there is an association between work and musculoskeletal disorders of the neck and limbs (Table 3.1).

Work relatedness is illustrated in the following sections by examples of specific tendon (section 3.2), nerve (section 3.3), muscle (section 3.4), joint (section 3.5), vascular (section 3.6) and bursa (section 3.7) disorders. Figure 3.1 illustrates the location of some of the disorders used as examples. There have also been a number of studies of upper-limb disorders as a whole (RSI, OCD, CTD), although in the absence of anatomic location or structure they are difficult to assess. However, since some of these studies provide important information on work related risk factors and upper-limb musculoskeletal symptoms, they are included and discussed in section 3.8. Tables presenting evidence on work relatedness are provided throughout sections 3.2 to 3.8; information to help interpret these tables is given in Box 3.1. The chapter ends with a summary of the evidence for the work relatedness of musculoskeletal disorders (section 3.9).

3.2 Evidence of the association between work and selected tendon disorders: shoulder tendinitis, epicondylitis, de Quervain's tendinitis, Dupuytren's contracture, Achilles tendinitis

3.2.1 Brief description of the disorders

A tendon is the part of a muscle that attaches the muscle to the bone or fascia, transferring force from the muscle to the bone or fascia to produce a joint motion. The tendon consists of collagen fibres arranged in a parallel fashion with a strength of about 50 per cent of that of cortical bones (Frankel and Nordin, 1980). Fibrous tissue surrounding the tendon forms a tendon sheath that protects the tendon against mechanical friction when passing over bony structure. The tendon sheath consists of a synovial membrane that reduces the friction against the bone (see Figure 3.2).

Tendinitis and tenosynovitis (the latter synonymous with tendovaginitis) are inflammations of the tendon and the synovial membrane of the tendon sheath, respectively. Common sites for these inflammations are (Figure 3.1):

- the tendons of the rotator cuff muscles (supraspinate, infraspinate, subscapularis, and teres minor muscles);
- the tendons of the long head of the biceps brachii;
- the tendons at the elbow (insertion of the finger extensor muscle which, when inflamed, is known as tennis elbow or lateral epicondylitis);
- the tendons of the abductor pollicis longus and the extensor pollicis brevis (de Quervain) at the wrist;
- the Achilles tendon at the ankle.

At these anatomical locations, the tendons move long distances while participating in a wide range of movements.

Peritendinitis is the inflammation of the tendon and adjacent tissue (most often muscle tissue); local swelling and edema may also be present (Adams, 1971; Kurppa *et al.*, 1979). Inflammation can be the result of a general inflammatory disease such as rheumatoid arthritis, but can also be due to mechanical irritation and friction, which may lead to local inflammation in tendons and in the tendon sheath (Kurppa *et al.*, 1979). Although tendinitis, peritendinitis, synovitis, myotendinitis, etc. are different disorders, they are often found together (Figure 3.2). It can be impossible to distinguish among them on a

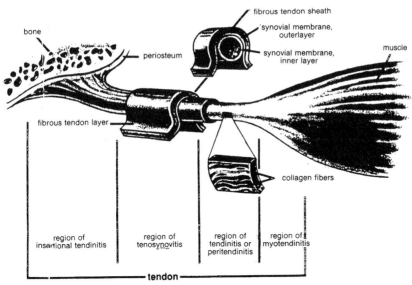

Figure 3.2 Structure of a tendon.

clinical level because they present the same signs and symptoms of inflammation. The 'tendinitis' studies considered here could therefore include any of the above-mentioned disorders. Dupuytren's contracture is a non-inflammatory degeneration of the tendons of the palmaris muscle in the hand (Figure 3.1).

3.2.2 Literature reviewed

Most articles described cross-sectional studies that examined the association between disorders and job title. Occupational exposures were approximated by using job titles; they were not measured in the study populations, except for one study. Furthermore, the description of ergonomic exposure in the job title groups was lacking in most of the scientific reports. Although the literature was sparse, it provided a diversity of tests of association ranging from experiments in animals and humans to observations of the events in sequence, i.e. that exposure precedes the development of the disorder. Risk evaluations with risk ratios and confidence intervals were lacking in most original articles. In these articles, whenever an attempt was made to perform epidemiologic analysis, there was no consideration of potential confounders, such as age, smoking, etc., except in a few studies. Furthermore, there was also no attempt to perform a cumulative exposure-response analysis in most studies.

Box 3.1 Help with interpreting the tables on work relatedness in chapter 3

Different types of study designs and variations on these designs, are mentioned in the tables of this chapter. The following explanations are provided in order to help readers without an epidemiologic background interpret the information in these tables.

The aim of epidemiologic studies is to examine the distribution and determinants of health-related events: for example the association between work factors and the development of work related musculoskeletal disorders. Two basic approaches are used: experimental and observational studies.

EXPERIMENTAL STUDIES

In *experimental studies* the researcher actively intervenes and changes one variable to see what happens to the other, while trying to prevent other variables from interfering and affecting the outcome of the experiment. *Intervention studies* are often regarded as quasi-experimental studies, since investigators have less control over implementation.

OBSERVATIONAL STUDIES

Because of the difficulties and ethics of conducting well-controlled experiments on human populations, and because there are abundant observational data, researchers have tended to concentrate on observational studies, in which they simply observe events that are occurring or have occurred; they do not play an active role in affecting the variables. Three main types of observational studies are discussed here.

The *cohort study* design follows up on a group of subjects (exposed and non-exposed) over time. It is based on a comparison of the number of cases that develop (the outcome which, here, is musculoskeletal disorders) in the exposed study subjects with those that develop in the comparison group (non-exposed group, also known as controls or referents). The exposure or study factor in the studies in this chapter would be the work factor examined (e.g. working overhead, repetition etc.). The proportion of new cases (i.e. the incidence) is measured in the exposed and the non-exposed groups; a comparison between these two incidences (in the exposed and non-exposed groups) is established (relative risk) and can be used to assess the observed association of the exposure with the development of cases. A relative risk (RR) of 1 implies that the incidence in the exposed and non-exposed groups is similar and therefore exposure is observed to have had no impact on the study subjects (i.e. no association has been observed between the exposure and the cases). A relative risk of greater than 1 implies that the incidence of cases was greater in the exposed group than in the non-exposed group and that an association has been observed between the exposure and the cases. For example RR = 2.5 means that subjects in the exposed group are 2.5 times more likely to become cases than those in the non-exposed group. Obviously the converse (RR less than 1) implies that the exposed group has fewer cases, and exposed subjects are therefore less likely to become cases than the comparison subjects (protective effect of the exposure).

The increased or decreased risk observed in the exposed group could, however, be due to mere chance events; two approaches are commonly used to gauge the possibility that the association observed between the exposure and the

Box 3.1 Help with the tables in chapter 3 (Cont.)

cases is 'true' rather than due to chance. Using various methods the *p* value can be calculated. Generally scientists interpret the *p* value obtained as follows: it is reasonable to conclude that the results observed are unlikely to have arisen by chance when the p value is less than 0.05 (meaning that there is less than a 5 per cent chance that the results observed are due to chance, in other words the results observed are due to chance in 1 out of 20 instances). Another approach used calculates the 95 per cent confidence intervals (CI); there is a 95 per cent chance that the true value is found between the lowest value and the highest value of the interval. Consider the following example: the observed value RR = 1.6 with 95 per cent CI of 0.8–2.4. The true value for the association lies between a RR = 0.8 and RR = 2.4 and this includes a RR = 1 (which implies no association between the exposure and the cases, as discussed above). Conversely, an observed value RR = 1.6 with 95 per cent CI of 1.4–2.0 is not compatible with a RR = 1; it is then reasonable to conclude that there is a true association between the exposure and the cases.

Another study design, the *case-control*, is based on the comparison of a group of subjects who are cases with a similar group of subjects (controls or referents) who are non-cases. The comparison group is therefore determined differently here than in the cohort study; here cases are compared with non-cases, versus exposed compared with non-exposed subjects in the cohort study. The presence of the exposure or study factor, i.e. the work factor under study (e.g. overhead work), is then assessed in both groups (cases and non-cases). Disease or case rates in this study design are meaningless, because the groups are determined by the researcher's selection criteria; therefore neither incidence or prevalence can be computed. What is meaningful, however, are the relative frequencies of subjects exposed to the study factor among the cases and non-cases; to determine the association of the exposure with the cases, the number of cases exposed to the study factor is compared with the number of non-cases who were also exposed to this same study factor using an odds ratio. The odds ratio (OR) can be seen as an estimate of the relative risk and can be interpreted as discussed above for the RR. For example, if the frequency of exposure is higher among cases, the odds ratio will exceed 1; if the OR is less than 1 then the frequency of exposure will be smaller among the cases than among the controls. As above, the *p* value or 95 per cent CI can be calculated for the OR.

The *cross-sectional* study design examines a subject population at one point in time; it does not follow up on the study population over time to look for new cases, as does the cohort study. Exposure to the study factor is established at the time of the investigation concurrently with the subjects who are cases. Those exposed can either be compared with non-exposed subjects from the same population or an external comparison population specifically selected for this purpose. The proportion of existing cases (prevalence) in the exposed and non-exposed groups can be calculated. Although a comparison could be established using the prevalence in the exposed and non-exposed groups, for reasons which will not be detailed here many epidemiologists prefer to use the odds ratio to estimate the risk associated with exposure in cross-sectional studies. The interpretation of the odds ratio, the *p* value and the 95 per cent confidence intervals has already been discussed above.

(Fletcher *et al.*, 1982; Friedman, 1987; Last, 1988; Checkoway *et al.*, 1989)

3.2.3 Results of epidemiologic studies

3.2.3.1 Shoulder tendinitis

The literature available was sparse. In the epidemiologic studies, shoulder tendinitis was defined as localized shoulder pain with tenderness over the humeral head (Hagberg and Wegman, 1987). Impingement syndrome was sometimes used as a synonym for shoulder tendinitis (Neer, 1983).

Of those studies selected and reviewed, there were cross-sectional studies (seven studies in total), case-control studies (one study only) and human experimental studies (one study only) reported (Table 3.2). Exposures were poorly defined. Cumulative exposure-response was not fully addressed by any study. A summary of the results of these studies is presented in Table 3.2 and some of these results are briefly discussed below.

Welders and plate workers had high prevalence rates (18 per cent and 16 per cent, respectively) of shoulder tendinitis (Table 3.2). The corresponding odds ratios were 13 and 11 when the welders and plate workers were contrasted with office workers with a prevalence rate of 2 per cent. In a case control study of male industrial workers, a similar odds ratio of 11 was found for exposure working with hands at or about shoulder level (Table 3.2). Assemblers with acute shoulder pain (myofascial syndrome and tendinitis) elevated their arms more frequently and for longer periods during work compared to controls (Bjelle *et al.*, 1981). Among exposed male and female industrial workers, Silverstein (1985) found a prevalence of 7.8 per cent of 'shoulder CTD' (shoulder tendinitis and degenerative joint disease) with an exposure association of 5.4 (Table 3.2). In the Silverstein study, exposure was defined as a work task with high demands on the wrist and hands in terms of force and/or repetition. Female students sustained transient shoulder tendinitis when they performed repetitive shoulder flexions. The flexion rate was 15 forward flexions per minute between 0 and 90 degrees during one hour (Table 3.2). A twofold risk of shoulder tendinitis was reported for folding, sewing, boarding and non-office workers when they were contrasted with knitting workers (Table 3.2), although whether knitting workers are suitable as referents is uncertain. In this report by McCormack *et al.* (1990) among the various exposed groups, the non-office workers consisting of cleaners and sweepers were regarded as the less exposed, but no ergonomic analysis of their work was conducted. Previous studies have shown that cleaners and sweepers have a hazardous exposure to the shoulder-neck region (Hagner and Hagberg, 1989).

Extensive work-like exposures may also occur in sports and thus there is a strong analogy between work related and sport-related tendinitis. Competitive swimmers with repetitive overhead arm movements are at risk for impingement syndrome (Kennedy *et al.*, 1978). Among professional athletes, an estimated 10 per cent of pitchers experienced shoulder tendinitis (Hill, 1983). The seasonal incidence of severe shoulder injuries among baseball players was 11 per cent (Hill, 1983). An estimated 16 per cent of these shoulder injuries were shoulder tendinitis (Hill, 1983). In a survey of 2496 swimmers in Canadian swimming clubs, 15 per cent reported significant shoulder disability, primarily due to impingement as it related to the butterfly and freestyle strokes (Hawkins and Kennedy, 1980). Tendinitis of the biceps was found in 6 per cent of the 84 best tennis players in the world (Priest and Nagel, 1976).

3.2.3.2 Epicondylitis

The literature selected and reviewed on lateral epicondylitis (tennis elbow) consisted of five cross-sectional studies and one cohort study. These are summarized in Table 3.3. Actual work exposure measurements for the finger extensor muscles have not been done in any study. In one study, there was an attempt to classify elbow stress at work into three categories, but the categorization was not based on an ergonomic analysis (Dimberg, 1987). Cumulative exposure-response was not addressed in any study. Lateral epicondylitis was defined as lateral elbow pain and tenderness over the lateral epicondyle, and in some reports pain at contraction (hand extensors) was also a required finding. Only the cohort study showed a clear increased risk for lateral epicondylitis among sausage makers, meat cutters and packers (Table 3.3). In one study (Dimberg, 1987) blue-collar workers had a lesser risk of epicondylitis than white-collar workers (Table 3.3).

There were two different studies showing cumulative exposure-response between tennis playing and lateral elbow pain, this lateral elbow pain being interpreted by the authors as tennis elbow. Out of 2633 tennis school participants, 694 had experienced elbow pain at some time in their playing history (Priest *et al.*, 1980). Factors associated with pain were age, and years and frequency of playing tennis (Priest *et al.*, 1980). Kitai *et al.* (1986) studied tennis elbow among 150 non-professional tennis players from local tennis clubs. They found that tennis elbow was associated with the number of playing hours per week even when controlling for playing years, age and age at the onset of pain.

3.2.3.3 De Quervain's tendinitis and hand-wrist tendinitis

Hand-wrist tendinitis is the inflammation of the tendons crossing the radiocarpal joint. Pain and tenderness over the tendons are the symptoms and signs required for diagnosis. De Quervain's tendinitis, a common hand-wrist tendinitis, is the inflammation of two of the tendons crossing the radiocarpal joint, namely the tendons to the long abductor and the short extensor muscle of the thumb. Sometimes the Finkelstein diagnostic test is required for diagnosis.

The literature on hand-wrist tendinitis comes from Finland and the USA. One cohort study and four cross-sectional studies were available for review. Most studies usually investigated hand-wrist tendinitis and de Quervain's tendinitis was generally not specified. In some of the cited studies, the authors did not distinguish between de Quervain's and other forms of hand-wrist tendinitis.

Table 3.2 Findings from selected studies on the work relatedness of shoulder tendinitis (Refer to Box 3.1 for help in interpreting the tables and Table 3.1 to find out how the studies were selected.)

Study population[1]	Outcome[2] and exposure information	Study design	Findings — On subjects (Prevalence)	Risk ratio OR	95% CI	Authors	Comments
Shipyard welders (n = 131 males) with more than five years' welding experience compared with office workers age 40 or above (n= 57 males).	• Workers with localized shoulder muscle fatigue were clinically investigated for tendinitis. • Welders worked with elevated arms, with hands at or above shoulder level. Exposure had been assessed for welders in another study by observation and EMG.	Cross-sectional	18% of the exposed were cases / 2% of the comparison group were cases	13	1.7–95	Herberts et al., 1981	• Adjustment for potential confounders not performed in the analysis. • The number of welding years was not associated with outcome. • No information provided on how representative the study population was of the welder population at the shipyard studied.
Shipyard plate workers (n = 188 males) compared with office workers age 40 or above (n = 57 males).	• Workers with localized shoulder muscle fatigue were clinically investigated for tendinitis. • Plate workers worked with elevated arms, with hands at or above shoulder level.	Cross-sectional	16% of the exposed were cases / 2% of the comparison group were cases	11	1.5–83	Herberts et al., 1984	• Adjustment for potential confounders not performed in the analysis. • No information provided on how representative the study population

Study population	Exposure / Outcome	Study design	Results (cases)	Results (comparison)	OR	95% CI	Reference	Comments
Male industrial workers belonging to an occupational health centre: 20 cases of degenerative shoulder tendinitis (3 cases of tendinitis were later excluded due to general inflammatory disorder) compared with 34 non-cases matched for age and workshop (2 non-cases for each case).	Exposure assessed for plate workers by observation and plant walk-through. • Exposure defined as work with hands at or above shoulder level. It was assessed by interview and observation of physician.	Case control	Of the cases, 11 had been exposed	Of the non-cases, 5 had been exposed	11	2.7–42	Bjelle et al., 1979	was of the plate worker population at the shipyard studied. • Adjustment for potential confounders not performed in the analysis.
Industrial workers: 212 males and 226 females exposed to high repetition and/or high force compared with	• Outcome defined as degenerative joint disease (acromioclavicular and glenohumeral joints), bicipital tenosynovitis, rotator cuff tendinitis, and frozen shoulder.	Cross-sectional	7.8% of the exposed were cases	1% of the comparison group were cases	5.4	1.3–23	Silverstein, 1985	• Adjustment for potential confounders (plant, not performed in the analysis. • Odds ratio adjusted for sex. • The study

Table 3.2 Work relatedness of shoulder tendinitis, (Cont.)

Study population[1]	Outcome[2] and exposure information	Study design	Findings — Prevalence — On subjects	Findings — OR — Risk ratio	Findings — OR — 95% CI	Authors	Comments
75 males and 61 females not exposed to high repetition or high force.	● Exposure categories based on repetition and force by hand/wrist (not shoulder), classified according to exposure measurements in a sample of workers.						population figures quoted represent a 90 per cent overall response rate from those originally selected.
Female students (n = 6) age 18–29 years.	● Outcome defined as tendinitis of acute stage 1 type (Neer, 1983). ● Exposure was repetitive shoulder flexions 0–90 degrees at a rate of 15 per min for 60 min with up to 3.1kg weight in hand.	Laboratory	2 cases (but all 6 subjects had tenderness of shoulder tendons)			Hagberg, 1981	● The 2 cases recovered within two weeks.
The exposed group comprised 163 female assembly line packers (representing 84% of the packaging	● Outcome consisted of humeral tendinitis. ● Packers performed repetitive arm work (repetitive motions up to 25 000 cycles per workday), had extreme work positions of the	Cross-sectional	9% of those exposed were cases — 4% of the comparison group were cases	2.6	0.91–7.4	Luopajärvi *et al.*, 1979	● Adjustment for potential confounders not performed in the analysis.

department in a food production firm). After the exclusion of 11 workers for previous trauma, arthritis or other pathologies, 152 females remained, who were compared with shop assistants in a department store, cashiers excluded ($n = 133$ females remaining after exclusion of 10 workers for trauma, arthritis and other pathologies). hands and arms, and shoulders were subject to static muscle work. Exposure assessed by observation, analysis of videos and interviews.

| Manufacturing workers: packaging/folding workers (41 males, 328 females) compared with knitting workers (203 males, 149 females). | • Outcome consisted of bursitis, bicipital tendinitis and impingement syndrome. • Exposure based on job categories, no actual exposure measurement or assessment. | Cross-sectional | 2.7% of those exposed were cases | 1.1% of the comparison group were cases | 2.4 | 0.8–7.8 | McCormack *et al.*, 1990 | • Adjustment for potential confounders (sex) not performed in the analysis. • All jobs combined, the study population figures quoted here represent 91% of the original sample |

Work related musculoskeletal disorders (WMSDs)

Table 3.2 Work relatedness of shoulder tendinitis, (Cont.)

Study population[1]	Outcome[2] and exposure information	Study design	Findings			Authors	Comments
			On subjects Prevalence	Risk ratio OR	95% CI		
							drawn. There was also a further 6.9% loss of subjects, from the population quoted here, for the clinical exam.
Manufacturing workers: sewing workers (28 males, 534 females) compared with knitting workers (203 males, 149 females).	• Outcome consisted of bursitis, bicipital tendinitis and impingement syndrome. • Exposure based on job categories, no actual exposure measurement or assessment.	Cross-sectional	2.5% of those exposed were cases 1.1% of the comparison group were cases	2.2	0.7–6.8	McCormack et al., 1990	• Adjustment for potential confounders (sex), not performed in the analysis. • All jobs combined, the study population figures quoted here represent 91% of the original sample drawn. There was also a further 6.9% loss of subjects, from the population quoted here, for the clinical exam.
Manufacturing workers: boarding	• Outcome consisted of bursitis, bicipital tendinitis and	Cross-sectional	2.4% of those exposed were cases 1.1% of the comparison group were	2.1	0.6–7.3	McCormack et al., 1990	• Adjustment for potential confounders (sex,

Population	Outcome/Exposure	Study design	Prevalence (exposed)	Prevalence (comparison)	OR	CI	Reference	Comments
workers (19 males, 277 females) compared with knitting workers (203 males, 149 females).	impingement syndrome. • Exposure based on job categories, no actual exposure measurement or assessment.		cases					not performed in the analysis. • All jobs combined, the study population figures quoted here represent 91% of the original sample drawn. There was also a further 6.9% loss of subjects, from the population quoted here, for the clinical exam.
Manufacturing workers: non-office workers (204 males, 264 females) compared with knitting workers (203 males, 149 females). Non-office workers were cleaners, sweepers, maintenance and transportation personnel.	• Outcome consisted of bursitis, bicipital tendinitis and impingement syndrome. • Exposure based on job categories, no actual exposure measurement or assessment.	Cross-sectional	2.1% of those exposed were cases	1.1% of the comparison group were cases	1.9	0.6–6.1	McCormack *et al.*, 1990	• Adjustment for potential confounders (sex), not performed in the analysis. • All jobs combined, the study population figures quoted here represent 91% of the original sample drawn. There was also a further 6.9% loss of subjects, from the population quoted here, for the clinical exam.
Female data entry workers (*n* = 104)	• Outcome consisted of humeral tendinitis. • Exposure based on job	Cross-sectional	1%	1%	0.54	0.03–8.9	Kukkonen *et al.*, 1983	

Table 3.2 Work relatedness of shoulder tendinitis, (Cont.)

Study population [1]	Outcome [2] and exposure information	Study design	Findings					Authors	Comments
			On subjects		Risk ratio	95% CI			
			Prevalence		OR				
compared with female workers in varying office tasks ($n = 57$).	categories, no actual exposure measurement or assessment.								
From 119 workers employed at a slaughterhouse, 113 workers (cutters $n = 52$, butchers $n = 38$, meat by-product workers $n = 23$) participated ($n = 82$ males, 31 females).	• Outcome consisted of supraspinous tendinitis. • Exposure based on job categories.	Cross-sectional	N/A [3]	3%	N/A [3]	N/A [3]		Viikari-Juntura, 1983	• No comparison group for shoulder tendinitis was used; only the prevalence was sought.

[1] Based on the information available in the article, no subject was lost from the original study poulation when only one set of numbers appears in the table under the 'study population' or when there is no relevant information in the 'comments' column.
[2] Unless otherwise specified, all outcomes in Table 3.2 were shoulder tendinitis and were identified by medical examination.
[3] N/A: Not available.

Table 3.3 Findings from selected studies on the work relatedness of lateral epicondylitis (tennis elbow) (Refer to Box 3.1 for help in interpreting the tables and Table 3.1 to find out how the studies were selected.)

Study population[1]	Outcome[2] and exposure information	Study design	Findings		Risk ratio RR	95% CI	Authors	Comments
			Incidence					
			Exposed subjects	Comparison group				
Sausage makers (n = 107 females) compared with employees in non-strenuous jobs, e.g. office workers and supervisors (n = 197 females).	• Exposure defined as having work tasks strenuous to the muscle-tendon structures of the upper limb. Exposure data obtained from previous published literature and plant walk-throughs.	Cohort: 31-month follow-up	11.1 cases per 100 person-years	1.1 cases per 100 person-years	10.3	N/A[3]	Kurppa et al., 1991	• Adjustment for potential confounders not performed in the analysis. • Loss to follow-up occurred through job transfers or termination of employment; however it is not clear how much, though it appears minor.
Meat cutters (n = 102 males) compared with employees in non-strenuous jobs, e.g. office workers, maintenance men and supervisors (n = 141 males).	• Exposure defined as having work tasks strenuous to the muscle-tendon structures of the upper limb. Exposure data obtained from previous published literature and plant walk-throughs.	Cohort: 31-month follow-up	6.4 cases per 100 person-years	0.9 cases per 100 person-years	7.1	N/A[3]	Kurppa et al., 1991	• Adjustment for potential confounders not performed in the analysis. • Loss to follow-up occurred through job transfers or termination of employment; however it is not

Table 3.3 Work relatedness of lateral epicondylitis (Cont.)

Study population[1]	Outcome[2] and exposure information	Study design	Findings				Authors	Comments
			Incidence		RR			
			Exposed subjects	Comparison group	Risk ratio	95% CI		
Packers (n = 118 females) compared with employees in non-strenuous jobs, eg office workers and supervisors (n = 197 females).	• Exposure defined as having work tasks strenuous to the muscle-tendon structures of the upper limb. Exposure data obtained from previous published literature and plant walk-throughs.	Cohort: 31-month follow-up	7.0 cases per 100 person-years	1.1 cases per 100 person-years	6.4	N/A	Kurppa *et al.*, 1991	clear how much, though it appears minor. • Adjustment for potential confounders not performed in the analysis. • Loss to follow-up occurred through job transfers or termination of employment; however it is not clear how much, though it appears minor.
			Prevalence		OR			
Meat cutters (n = 90 males) compared with construction foremen not exposed to repetitive	• Exposure defined as overstrain of the extensors and flexors of the wrist and fingers while cutting frozen meat. Exposure data obtained from previous	Cross-sectional	8.9%	1.4%	6.9	0.85–57	Roto and Kivi, 1984	

movement of the upper limbs (n = 77 males). All meat cutters and 72 of the 77 non-exposed subjects participated. | published literature and plant walk-throughs. | | | | | |

Study population	Exposure assessment	Design			OR	CI	Reference	Comments
Manufacturing workers: packaging/folding workers (41 males, 328 females) compared with knitting workers (203 males, 149 females).	• Exposure based on job categories, no actual exposure measurement or assessment.	Cross-sectional	2.2%	1.4%	1.5	0.5–4.7	McCormack et al., 1990	• Adjustment for potential confounders (sex), not performed in the analysis. • All jobs combined, the study population figures quoted here represent 91% of the original sample drawn. There was also a further 6.9% loss of subjects, from the population quoted here, for the clinical exam.
Manufacturing workers: sewing workers (28 males, 534 females) compared with knitting workers (203 males, 149 females).	• Exposure based on job categories, no actual exposure measurement or assessment.	Cross-sectional	2.1%	1.4%	1.5	0.5–4.3	McCormack et al., 1990	• Adjustment for potential confounders (sex), not performed in the analysis. • All jobs combined, the study population figures quoted here represent 91% of the

Table 3.3 Work-relatedness of lateral epicondylitis (Cont.)

Study population[1]	Outcome[2] and exposure information	Study design	Findings		Risk ratio	95% CI	Authors	Comments
			Prevalence		OR			
			Exposed subjects	Comparison group				
Manufacturing workers: non-office workers (204 males, 264 females) compared with knitting workers (203 males, 149 females). Non-office workers were cleaners, sweepers, maintenance and transportation personnel.	• Exposure based on job categories, no actual exposure measurement or assessment.	Cross-sectional	1.9%	1.4%	1.4	0.5-4.1	McCormack *et al.*, 1990	original sample drawn. There was also a further 6.9% loss of subjects, from the population quoted here, for the clinical exam. • Adjustment for potential confounders (sex), not performed in the analysis. • All jobs combined, the study population figures quoted here represent 91% of the original sample drawn. There was also a further 6.9% loss of subjects, from the population quoted here, for the clinical exam.

Study population	Exposure assessment	Study design					Reference	Comments
Manufacturing workers: boarding workers (19 males, 277 females) compared with knitting workers (203 males, 149 females).	Exposure based on job categories, no actual exposure measurement or assessment.	Cross-sectional	1.0%	1.4%	0.7	0.2–3.0	McCormack et al., 1990	Adjustment for potential confounders (sex), not performed in the analysis. All jobs combined, the study population figures quoted here represent 91% of the original sample drawn. There was also a further 6.9% loss of subjects, from the population quoted here, for the clinical exam.
Meat cutters (n = 91 males), packers (n = 97 females, 22 males), sausage makers (n = 95 females, 17 males) compared with employees in non-strenuous jobs, e.g. office workers, maintenance men and supervisors (n = 124 males,	Exposure defined as highly repetitive work and tasks strenuous to the muscle-tendon structures of the upper limb. Exposure data obtained from previous published literature and plant walk-throughs.	Cross-sectional	0.6%	0.5%	1.2	N/A[3]	Viikari-Juntura et al., 1991	The diagnosis of epicondylitis was established in three cross-sectional investigations within 19 months.

Table 3.3 Work relatedness of lateral epicondylitis (Cont.)

Study population[1]	Outcome[2] and exposure information	Study design	Findings Prevalence Exposed subjects	Findings Prevalence Comparison group	Risk ratio OR	95% CI	Authors	Comments
164 females). The above subjects participated in the clinical examination and represent 94% of the original study population.								
The exposed group comprised 163 female assembly line packers (representing 84% of the packaging department in a food production firm). After the exclusion of 11 workers for previous trauma, arthritis or other	● Packers performed repetitive arm work (repetitive motions up to 25 000 cycles per workday), had extreme work positions of the hands and arms, and shoulders were subject to static muscle work. Exposure assessed by observation, analysis of videos and interviews.	Cross-sectional	2.6%	2.3%	1.2	0.3–5.3	Luopajärvi et al., 1979	● Adjustment for potential confounders not performed in the analysis.

Study population[1]	Exposure	Study design	Prevalence	OR/RR	95% CI	Reference	Comments
pathologies, 152 females remained, who were compared with shop assistants in a department store, cashiers excluded ($n = 133$ females remaining after exclusion of 10 workers for trauma, arthritis and other pathologies).							
Engineering industrial blue-collar workers ($n = 340$ males and females) compared with white-collar workers ($n = 200$ males and females). The numbers quoted represent a 99% response rate from the original sample.	• Elbow stress classification done by occupational health personnel. No actual measurements or observations.	Cross-sectional	5.3%	0.7	0.3–1.2	Dimberg, 1987	• No association was found between epicondylitis and elbow stress.
			11%				• Adjustment for potential confounders (gender, tennis playing) not performed in the analysis.

[1]Based on the information available in the article, no subject was lost from the original study population when only one set of numbers appears in the table under the 'study population' or when there is no relevant information in the 'comments' column.
[2]Unless otherwise specified, all outcomes in Table 3.3 were epicondylitis and were identified by medical examination.
[3]N/A = Not available

Tendinitis of tendons other than those engaged in de Quervain's disease is uncommon. Also, the term 'muscle tendon syndrome at the wrist' used in one study may be regarded as hand-wrist tendinitis (Kuorinka and Koskinen, 1979).

Results of the epidemiologic studies are shown in Table 3.4 and are, in part, discussed hereafter. High risks were detected among meat processing workers (risk ratio 36 for packers and 24 for sausage makers) in the cohort study (Table 3.4). Cross-sectional studies also showed increased risks of hand-wrist tendinitis in the meat industry and in other manufacturing industries (Table 3.4). In a study of the manufacturing workforce in different plants in the United States, an odds ratio of between three and eight was observed, with a prevalence rate of between 0.9 per cent and 6.4 per cent for different occupational groups (McCormack *et al.*, 1990). An indication of exposure-response was seen in a study of industrial workers where exposure in terms of repetition and force was categorized into either high or low (Silverstein, 1985). Exposure to high repetition gave an odds ratio of 3.3, high force gave an odds ratio of 6.1, and if the exposure consisted of both high repetition and high force the odds ratio for hand-wrist tendinitis was 29 (Table 3.4). Although no epidemiologic studies are available, hand-wrist tendinitis and de Quervain's tendinitis have been associated with tennis, racquetball, squash and badminton. For a review see Osterman *et al.* (1988).

3.2.3.4 Dupuytren's contracture

Dupuytren's contracture is a degeneration of the palmar fascia in the hand, which is the distal part of the tendon to the palmaris longus muscle (Figure 3.1).

Male industrial workers (n = 216) whose work involved bagging and packing were examined for the occurrence of Dupuytren's contracture, in addition to a group of control workers at the same plant whose work did not involve bagging and packing (n = 84) (Bennett, 1982). The exposed workers had a 7.4 per cent prevalence of Dupuytren's contracture. In the control group, only one case was seen. Age-adjusted standardized morbidity ratios (SMRs) for the bagging and packing workers were calculated, using Dupuytren's contracture frequency data obtained from 4374 manual workers engaged in light to heavy manual work. The observed and expected figures for the exposed subjects were 16 and 8.08, respectively, yielding an SMR of 198. This is the only study indicating that Dupuytren's contracture may be related to work exposure. Hueston (1963), in a review of studies done between 1912 and 1962, concluded that occupation was not related to Dupuytren's contracture. In this review, two studies indicated an association between Dupuytren's contracture and the job title 'brewery worker'; however, this association was explained by a greater frequency of alcoholism among brewery workers. Furthermore, two other studies, in 1951 and 1962, with large study populations (5000 locomotive works workers, 1000 steel workers, 1000 miners and 1000 clerks), failed to show any relationship between Dupuytren's contracture and job title, according to Hueston (1963). Evidence for or against the work relatedness of Dupuytren's contracture is sparse. This disorder will therefore not be examined for causality in the next subsection (3.2.4), and will be only briefly mentioned in the conclusion (3.2.5).

3.2.3.5 Achilles tendinitis

Epidemiologic studies on Achilles tendinitis are rare. Most scientific articles about Achilles tendinitis concern surgical procedures, case reports and treatment. Epidemiologic reports about Achilles tendinitis are also lacking in sports medicine.

Ballet dancers have been singled out as a risk group for Achilles tendinitis (Fernandez-Palazzi *et al.*, 1990). Three professional dance companies (*n* = 42) were studied for three years in Caracas (Fernandez-Palazzi *et al.*, 1990). Achilles tendinitis was found in 13 dancers (31 per cent). The exposure factors identified were repetitive stretching and contraction of the tendon, and impact force on the tendon when jumping. Among athletes seeking medical care, Achilles tendinitis is a common problem with a high incidence (Leppilahti *et al.*, 1991). Tendinitis has been induced in rabbits' Achilles tendons by electrical stimulus leading to repetitive contractions (Rais, 1961). Evidence for or against the work relatedness of Achilles tendinitis is sparse. This disorder will therefore not be examined for causality in the next subsection (3.2.4), and will be only briefly mentioned in the conclusion (3.2.5).

3.2.4 Results and causality assessment

3.2.4.1 Strength of the association

The strength of the association in the reviewed studies was generally high for *shoulder* and *hand-wrist tendinitis*. There were many different studies reporting odds ratios of greater than five with confidence intervals above one, indicating effects not explained by chance. Such a large effect cannot be easily explained by any sources of bias (e.g. a confounder), especially biases that were not evident to the investigators. It should also be noted that the confidence intervals for these two disorders were generally wide, indicating that the risk ratios were approximations.

The magnitude of risk ratio observed implies that a high percentage of tendinitis may be attributable to the exposure factor among the exposed in working populations. For example, for shoulder tendinitis in the Silverstein study (1985) an attributable fraction of 0.81 (95 per cent CI 0.23–0.96) can be calculated, in the Bjelle *et al.* study (1979) the attributable fraction is 0.90 (95 per cent CI 0.62–0.98) and for the Herberts *et al.* study (1981) this value is 0.92 (95 per cent CI 0.41–0.99). See appendix III for more information on attributable fractions.

The epidemiologic literature is not convincing on the work relatedness of *lateral epicondylitis*. The cohort study and the literature on tennis elbow among tennis players do show results in favour of such a relationship. In general, cohort

Table 3.4 *Findings from selected studies on the work relatedness of hand-wrist tendinitis (Refer to Box 3.1 for help in interpreting the tables and Table 3.1 to find out how the studies were selected.)*

Study population [1]	Outcome [2] and exposure information	Study design	Findings						Authors	Comments
			Incidence		Risk ratio	95% CI				
			Exposed subjects	Comparison group	RR					
Packers (n = 118 females) compared with employees in non-strenuous jobs, e.g. office workers and supervisors (n = 197 females).	• Exposure defined as having work tasks strenuous to the muscle-tendon structures of the upper limb. Exposure data obtained from previous published literature and plant walk-throughs.	Cohort: 31-month follow-up	25.3 cases per 100 person-years	0.7 cases per 100 person-years	36		N/A [3]		Kurppa *et al.*, 1991	• Adjustment for potential confounders not performed in the analysis. • Loss to follow-up occurred through job transfers or termination of employment; however it is not clear how much, though it appears minor.
Sausage makers (n = 107 females) compared with employees in non-strenuous jobs, e.g. office workers and supervisors (n = 197 females).	• Exposure defined as having work tasks strenuous to the muscle-tendon structures of the upper limb. Exposure data obtained from previous published literature and plant walk-throughs.	Cohort: 31-month follow-up	16.8 cases per 100 person-years	0.7 cases per 100 person-years	24		N/A [3]		Kurppa *et al.*, 1991	• Adjustment for potential confounders not performed in the analysis. • Loss to follow-up occurred through job transfers or termination of employment; however it is not

Population	Exposure assessment	Study design	Prevalence		OR		Reference	Comments
Meat cutters (n = 102 males) compared with employees in non-strenuous jobs, e.g. office workers, maintenance men and supervisors (n = 141 males).	• Exposure defined as having work tasks strenuous to the muscle-tendon structures of the upper limb. Exposure data obtained from previous published literature and plant walk-throughs.	Cohort: 31-month follow-up	12.5 cases per 100 person-years	0.9 cases per 100 person-years	14	N/A[3]	Kurppa et al., 1991	clear how much, though it appears minor. • Adjustment for potential confounders not performed in the analysis. • Loss to follow-up occurred through job transfers or termination of employment; however it is not clear how much, though it appears minor.
Industrial workers with high force and high repetition (n = 68 males, 74 females) compared with industrial workers with low force and low repetition (n = 75 males, 61 females).	• Repetition and force used by hands, classified according to exposure measurements in a sample of workers.	Cross-sectional	12%	1%	29	N/A[3]	Silverstein, 1985	• Plant-adjusted odds ratio. • The study population figures quoted represent a 90% overall response rate from those originally selected.

Table 3.4 Work-relatedness of hand-wrist tendinitis (Cont.)

Study population[1]	Outcome[2] and exposure information	Study design	Findings					Authors	Comments
			Prevalence		Risk ratio	95% CI			
			Exposed subjects	Comparison group	OR				
Industrial workers with high force and low repetition (n = 101 males, 52 females) compared with industrial workers with low force and low repetition (n = 75 males, 61 females).	• Repetition and force used by hands, classified according to exposure measurements in a sample of workers.	Cross-sectional	4%	1%	6.1	N/A[3]	Silverstein, 1985	• Plant-adjusted odds ratio. • The study population figures quoted represent a 90% overall response rate from those originally selected.
Industrial workers with low force and high repetition (n = 43 males, 100 females) compared with industrial workers with low force and low repetition (n = 75 males, 61 females).	• Repetition and force used by hands, classified according to exposure measurements in a sample of workers.	Cross-sectional	3%	1%	3.3	N/A	Silverstein, 1985	• Plant-adjusted odds ratio. • The study population figures quoted represent a 90% overall response rate from those originally selected.

The exposed group comprised 163 female assembly line packers (representing 84% of the packaging department in a food production firm). After the exclusion of 11 workers for previous trauma, arthritis or other pathologies, 152 females remained, who were compared with shop assistants in a department store, cashiers excluded (*n* = 133 females remaining after exclusion of 10 workers for trauma,

• Packers performed repetitive arm work (repetitive motions up to 25 000 cycles per workday), had extreme work positions of the hands and arms, and shoulders were subject to static muscle work. Exposure assessed by observation, analysis of videos and interviews.

Cross-sectional

53%

14%

7.1

3.9–12.8 Luopajärvi *et al.*, 1979

• Adjustment for potential confounders not performed in the analysis.

Table 3.4 Work relatedness of hand-wrist tendinitis (Cont.)

Study population[1]	Outcome[2] and exposure information	Study design	Findings Prevalence Exposed subjects	Comparison group	Risk ratio OR	95% CI	Authors	Comments
arthritis and other pathologies).								
Manufacturing workers: boarding workers (19 males, 277 females) compared with knitting workers (203 males, 149 females).	● Exposure based on job categories, no actual exposure measurement or assessment.	Cross-sectional	6.4%	0.9%	8.0	2.3–27	McCormack et al., 1990	● Adjustment for potential confounders (sex), not performed in the analysis. ● All jobs combined, the study population figures quoted here represent 91% of the original sample drawn. There was also a further 6.9% loss of subjects, from the population quoted here for the clinical exam.
Manufacturing workers: sewing workers (28 males, 534 females)	● Exposure based on job categories, no actual exposure measurement or assessment.	Cross-sectional	4.4%	0.9%	5.4	1.6–18	McCormack et al., 1990	● Adjustment for potential confounders (sex), not performed in the analysis.

compared with knitting workers (203 males, 149 females).					• All jobs combined, the study population figures quoted here represent 91% of the original sample drawn. There was also a further 6.9% loss of subjects, from the population quoted here, for the clinical exam.
Manufacturing workers: packaging/folding workers (41 males, 328 females) compared with knitting workers (203 males, 149 females).	• Exposure based on job categories, no actual exposure measurement or assessment. Cross-sectional	3.3%	0.9%	3.9	1.1–14	McCormack *et al.*, 1990	• Adjustment for potential confounders (sex), not performed in the analysis. • All jobs combined, the study population figures quoted here represent 91% of the original sample drawn. There was also a further 6.9% loss of subjects, from the population quoted here, for the clinical exam.

Table 3.4 Work relatedness of hand-wrist tendinitis (Cont.)

Study population[1]	Outcome[2] and exposure information	Study design	Findings		Risk ratio OR	95% CI	Authors	Comments
			Exposed subjects	Comparison group				
			Prevalence					
Manufacturing workers: non-office workers (204 males, 264 females) compared with knitting workers (203 males, 149 females). Non-office workers were cleaners, sweepers, maintenance and transportation personnel.	• Exposure based on job categories, no exposure measurement or assessment.	Cross-sectional	2.1%	0.9%	2.5	0.7–9.3	McCormack et al., 1990	• Adjustment for potential confounders (sex), not performed in the analysis. • All jobs combined, the study population figures quoted here represent 91% of the original sample drawn. There was also a further 6.9% loss of subjects, from the population quoted here, for the clinical exam.
Out of 115 industrial workers: scissor makers, 93 (n = 90 females, 3 males) were examined and	• Exposure determined by job analysis: a work history (including calculating number of parts handled) and a work method analysis were	Cross-sectional	18%	14%	1.4	0.7–2.9	Kuorinka and Koskinen, 1979	• Tendinitis related to number of parts handled. • Adjustment for potential confounders (sex), not performed

in the analysis.

also completed the work history and production data collection. These subjects were compared with shop assistants (*n* = 133 females) in a large department store.

done. Grasping with the fingers wide open and the use of force were considered important.

[1] Based on the information available in the article, no subject was lost from the original study population when only one set of numbers appears in the table under the 'study population' or when there is no relevant information in the 'comments' column.

[2] Unless otherwise specified, all outcomes in Table 3.4 were tendinitis and were identified by medical examination.

[3] N/A = Not available

Box 3.2 Specificity of the association for all the musculoskeletal disorders discussed in chapter 3

Specificity in the cause is the precision with which a given outcome (musculoskeletal disorder of interest here) will always result from a given cause under investigation (work related factor); specificity in the effect is the precision with which a given cause (work related factor) will predict a given outcome (musculoskeletal disorder).

The specificity in the cause is expected to be low for many of the musculo-skeletal disorders discussed in this chapter because of their multi-factorial nature. For many of these disorders, there is an underlying pathological process (e.g. compression or ischemia) that could be caused by work as well as by acute trauma and/or systemic disease. For example, rheumatic and general inflammatory diseases, as well as ergonomic exposures, may trigger, exacerbate or aggravate tendinitis. Both anatomical malformations and occupational exposure may cause compression and induce the thoracic outlet syndrome. Bursitis may be caused by general inflammatory diseases, such as rheumatoid arthritis, as well as by work related factors. Besides occupational exposures, sports exposure (Vingård, 1991), high body mass index (Rissanen *et al.*, 1990), age (Danielsson *et al.*, 1984) and previous injury (Kohatsu and Schurman, 1991), as well as systemic factors, may all result in the underlying pathogenic process leading to polyarthroses or generalized arthroses. Therefore work would not be expected to be the only initiator of the pathological processes involved in the development of the musculoskeletal disorders discussed in this chapter.

The specificity of the effect is low for all musculoskeletal disorders examined in this chapter. A specific work related factor can cause a number of different musculoskeletal disorders. For example, work factors found associated with the thoracic outlet syndrome can also be associated with shoulder tendinitis and tension neck syndrome.

---

studies are better designed than cross-sectional studies and provide stronger evidence. On the other hand, the strength of the association found in the cross-sectional studies on lateral epicondylitis was weak.

3.2.4.2 Specificity of the association

See Box 3.2.

3.2.4.3 Temporal association

The criterion of temporality, which means that the exposure precedes the onset of the disorder, was not demonstrated in all studies. The latency between onset of exposure and tendinitis has not been addressed by any researcher.

3.2.4.4 Consistency of the association

There is consistency (replicability) across studies from a range of industries and occupations showing increased risk of *shoulder tendinitis* with repetitive and overhead work. Overhead work results in a high load on the rotator cuff tendons in the shoulder. Work relatedness of shoulder tendinitis has been shown by a variety of different study designs: cross-sectional, case-control and experimental studies. For *hand-wrist tendinitis*, forceful and repetitive gripping is described as a hazardous factor across studies from a range of industries and occupations. Consistency is not found for *lateral epicondylitis*; numerous studies showing no relationship to work have been published.

3.2.4.5 Predictive performance of the association

Whether observed associations can predict unknown facts, consequent to the associations, has not yet been tested for tendon disorders. For *hand-wrist tendinitis*, different studies have shown a cumulative exposure-response relationship where exposure was measured as duration of employment. The number of parts handled was found to be associated with *hand-wrist tendinitis* (Kuorinka and Koskinen, 1979). High repetition and high force, in combination, were shown to be more important as risk factors for *hand-wrist tendinitis* than the two exposure factors separately (Silverstein, 1985). Cumulative exposure-response has not been shown for either *shoulder tendinitis* or *lateral epicondylitis*.

3.2.4.6 Coherence of the association observed with current knowledge

COHERENCE OF THE ASSOCIATION: SHOULDER TENDINITIS
The pathogenesis of shoulder tendinitis is fairly well understood. The predisposing factor for shoulder tendinitis is often degeneration. The initial form of degeneration is cell death within the tendon, forming debris in which calcium may deposit. Degeneration of the tendon is caused by impairment of the blood perfusion and nutrition, in addition to mechanical stress. The tendons to the supraspinate, the biceps brachii (long head), and the upper parts of the infraspinate muscle have a zone of avascularity. Signs of degeneration such as cell death, calcium deposits and microruptures are located predominantly in this area of avascularity.

How can degeneration occur? Compression and static tension of the shoulder tendons can cause impairment of circulation, and thus accelerated degeneration. Compression of the tendons occurs when the arm is elevated (e.g. while work is done above shoulder level). During elevation of the arm, the rotator cuff tendons and the insertions on the greater tuberosity are forced under the coracoacromial arch; this process of compression is often referred to as impingement. Compression of the rotator cuff tendons (especially the supraspinate tendon) then results, because of the narrow space between the humeral head and the tight coracoacromial arch (Figure 3.3). Patients with long-term disability due to chronic bursitis or complete or partial tears of the rotator cuff tendons or biceps brachii usually have the impingement syndrome. Besides

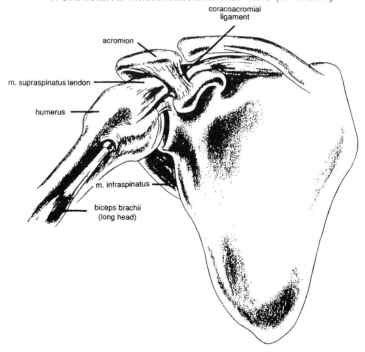

FRONTAL VIEW

Figure 3.3 Compression of the rotator cuff tendon.

being affected by impingement, circulation in the tendon is also inversely proportional to muscle tension and ceases at greater tensions. Studies (Järvholm *et al.*, 1988) have shown that, at 30 degrees of forward flexion or abduction in the shoulder joint, the intramuscular pressure in the supraspinate muscle exceeds 30 mm mercury (Figure 3.4). At this pressure level, impairment of blood circulation will occur. Since the major blood vessel supplying the supraspinate tendon runs through the supraspinate muscle, it is likely that the circulation of the tendon may already be disturbed at 30 degrees of forward flexion or abduction in the shoulder joint. Many workers in industry have tasks that involve such flexion or abduction.

The degenerated tendon. Once the tendon is degenerated, exertion may trigger an inflammatory 'foreign body' response to the debris of dead cells, resulting in active tendinitis. Also, an infection (viral or bacterial, e.g. urogenital: urinary tract infection, gonorrhoea etc.) or systemic inflammation may predispose a subject to reactive tendinitis in the shoulder. One hypothesis to explain this predisposition is that an infection that activates the immune system also increases the possibility of a 'foreign response' to the degenerative structures in the tendon.

It can therefore be seen how work related exposures could trigger mechanisms leading to shoulder tendinitis. Thus an association between tendinitis and work exposure is coherent with our current knowledge.

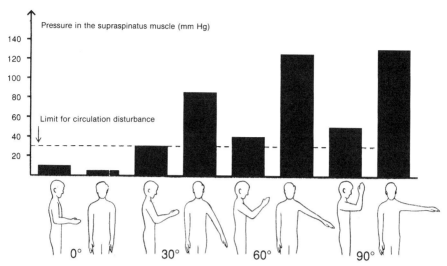

Figure 3.4 Pressure in the supraspinatus muscle according to arm and shoulder position.

COHERENCE OF THE ASSOCIATION: LATERAL EPICONDYLITIS

The pathogenesis of lateral epicondylitis is still obscure. The predominant hypothesis is that microruptures occur at the attachment of the muscle to the bone, more specifically between the tendon attachment (insertion) and the periosteum of the bone, causing inflammation (Figure 3.2). The microruptures could be due to repetitive high force, such as that found in some work situations, which exceeds the strength of the collagen fibres of the tendon attachment. As a consequence of repetitive trauma, these microtears, usually located in the origin of the extensor carpi radialis brevis, lead to the formation of fibrosis and granulation tissue (Leach and Miller, 1987).

COHERENCE OF THE ASSOCIATION: HAND-WRIST TENDINITIS

Ligaments hold tendons in place at the wrist either by securing the tendon sheath in place, or by forming compartments for the tendons to run through. The first dorsal wrist compartment may be narrow and friction may result, causing inflammation and swelling of the tendons. For example, work tasks that require forceful repetitive movements may result in friction in the first dorsal wrist compartment. Thus an association with work would be coherent with current knowledge regarding the development of hand-wrist tendinitis.

3.2.5 Conclusion

Tendinitis at the shoulder, elbow (lateral epicondyle), wrist and, to a lesser extent, hand (Dupuytren's contracture) and ankle (Achilles tendinitis) have been associated with occupational exposure. However, the epidemiologic literature is most convincing on the work relatedness of *shoulder* and *hand-wrist tendinitis*.

The strength of the association in the reviewed studies was generally high for *shoulder* and *hand-wrist tendinitis*. There were many different studies reporting odds ratios of greater than five. There is consistency across studies showing increased risk of *shoulder tendinitis* with repetitive and overhead work. Increased risk of *hand-wrist tendinitis* with repetitive and forceful gripping has also been consistently observed across studies. The pathogenesis of *shoulder tendinitis* is fairly well understood, and shows coherence with work related exposures. The possibility of an association between work and *hand-wrist tenditinis* is also coherent with current knowledge.

The epidemiologic literature does not make a convincing case for *lateral epicondylitis* (tennis elbow) being work related. The cohort study and the literature on tennis elbow among tennis players are in favour of such a relationship, and it has been mentioned that cohort studies generally provide stronger evidence than cross-sectional designs. In the selected and reviewed cross-sectional studies, however, the strength of the relationship was weak. Whether *Dupuytren's contracture* is work related is uncertain. There is one recent study showing work relatedness, but many other studies have failed to show any relationship between occupation and the disease. It is likely that *Achilles tendinitis* is work related in specific jobs where there is high exposure, such as in ballet dancing.

Occupational hazards for tendinitis show similarities to hazards in sports activities. Sport-related tendinitis has also been reported among swimmers (shoulder), baseball players (shoulder) and tennis players (elbow).

3.3 Evidence of the association between work and selected peripheral nerve disorders: carpal tunnel syndrome, thoracic outlet syndrome, radiculopathy, vibration neuropathy

3.3.1 Introduction to nerve entrapments and vibration neuropathy

The peripheral nerves that carry signals to and from the central nervous system are nerve trunks consisting of nerve fibres and connective tissue (Figure 3.5). The nerve fibres are packed within the endoneural connective tissue inside fascicles. The fascicles are surrounded by connective tissue (perineurium), a sheath of considerable mechanical strength, and are embedded in an epineurium that supports and protects the nerve fibres. The epineurium carries the intraneural microvascular system – the epineurial vessels. Several fascicles are usually grouped together in bundles constituting well-defined sub-units of the nerve trunk (Figure 3.5). The nerve fibres consist of a nerve cell body with one long extension: the axon. Motor nerve cells in the spinal cord may have axons longer than one metre to innervate muscle fibres in the foot. The axons are

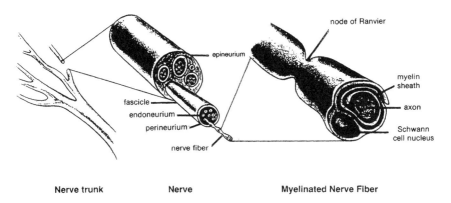

Figure 3.5 Structure of a nerve.

surrounded by Schwann cells. In myelinated nerve fibre, only one Schwann cell is wrapped around the axon in contrast to non-myelinated nerve fibre, where the Schwann cell is wrapped around several axons. In the axons, both anterograde as well as retrograde intercellar transport of materials occurs (Lundborg, 1988).

Disturbance of the axonal transport system is probably one of the mechanisms for entrapment neuropathies. The term 'nerve entrapment' refers to a pathological condition caused by an incompatibility between the volume of a peripheral nerve structure and the anatomical space available to the nerve structure (Lundborg, 1988).

In acute injuries, the nerve fibre may be damaged, resulting in symptoms observed in the nerve and usually loss of both motor and sensory functions. An example of this type of acute injury is a cut at the palmar side of the wrist where the median nerve is superficially located. Mechanical trauma may also cause acute structural damage to small nerves and receptors.

3.3.1.1 Entrapment mechanism

Nerve entrapments may be due to cumulative trauma over a prolonged period. While entrapment is occurring, the nerve is injured because there is increased pressure on it. In the early stages, this increased pressure will impair blood flow, oxygenation and the axonal transport system. If the pressure is high, mechanical blocking of the depolarization process in the nerve will occur (Lundborg, 1988). Increased pressure on the nerve can be caused by edema of the surrounding tissues (i.e. swelling) in the canals through which a nerve runs, such as the carpal tunnel. Edema may result from mechanical irritation of surrounding structures such as tendons or muscles. If a nerve is entrapped at one point there may be increased susceptibility to mechanical pressure both distal and proximal to the entrapment point. This increased distal and proximal susceptibility is caused by the impairment in the retrograde and anterograde axonal flows. Thus multiple entrapments along the same nerve are somewhat common; for example, it is not uncommon to suffer from both radiculopathy and carpal tunnel syndrome at the same time. This phenomenon of multiple entrapments along the same nerve is called double crush syndrome (Lundborg, 1988).

3.3.1.2 Vibration neuropathy

Workers exposed to vibration are at risk for carpal tunnel syndrome (see subsection 3.3.2 on carpal tunnel syndrome); however, only part of the neuropathy observed in the hands among workers exposed to vibration can be explained by entrapment of the median nerve at the wrist (Hagberg *et al.*, 1991). Loss of sensory function and tactile gnosis with paresthesia in the hands at night have been found in vibration-exposed workers (Lundström and Johansson, 1986). It may be that vibration causes intraneural edema in addition to structural damage of small nerves and damage to the tactile receptors in the hand (Lundborg *et al.*, 1990). Although there are possible links to be explored between work and vibration neuropathy, this disorder is mentioned only briefly here in the introduction.

Section 3.3 examines selected nerve entrapment disorders as examples of the relationship between work and peripheral nerve disorders. Evidence for the work relatedness of carpal tunnel syndrome, thoracic outlet syndrome and radiculo-pathy is reviewed in subsections 3.3.2, 3.3.3 and 3.3.4, respectively (Figures 3.1 and 3.6). There are a number of other nerve entrapment disorders that may be work related, but studies providing evidence of such work relatedness are currently lacking. Examples of such nerve entrapments are listed below:

- Pronator syndrome, a compression of the median nerve in the elbow region by a tendinous band of the pronator teres muscle.
- Ulnar nerve entrapment in the elbow region, the second most frequent upper-extremity entrapment neuropathy. This cubital tunnel syndrome is caused by constriction of the aponeurosis of the flexor carpi ulnaris muscle.
- Another ulnar nerve entrapment, this time at the wrist in the Guyon canal.
- The posterior interosseal nerve, which is a branch of the radial nerve, may be entrapped at the Frohse's arcade, just distal to the elbow.
- Digital nerves may be entrapped in the hand. The entrapment may be caused by trauma, cyst osteophytes or tumours.
- The sciatic nerve may be entrapped at the spine (Spitzer *et al.*, 1987).
- The peroneal nerve may be entrapped as it passes through a fibrosis tunnel between the edge of the peroneus longus muscle and the fibula.
- Tarsal tunnel syndrome, an entrapment of the posterior tibial nerve.
- There is also anterior tarsal tunnel syndrome, caused by an entrapment of the deep peroneal nerve in the foot.

There are a number of other entrapment syndromes in addition to the above, but they are rare (Dawson *et al.*, 1983). For those nerve disorders selected as examples for the discussion on work relatedness (carpal tunnel syndrome, thoracic outlet syndrome and radiculopathy), a general conclusion can be found in subsection 3.3.5.

3.3.2 Carpal tunnel syndrome (CTS)

3.3.2.1 Brief description of the disorder

Carpal tunnel syndrome (CTS) is the compression of the median nerve at the wrist (Figure 3.6). It is characterized by symptoms of pain, numbness and tingling in the median nerve distribution of the hand, often worse at night. Motor branch involvement can result in thenar atrophy. Physical findings have varying degrees of sensitivity and specificity when compared with electrodiagnostic tests (Katz *et al.*, 1991). Electrodiagnostic tests are currently considered the 'gold standard', but their sensitivity and specificity have been estimated by comparison with 'classical symptoms' (Nathan *et al.*, 1988; Stevens *et al.*, 1988) and 'normal' values depend on electrode placement, method, temperature, age, height, finger circumference and wrist ratio (Stetson *et al.*, 1992).

3.3.2.2 Literature reviewed

There have been a number of cross-sectional and case-control studies of carpal tunnel syndrome using symptoms and physical findings (positive Phalen's wrist flexion test or Tinel's sign) or electrodiagnostic tests. Some studies relied on workplace medical records or worker's compensation claims data for case definition. A number of other studies relied solely on questionnaire-based symptom complexes. Additionally, there are a number of clinical and laboratory-based studies that address mechanisms by which various work related and non-work related risk factors may contribute to CTS. These non-work related risk factors include diseases and conditions that contribute to reducing carpal tunnel volume (diabetes mellitus, rheumatoid arthritis, Colles fracture, acromegalia, thyroid disorders, pregnancy). Other non-work related risk factors have also been postulated, although there has been conflicting evidence with respect to hysterectomy, bilateral oophorectomy, age at menopause and obesity (Punnett and Robins, 1985; Punnett *et al.*, 1985; Silverstein, 1985; Nathan *et al.*, 1992).

3.3.2.3 Results of epidemiologic studies

The results of the studies reviewed are presented in Table 3.5; some of these are discussed below. Population-based estimates of incidence (Stevens *et al.*, 1988) were 0.1 cases per 100 person/years for 1961 to 1980, and were increasing at that time. Incidence was approximately three times greater in women than men, but this may reflect gender differences in treatment-seeking behaviours (cases were those seeking treatment at the Mayo Clinic). By contrast, there was a 1.2:1

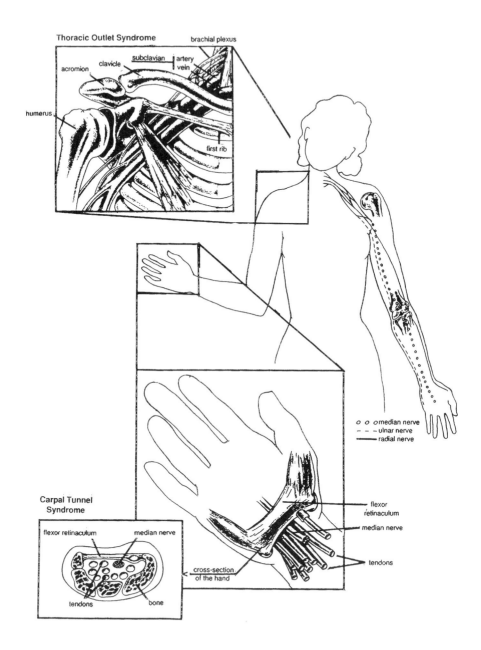

Figure 3.6 Median, ulnar and radial nerves and the carpal tunnel and thoracic outlet syndromes.

female-to-male ratio based on worker's compensation claims for work related CTS, and the mean age of those filing CTS claims was 35, compared with 51 for general-population CTS cases (Franklin *et al.*, 1991). Industry-wide incidence was approximately 0.2/100 worker/years, while certain industrial classifications had relative risks of 10–15 (oyster, crab and clam packing, meat and poultry dealers, slaughterhouses, fish canneries and processing). In cross-sectional studies, prevalence of CTS ranged from 0.6 per cent in low-force/low-repetition jobs of industrial workers (Silverstein *et al.*, 1987) to 61 per cent in grinders and other occupations demanding similar hand activities (Nathan *et al.*, 1988). In some studies, there was an association with increase in age, but not in others.

3.3.2.4 Results and causality assessment

STRENGTH OF THE ASSOCIATION

There are strong associations between exposure to highly repetitive and forceful work and carpal tunnel syndrome. The risk of CTS associated with high-force/high-repetition jobs was 4 to 15, while exposure to repetitiveness alone was around two to five (Silverstein *et al.*, 1987) (Table 3.5). Repetitiveness combined with cold increased the risk (Chiang *et al.*, 1990). Strong associations were demonstrated with flexed postures that increased with hours of exposure (de Krom *et al.*, 1990). Although use of vibrating tools is often correlated with forceful repetitive work, Wieslander *et al.* (1989) reported an odds ratio of six for vibration exposure.

SPECIFICITY OF THE ASSOCIATION
See Box 3.2

TEMPORAL ASSOCIATION

With the exception of the Wieslander *et al.* (1989), Cannon *et al.* (1981) and de Krom *et al.* (1990) case-control studies, the studies cited are cross-sectional; in these latter studies some evidence does exist that work exposure preceded the onset of the disorder. The Silverstein (1985) and Silverstein *et al.* (1986; 1987) studies include onset on current job in the case definition, and subject selection was based on being in the study job for at least one year prior to data collection. Although there may be some misclassification with respect to onset, this suggests a temporal relationship. The Franklin *et al.* (1991) study makes the assumption that CTS workers' compensation claims were filed based on the onset being in the job/industry at the time of filing. The studies based solely on electro-diagnosis do not demonstrate a temporal relationship because abnormality may have occurred prior to exposure.

Table 3.5 Findings from selected studies on the work relatedness of carpal tunnel syndrome (Refer to Box 3.1 for help in interpreting the tables and Table 3.1 to find out how the studies were selected.)

Study population[1]	Outcome[2] and exposure information	Study design	Findings	Authors	Comments
Washington State workers (approximately 1.3 million full-time workers in 1988).	• Outcome assessed using worker's compensation claims for CTS. • Exposure based on industry codes.	Cohort: from 1984 to 1988	• Industry-wide incidence rate of 2 claims/1000. • RR values of 14.8 (95% CI 11.2–19.5) for oyster and crab packers, and 13.8 (95% CI 11.6–16.4) for meat and poultry industries compared with industry-wide.	Franklin *et al.*, 1991	• Among claimants, the female-to-male ratio was 1.2:1. • Mean age of claimants was 37.4.
652 industrial workers who were categorized into 4 groups: (1) low-force/low-rep (comparison group); (2) high-force/low-rep; (3) low-force/high-rep; (4) high-force/high-rep.	• CTS determined by medical examination and interviews. • Exposure assessed by EMG and video analysis of jobs.	Cross-sectional	• High-force/high-repetition jobs: OR 15.5 ($p < 0.001$). • High-repetition jobs: OR 5.5 ($p < 0.5$).	Silverstein *et al.*, 1987	• Controlled for age, gender, plant, years on job. • The study population numbers quoted represent a 90% response rate from the original population.
Dutch population: 28 CTS cases from a 501-subject community sample and 128 CTS cases from the local hospital (total n = 156 CTS cases, i.e. 131 women, 25 men) compared with community non-cases: 310 women and 163 men.	• CTS diagnosed by clinical examination and neurophysiological tests. • Questionnaire on exposure.	Nested case control	• 5.6% prevalence in the general population (the 28 cases from 501 subjects in the community sample). • For 20–40 hours/week exposure compared with no exposure: Flexed wrist: OR 8.7, 95% CI 3.1–24.1; Extended wrist: OR 5.4, 95% CI 1.1–27.4	de Krom *et al.*, 1990	• Adjusted for age and gender. • The figures quoted for the study population represent an approximately 70% response rate from the original population sampled (for both the hospital and community).

Study population	Methods	Study type	Results	Reference	Comments
38 male patients with CTS release surgery compared with, for each case, 2 surgical cases (i.e. 1 gall bladder surgery and varicose vein surgery) and 2 other referents from the community (total comparison group = 152 males). With losses, 34 cases and 143 non-cases were left.	• Nerve conduction velocity (NCV) used to diagnose cases. • Telephone interviews for exposures using standard questionnaire.	Case controls	• Hand-held vibrating tool > 20 yrs: OR 4.8, 95% CI 1.5–15.6. • Repetition of wrist movement > 20 yrs: OR 4.6, 95% CI 1.8–11.9.	Wieslander *et al.*, 1989	• Controlled for age and sex.
Ski manufacturing workers: 106 with repetitive jobs compared with 67 with non-repetitive jobs. These numbers represent 70% and 64% participation rates for the original population for repetitive and non-repetitive jobs, respectively.	• CTS determined by electrodiagnosis of median-ulnar difference (latency on response time) and signs (Phalen's or Tinel's). • Exposure based on observation of jobs.	Cross-sectional	• Prevalence 15.4% repetitive group vs 3.1% non-repetitive group: for either hand OR 4.0, 95% CI 1.0–15.8.	Barnhart *et al.*, 1991	• Difference due to a fast response time in the ulnar nerves rather than slow in the median nerves. • Minimal control of confounding, but control for age and gender done.
471 industrial workers.	• Outcome defined as NCV-determined cases of impaired sensory conduction (sensory latency). • Jobs were grouped into 5 occupational classes according to amount of resistance and repetition (from very light to very heavy).	Cross-sectional	• High-force/high-repetition jobs compared with the very light group: OR 4.0, 95% CI 1.5–11.0.	Nathan *et al.*, 1988	• No description of symptom status; controlled for age and gender. • Information on subjects lost from the original study population is unclear.

Table 3.5 Work relatedness of carpal tunnel syndrome (Cont.)

Study population[1]	Outcome[2] and exposure information	Study design	Findings	Authors	Comments
207 frozen food factory workers.	• CTS based on clinical findings and electro-physiological tests. • Job analysis to create 3 groups: low-cold/low-repetition (group I), low-cold/high-repetition (group II), high-cold/high-repetition (group III) using the Silverstein force/repetition criteria.	Cross-sectional	• Prevalence: group I = 4%; group II = 41%; group III = 37%. • Compared with group I: repetition and cold: OR 9.4, 95% CI 2.4–37.2 and repetition only: OR 2.2, 95% CI 0.2–21.1.	Chiang et al., 1990	• Controlled for sex, age and employment time. • No comments on the number of non-participants.
Aircraft engine workers: 30 CTS cases (3 males, 27 females) compared with 3 non-cases per case (sex matched) from the same plant.	• 16 CTS cases came from worker's compensation claims and 14 CTS cases were diagnosed by medical dept (but had no claim). • Exposure based on job category.	Case control	• Job in which vibrating tools are used: OR 7.0 ($p < 0.01$). • Jobs with repetitive wrist movement: OR 2.1 ($p = 0.05$).	Cannon et al., 1981	• Controlled for gynecology surgery and years on job.
Out of 214 active garment workers, 205 were physically examined and sufficient information was obtained by questionnaire on 179 workers (162 women and 17 men). Men were excluded and the 162 women were compared with 73 female	• A physical exam was performed; however, for the outcome CTS, a questionnaire was used. • Exposure assessed by questionnaire for work history and video analysis.	Cross-sectional	• 18% prevalence in garment workers with a prevalence odds ratio of 3.0, 95% CI 1.2–7.6 compared with hospital workers.	Punnett et al., 1985 Punnett and Robins, 1985	• Controlled for age. • CTS symptoms modified by hormonal status and native language.

Study population	Methods	Design	Outcomes	Reference	Comments
full-time workers in a hospital (excluding typists). This latter figure of 73 represented 34% of all the full-time and part-time hospital workers.					
Out of 19 butchers from two slaughterhouses, 17 male butchers participated.	• Cases determined by clinical examination and electrodiagnosis. • Exposure based on questionnaires and job observation.	Cross-sectional	• 53% prevalence (subjective symptoms and electrophysiological evidence). • CTS greater in non-dominant hand that pulls meat.	Falck and Aarnio, 1983	• Age and obesity important.
Poultry workers (n = 27 men, 66 women) compared with applicants for positions (n = 44 men, 41 women).	• Electrodiagnosis of cases. • Exposure based on current workers vs pre-employed.	Cross-sectional	• Midpalmar sensory latency greater in poultry workers.	Schottland et al., 1991	• Controlled for age and gender. • Results mentioned in this table are based on the actual data reported in this article, not on the authors' abstracts.

[1]Based on the information available in the article, no subject was lost from the original study population when only one set of numbers appears in the table under the 'study population' or when there is no relevant information in the 'comments' column.
[2]Unless otherwise specified, all outcomes in Table 3.5 were carpal tunnel syndrome and were identified by medical examination.

CONSISTENCY OF THE ASSOCIATION

There is evidence for consistency of the association; the association is found across different studies (i.e. replicability) and survives the test of alternative hypotheses (i.e. survivability). While there are differences in case definition among the studies cited (ranging from symptoms to electrodiagnostic findings), these studies were conducted in a wide range of industries in which there were highly repetitive and forceful jobs. The Franklin *et al.* (1991) study included two-thirds of all workers in Washington State. The other studies included workers in appliance manufacturing, electronics, foundries, bearing manufacturing, apparel sewing, investment casting, steel, meat and food packaging, electronics and plastics, ski manufacturing and grocery checking. None of these studies looked at the office environment except as mixed exposure for the comparison group (controls).

Virtually all these studies recognized age as a possible effect modifier in the analysis of the data. With the exception of the Barnhart *et al.* (1991) study, gender was recognized as a potential confounder in all studies and was controlled by restriction (i.e. studies of only one gender: for example, Wieslander *et al.*, 1989 – males; Morgenstern *et al.*, 1991 – females) or stratification (rates for females calculated separately from those for males). Females were not at significantly greater risk for CTS in the Silverstein *et al.* (1987) study. There was a 1.2:1 female-to-male ratio in the Franklin *et al.* (1991) study. Misclassification of work exposure was possible in all studies, particularly those in which job title or industry was used as a surrogate measure for exposure. In the de Krom *et al.* (1990) study, recall of number of hours spent in a particular posture is subject to considerable misclassification. The effect is likely to minimize risk estimates. Participation rates were not described in the Nathan *et al.* (1988) or Schottland *et al.* (1991) studies, making it difficult to evaluate the representativeness of the findings. Reproductive status (bilateral oophorectomy, hysterectomy, use of birth control pills and menopause) were not significant confounders in either the Silverstein (1985) or Morgenstern *et al.* (1991) studies. Hysterectomy was a significant predictor in the de Krom *et al.* (1990) population-based study. Nathan *et al.* (1992) found body mass index (height and weight) predictive of decrease in median nerve function electrodiagnostically five years later. Stetson *et al.* (1992), controlling for exposure and other anthropometric variables, found that height was negatively associated with amplitude and positively associated with sensory latency in non-symptomatic, unexposed adults (*n* = 105). Gender was not a significant factor when taking these anthropometric measurements into account. Findings should be adjusted for age, height and finger circumference. Using these adjustments, Stetson *et al.* (1993) found significant decreases in electrodiagnostic test functioning for median nerves of long-term auto workers compared with those who had never done high-force or high-repetition work. In general, the magnitude of the associations with work, particularly in the Silverstein *et al.* (1987) study (while controlling for a variety of potentially intervening factors), makes it unlikely that these associations can be explained by other, unaccounted for, risk factors.

PREDICTIVE PERFORMANCE OF THE ASSOCIATION

No epidemiologic studies have reported a reduction in CTS prevalence/ incidence as a consequence of reduction in work related exposure. However, in a small study of 22 auto workers with relatively acute onset of CTS diagnosed electrodiagnostically, Gordon *et al.* (1987) reported that 18 who had changed jobs (17 treated conservatively) had symptom improvement, and one with follow-up studies was normal. The Silverstein *et al.* (1987) study suggests an exposure-response relationship, in that those workers in high-force/high-repetition jobs had a higher risk of CTS than those in just high-repetition jobs who, in turn, had a higher risk than those in low-force/low-repetition jobs. On the other hand, in a follow-up study at one of these plants, when there had been no change in force or repetitiveness characteristics of the study jobs, there was no change in prevalence, but those who were originally symptomatic in the high-force/high-repetition category were no longer in that job category three years later (Silverstein *et al.*, 1988).

COHERENCE OF THE ASSOCIATION OBSERVED WITH CURRENT KNOWLEDGE

Mechanical stress (stretching or compression of the median nerve at the wrist) and ischemia are hypothesized to be the mechanisms of CTS. Extreme wrist postures (either prolonged or repetitive) increase the carpal canal pressure, resulting in paresthesia (Gelberman *et al.*, 1981; Werner *et al.*, 1983; Szabo and Chidgey, 1989). Repetitive stretching of nerves and tendons in the carpal canal can produce an inflammatory response, which in turn leads to tissue scarring and decreased effective canal size, thereby compressing the nerve (Armstrong *et al.*, 1984). Thus it can be shown that work related exposures are coherent with the development of CTS. Effects of repetitive or forceful gripping on carpal canal pressures may be modified by individual capacity, including the presence of systemic disorders, conditions or anomalies that effectively reduce the size of the carpal canal (diabetes, rheumatoid arthritis, myxedema, pregnancy [transient CTS], Colles fracture, muscle in the carpal canal). Some of these are discussed in chapter 9.

3.3.2.5 Conclusion

There is strong evidence supporting the contribution of work related factors to the development of carpal tunnel syndrome. The strength of the association between work and CTS was found to be high, consistency of the association has been shown, there is some evidence of a cumulative exposure-response and of a temporal association, and the possibility that such an association exists is coherent with current knowledge.

3.3.3 Thoracic outlet syndrome (TOS)

3.3.3.1 Brief description of the disorder

There is some debate about the definition and treatment of thoracic outlet syndrome (TOS). Some of the confusion stems from the fact that TOS may be regarded as a catch-all for many different syndromes with different etiologies and clinical courses. TOS (Figure 3.6) is a neurovascular impingement syndrome at different anatomical levels where the brachial plexus and the subclavian vessels may be entrapped as they pass through, en route from the cervical spine to the arm (Cuetter and Bartoszek, 1989). TOS syndrome can be divided into neurogenic TOS and vascular TOS (Hall, 1987). Common neurogenic TOS is a result of compression of the brachial plexus at different levels (Figure 3.6). The anatomical site of compression also gives rise to the names of the subtypes of thoracic outlet syndrome. These subtypes are cervical rib compression or fibrous band compression (from the cervical spine to the first rib), scalenus syndrome, costo-clavicular syndrome and hyperabduction syndrome or pectoral minor syndrome. The symptoms reflect the degree to which particular nerves and vascular structures are compressed (Karas, 1990).

A common type of neurogenic TOS is caused by a cervical rib or a fibrous band; these are malformations. The cervical rib or fibrous band can impinge the lowest part of the brachial plexus, causing pain down the ulnar side of the arm and forearm, and sometimes in the hand. This pain may be exacerbated by strenuous use of the hand (Hall, 1987). In epidemiologic studies, positive outcomes of the abduction external rotation test (Roos test) and shoulder-arm pain are used as the diagnostic criteria for neurogenic TOS. The form of TOS based on these diagnostic criteria, which do not require objective findings such as electromyography (EMG), denervation or muscle wasting, is referred to by Hall (1987) as 'the disputed, symptoms only, form of TOS'. Vascular forms of TOS are rare. There is no occupational epidemiologic study of arterial or venous TOS syndromes; the rest of this subsection will therefore be concerned only with neurogenic TOS.

3.3.3.2 Literature reviewed

There are only a few studies reported in which the prevalence of thoracic outlet syndrome has been studied in relation to the occupational group. The lack of consensus on diagnostic criteria for thoracic outlet syndrome is probably one reason why so few studies are available.

3.3.3.3 Results of epidemiologic studies

The results of the studies selected and reviewed are presented in Table 3.6. Some of these results are discussed below. The prevalence rate of neurogenic TOS varied considerably among industrial workers (Table 3.6). The rates ranged from 0.3 per cent for industrial workers in the US (Silverstein, 1985, study not shown in the table) to 44 per cent for female assembly-line workers (Sällström and Schmidt, 1984). When occupational groups were pooled according to exposure to repetitive arm movements, the odds ratio was 4.0 with a 95 per cent confidence interval of 1.2–13 (Hagberg and Wegman, 1987, study not shown in table). Vibration exposure (measured as vibration level) was associated with a risk of 1.4 and a 95 per cent CI of 1.1–1.7 (these figures are not shown in the table) in a study of plate workers and assemblers compared with white-collar workers (Toomingas *et al.*, 1991).

3.3.3.4 Results and causality assessment

STRENGTH OF THE ASSOCIATION
The strength of the association between work related exposure and thoracic outlet syndrome (neurogenic TOS) in individual studies was weak. Although odds ratios of up to 10 were found, the confidence intervals were wide, thus reducing the strength of the association between TOS and work related exposure in each individual study. One study revealed an odds ratio of 4.0 for occupational groups pooled according to exposure to repetitive arm movements (Hagberg and Wegman, 1987).

SPECIFICITY OF THE ASSOCIATION
See Box 3.2

TEMPORAL ASSOCIATION
No study has addressed the problem of temporality (temporal association). That is, we are not sure whether the onset of exposure preceded the onset of thoracic outlet symptoms, since all studies were cross-sectional and all were based on prevalent cases.

CONSISTENCY OF THE ASSOCIATION
There was consistency (replicability) across studies for a few occupations showing repetitive arm movements as a risk factor for neurogenic TOS. In the literature, several case reports point to the relationship between thoracic outlet syndrome and manual work (Sällström and Schmidt, 1984). One contradictory finding was observed; one group of assemblers showed a decreased risk for neurogenic TOS compared with controls. However, in this study, the assemblers were considerably younger than the controls and, since age was associated with TOS, the difference in age may explain this contradictory finding.

PREDICTIVE PERFORMANCE OF THE ASSOCIATION
Evidence for the predictive performance of the associations shown is lacking.

Table 3.6 Findings from selected studies on the work relatedness of thoracic outlet syndrome (Refer to Box 3.1 for help in interpreting the tables and Table 3.1 to find out how the studies were selected.)

Study population[1]	Outcome[2] and exposure information	Study design	Findings					Authors	Comments
			Prevalence		OR	95% CI			
			Exposed subjects	Comparison group					
The exposed group comprised 163 female assembly line packers (representing 84% of the packaging department in a food production firm). After the exclusion of 11 workers for previous trauma, arthritis or other pathologies, 152 females remained, who were compared with 143 shop assistants in a department store, cashiers excluded (*n* = 133 females remaining after exclusion of 10 workers for trauma, arthritis	● Packers performed repetitive arm work (repetitive motions up to 25 000 cycles per workday), had extreme work positions of the hands and arms, and shoulders were subject to static muscle work. Exposure assessed by observation, analysis of videos and interviews.	Cross-sectional	3%	0%	10	0.54-182		Luopajärvi *et al.*, 1979	● Adjustment for potential confounders not performed in the analysis.

and other pathologies).

Male assembly line workers (*n* = 83) at a truck manufacturing company compared with male office workers (*n* = 27).	• Exposure based on job category. Work tasks were: (1) assembly line work at increased rate, welding, drilling and turning. Tasks were often performed in awkward positions; (2) office work consisting of typing and editing texts.	Cross-sectional	14%	0%	9.6	0.55–168	Sällström and Schmidt, 1984	• Adjustment for potential confounders not performed in the analysis. • The numbers quoted for the study population represent, overall for all jobs, a participation rate of 96% from the original population.
Female assembly line workers (*n* = 9) at a truck manufacturing company compared with female office workers (*n* = 35).	• Exposure based on job category. Work tasks were: (1) assembly line work at increased rate, welding, drilling and turning. Tasks were often performed in awkward positions; (2) office work consisting of typing and editing texts.	Cross-sectional	44%	17%	3.9	0.80–19	Sällström and Schmidt, 1984	• Adjustment for potential confounders not performed in the analysis. • The numbers quoted for the study population represent, overall for all jobs, a participation rate of 96% from the original population.
Female cash register operators (*n* = 37) at three supermarkets compared with female office workers (*n* = 35).	• Exposure based on job category. Cash operators had the cash registers placed to the right of them. The operators had to keep their right arms in half abduction and elevation, demanding	Cross-sectional	32%	17%	1.7	0.54–5.3	Sällström and Schmidt, 1984	• Adjustment for potential confounders not performed in the analysis. • The numbers quoted for the study population represent, overall for all jobs, a

Table 3.6 Work relatedness of thoracic outlet syndrome (Cont.)

Study population[1]	Outcome[2] and exposure information	Study design	Findings					Authors	Comments
			Prevalence		OR	95% CI			
			Exposed subjects	Comparison group					
	continuous contraction of the muscles of the shoulders, neck, and back. Office work consisted of typing and editing texts.								participation rate of 96% from the original population.
Out of a population of 112 plate workers at a factory, 71 were compared with white-collar workers who were engineers, supervisors and managers – no VDU workers or clerks (*n* = 45 males randomly selected from 500 white-collar workers at an engineering factory).	• Outcome assessed by a positive abduction external rotation test. • Exposure measured by job category. Welding and grinding were the tasks performed by the plate workers. Tasks were often performed in awkward positions.	Cross-sectional	31%	16%	2.4	0.94-6.3		Toomingas *et al.*, 1991	
A population of assemblers (*n* = 70	• Outcome assessed by a positive abduction	Cross-sectional	6%	16%	0.33	0.1-1.2		Toomingas *et al.*, 1991	

males) at a truck assembly plant was compared with office white-collar workers who were engineers, supervisors and managers – no VDU workers or clerks ($n = 45$ males randomly selected from 500 white-collar workers at an engineering factory).

external rotation test.
● Exposure measured by job category. Assembly line work at increased rate. Assembling was done using impact wrenches, screwdrivers.

1 Based on the information available in the article, no subject was lost from the original study population when only one set of numbers appears in the table under the 'study population' or when there is no relevant information in the 'comments' column.
2 Unless otherwise specified, all outcomes in Table 3.6 were thoracic outlet syndromes and were determined by clinical examination.

No study has examined the cumulative exposure-response relationship between exposure to repetitive arm movements and the outcome of neurogenic TOS. However, the association between vibration-level exposure at work and TOS, when controlling for age, is an indicator of an exposure-response relationship (Toomingas *et al.*, 1991).

COHERENCE OF THE ASSOCIATION OBSERVED WITH CURRENT KNOWLEDGE

There is coherence between symptoms of TOS and experimental studies. Experimentally induced pressure on nerves was shown to cause symptoms of paresthesia, motor loss, and sensory perception disturbances (Szabo and Gelberman, 1987; Lundborg, 1988). Further, there are similarities between carpal tunnel syndrome and TOS. CTS and neurogenic TOS are both entrapment syndromes. In CTS the median nerve is entrapped in the carpal tunnel, while in neurogenic TOS the brachial plexus is entrapped when passing between muscles in the thoracic outlet on the way to the arm (Figure 3.6). Both syndromes result in the same type of symptoms as seen in experimentally induced pressure on nerves. Since the association between work related exposures and CTS is thought to be coherent with current knowledge, the similarities between CTS and TOS further support the coherence of TOS with physical exposures that can cause increased pressure on nerves and result in symptoms. Whether TOS is work related in individuals without anatomic malformations is unclear. Thoracic outlet syndrome may be work related in individuals with constitutional factors such as cervical rib or fibrous bands. Repetitive work may promote the symptoms of TOS syndrome in such individuals.

3.3.3.5 Conclusion

Epidemiologic studies indicate an exposure-response relationship between manual work and TOS. There is consistency in this association between repetitive arm movements, manual work and TOS, and this relationship is coherent with current knowledge. However, there are few data to support a cumulative exposure-response relationship.

3.3.4 Radiculopathy (or cervical syndrome)

3.3.4.1 Brief description of the disorder

Radiculopathy is the compression of the nerve root by a herniated disc or by a narrowed intervertebral foramen (Figure 3.1). The etiology of cervical herniation is obscure, as is the etiology of narrowed intervertebral foramina. The symptoms are severe pain with distinct radiation of the pain to specific dermatomes. Tingling and numbness may be present. Wasting of muscles can sometimes be seen, as in a C7 syndrome, where atrophy of the first dorsal interosseus muscle may occur. Loss of muscle power may impair the patient. The diagnosis may be established by clinical examination, where the findings are specific loss of sensitivity in affected dermatomes, diminished motor reflexes and a positive test of compression of the cervical spine or the intervertebral foramina. Electromyography (EMG) may show muscle degeneration. Radiographic examination with a CAT-scan or MRI may show a hernia or a crushed spinal nerve. A synonym for cervical radiculopathy is cervical syndrome (Waris, 1979).

3.3.4.2 Literature reviewed

There are only a few cross-sectional studies available addressing the problem of radiculopathy. The main aim of these studies was not to assess the prevalence and exposure associations of cervical radiculopathy; rather, radiculopathy happens to have been registered as a diagnostic entity in the occupational groups studied. Since cervical radiculopathy appears to have a low incidence and low prevalence (prevalence: 1 to 5 per cent, Table 3.7) a case-control study design would be more appropriate to evaluate whether there is an association with work exposure. However, only cross-sectional studies are currently available.

3.3.4.3 Results of epidemiologic studies

Results of the studies selected and reviewed are shown in Table 3.7. Some of these results are discussed below. In studies of data entry operators, dockers and assembly-line packers, risk ratios below 1.0 were found (Table 3.7). The confidence intervals were wide and the results are difficult to interpret. It is likely that the low prevalences observed in these studies, especially among the exposed workers, represent a primary and secondary healthy worker selection (Hagberg and Wegman, 1987).

3.3.4.4 Conclusion

In the few studies published where radiculopathy was addressed, there was no evidence of an exposure-response relationship between work and cervical radiculopathy. However, since the power of these studies was low, the existence of an exposure-response relationship cannot be excluded.

3.3.5 Conclusion

The peripheral nerve disorders most commonly thought of as being associated with work were used as examples to examine evidence for or against work relatedness. These were carpal tunnel syndrome (CTS), thoracic outlet

Table 3.7 Findings from selected studies on the work relatedness of radiculopathy (cervical syndrome) (Refer to Box 3.1 for help in interpreting the tables and Table 3.1 to find out how the studies were selected.)

Study population[1]	Outcome[2] and exposure information	Study design	Findings Prevalence Exposed subjects	Findings Prevalence Comparison group	OR	95% CI	Authors	Comments
Female data entry workers (n = 104) compared with female office workers in varying office tasks (n = 57).	• Exposure based on job categories, no actual exposure measurement or assessment.	Cross-sectional	1%	2%	0.54	0.03–8.9	Kukkonen et al., 1983	• Adjustment for potential confounders not performed in the analysis.
Dockers (n = 215 males) compared with male civil servants (n = 188).	• Outcome based on a diagnosis of cervical disc disease. • Exposure based on job category: dockers performed manual work whereas civil servants did not.	Cross-sectional	2%	5%	0.47	0.14–1.5	Partridge and Duthie, 1968	• The figures quoted for the study population represent a 96% response rate for dockers and 91% for civil servants.
The exposed group comprised 163 female assembly line packers (representing 84% of the packaging department in a food production firm). After the	• Packers performed repetitive arm work (repetitive motions up to 25 000 cycles per workday), had extreme work positions of the hands and arms, and shoulders were subject to static muscle work.	Cross-sectional	1%	2%	0.27	0.03–2.8	Luopajärvi et al., 1979	• Adjustment for potential confounders not performed in the analysis.

exclusion of 11 workers for previous trauma, arthritis or other pathologies, 152 females remained, who were compared with 143 shop assistants in a department store, cashiers excluded ($n = 133$ females remaining after exclusion of 10 workers for trauma, arthritis and other pathologies).

Exposure assessed by observation, analysis of videos and interviews.

[1]Based on the information available in the article, no subject was lost from the original study population when only one set of numbers appears in the table under the 'study population' or when there is no relevant information in the 'comments' column.

[2]Unless otherwise specified all outcomes in Table 3.7 were cervical syndromes and were determined by clinical examination.

syndrome (TOS) and radiculopathy. The work relatedness of each peripheral nerve disorder was considered individually. Although all the disorders discussed have a similar etiologic basis (nerve entrapment or compression), there are differences in their patho-physiological mechanisms, in the numbers of epidemiologic studies available for each disorder and in the results of these studies.

There is strong evidence supporting the contribution of work related factors to the development of carpal tunnel syndrome. The strength of the association between work and CTS was found to be high, consistency of the association has been shown, there is some evidence of an exposure-response relationship and a temporal association, and the possibility that such an association exists is coherent with current knowledge.

There is epidemiologic evidence that work related exposures are associated with TOS; however, with the few studies currently available for review, the strength of this evidence is weak. There is also a possibility that some personal characteristics (anatomic variations) are mediating factors in the relationship between work and TOS, although much of this is unclear at this stage. Consistency is observed across studies for the risk factors postulated (manual work, repetitive arm movements) for this disorder. The observed association between work and TOS is coherent with pre-existing theories and knowledge.

In the three cross-sectional studies, there was no evidence of an association between work and radiculopathy. However, since the power of these studies was low, the existence of an exposure-response relationship cannot be ruled out.

3.4 Evidence of the association between work and selected muscle disorders: tension neck syndrome

3.4.1 Brief description of the disorder

3.4.1.1 General background

A muscle is an organ with the ability to contract. It consists of muscle fibres (muscle cells), nerve elements (motor neurons, afferent neurons, receptors of different types), and connective tissue and vascular elements (tendon and blood vessels) that keep the muscle fibres together (Figure 3.7). Different methods are used to classify muscle fibres. Type I fibre is sometimes called slow-twitch or red-muscle fibres. Type II is often referred to as fast-twitch or white-muscle fibres. The type II fibres can be classified into different sub-types. In the muscle fibre, the smallest *morphological* contractile unit is the sarcomere, built of actin and myosin filaments. The smallest *functional* unit in the muscle is the motor unit, which consists of a motor neuron cell and the muscle fibres its branches supply. Myalgia is the medical term for the symptom of muscle pain and myopathy is the term for measurable pathological changes in a muscle with or

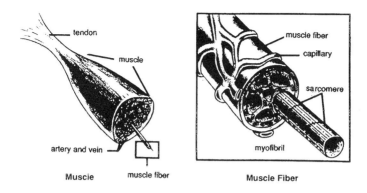

Figure 3.7 Structure of a muscle.

Table 3.8 Painful myopathies (according to Morgan-Hughes, 1979)

- Inflammatory myopathies
- Polymyalgia rheumatica
- Acute alcoholic myopathy
- Acute and subacute drug-induced myopathies
- Myopathies of metabolic bone disease
- Myopathies due to specific defects in muscle energy metabolism
- Idiopathic paroxysmal myoglobinurias
- **Muscle pain syndromes of unknown etiology**

without symptoms. Muscle disorders (myopathies) can be congenital, such as muscular dystrophies, or secondary to a general inflammatory disease, such as rheumatoid arthritis (Brooke, 1977). Muscle pain (myalgia) is a common experience in healthy people after vigorous or unaccustomed exercise. There are many types of painful conditions in muscle (Table 3.8). Although the last decades have brought advanced diagnostic procedures, most painful myopathies cannot be diagnosed as pathogenetic, i.e. the mechanisms behind the myopathies are unknown (Mills and Edwards, 1983).

Muscle pain syndromes of unknown etiology can be general or regional. General muscle pain localized in the four quadrants of the body is called primary fibromyalgia. Primary fibromyalgia is not work related because, by definition, trauma-induced myalgia is excluded by the specific diagnostic criteria set by the American College of Rheumatology (Wolfe *et al.*, 1990). Regional muscle pain syndromes often fall under the term myofascial syndrome. Grosshandler and Burney (1979) defined this as a painful condition of skeletal muscle characterized by the presence of one or more discrete areas (trigger points) that are tender and hypersensitive and from which pain may radiate when pressure is applied. These are the muscle-related syndromes that could be associated with work exposures.

3.4.1.2 *Tension neck syndrome*

Tension neck syndrome (Figure 3.1) is a term used for myofascial syndrome localized in the shoulder and neck region (Viikari-Juntura, 1983). This term has been widely used in epidemiologic studies of work related myofascial pain. The diagnostic criteria for tension neck syndrome are usually pain in the shoulder or neck, in addition to tenderness over the descending part of the trapezius muscle (shoulder-neck area) (Waris *et al.*, 1979; Hagberg and Wegman, 1987). Myofascial syndrome may also be prevalent in the low back, and in such cases is called 'pain without radiation' or 'chronic pain syndrome' (Spitzer *et al.*, 1987).

3.4.2 Literature reviewed

The literature available concerning work related muscle disorders was incomplete. Most articles spoke of cross-sectional studies that had examined the association between disorders and job title. Even if the exposure was assessed only by using the job title as a surrogate, in many studies it was fairly well described according to observational analysis among exposed subjects. Risk evaluation with odds ratios and confidence intervals was lacking in most original articles. The odds ratios with confidence intervals presented in Table 3.9 were computed for our purposes from prevalence rates given in the original articles. The computed confidence intervals were based on Taylor series (exact). In all but a few of the studies, whenever attempts were made to conduct epidemiologic analyses, there was no consideration of potential confounders, such as age or smoking. Furthermore, there was no attempt to perform cumulative exposure-response analysis in the studies.

3.4.3 Results of epidemiologic studies

Results of the studies selected and reviewed on the work relatedness of tension neck syndrome are shown in Table 3.9. Some of these results are discussed below.

Tension neck syndrome is more common among females than males. The odds ratio for selected female industrial workers in the United States compared with male industrial workers was 5.9 (Silverstein, 1985); however, whether this difference in gender is due to genetic or to gender difference in exposure both at work and at home is unknown. Occupational groups with work tasks requiring repetitive arm movements and constrained work postures showed high rates of tension neck syndrome (Table 3.9). There were also high rates of tension neck syndrome in the referents, although not as high as in each of their corresponding exposed groups. The odds ratios for tension neck syndrome varied from 1.5 to 7.3 for nine exposed groups (teachers and nurses, data entry operators, typists, conversational terminal operators, industrial scissor makers, film-rolling workers, lamp assemblers, industrial workers and assembly line workers) when compared with their respective comparison groups.

Among assembly line workers with shoulder pain, muscle tenderness of the trapezius muscle was common (Bjelle *et al.*, 1981, study not shown in table). The assemblers in this study were exposed to repetitive arm elevations and were required to keep their arms elevated for long periods throughout the workday. All epidemiologic studies were cross-sectional. There was one experimental study in which females performing repetitive shoulder flexions developed transient shoulder pain and tenderness (Hagberg, 1981). In this study, the flexion rate used was 15 forward flexions per minute between 0 and 90 degrees, for one hour. The pain and tenderness observed was probably muscle soreness. No experimental studies on chronic muscle pain have been found.

3.4.4 Results and causality assessment

3.4.4.1 Strength of the association

The magnitude of associations between exposure (measured using the occupational group) and tension neck syndrome were moderate. The risks varied between three and seven for eight exposed groups (Table 3.9). Generally, the risk ratios were high enough that sources of bias were less likely, even if these had not been apparent to the investigators. For some studies, attributable fractions (exposed) have been calculated to indicate the portion of tension neck syndrome that may be attributable to the work exposure factor examined in the specific context of the studies quoted. For example, the attributable fraction (exposed) in the Kuorinka and Koskinen (1979) study would be 76 per cent (95 per cent CI 57–86 per cent); for the Hünting *et al.* (1981) study it would be 80 per cent (95 per cent CI 44–92 per cent). More information on attributable fractions may be found in appendix III.

3.4.4.2 Specificity of the association

See Box 3.2

3.4.4.3 Temporal association

In studies of tension neck syndrome, it was not clearly demonstrated that work exposure preceded the onset of symptoms (i.e. temporal association). The latency between onset of work exposure and tension neck syndrome is unknown.

3.4.4.4 Consistency of the association

For data entry workers and typists there was more than one study available.

Table 3.9 Findings from selected studies on the work relatedness of tension neck syndrome (Refer to Box 3.1 for help in interpreting the tables and Table 3.1 to find out how the studies were selected.)

Study population[1]	Outcome[2] and exposure information	Study design	Findings		Risk ratio	95% CI	Authors	Comments
			Exposed subjects	Comparison group				
			Prevalence		OR			
Assembly line workers on a shoe manufacturing line (*n* = 102 females) compared with non-assembly workers (*n* = 102 females).	• Outcome assessed as rhomboid muscle tenderness in the clinical examination. • Exposure based on job group. Observations made from films of the work tasks and showed each assembly line worker handled 3400 shoes per day (repetitive arm movements).	Cross-sectional	14%	1%	7.3	1.6–33	Amano *et al.*, 1988	• Adjustment for potential confounders not performed in the analysis. • Findings shown are for the left side. Figures for the right side are almost identical. • No information on participation rate from the original study population.
Data entry operators (*n* = 50 females, 3 males) compared with traditional office workers (*n* = 33 females, 22 males).	• Outcome determined by medical examination (tendomyotic pressure pains in shoulder and neck). • Exposure based on job category. Constrained head and arm posture for the exposed group.	Cross-sectional	38%	11%	4.9	1.8–13	Hünting *et al.*, 1981	• Adjustment for potential confounders (sex) not performed in the analysis.

Population	Methods	Design					Reference	Comments
Typists (*n* = 74 females and 4 males) compared with traditional office workers (*n* = 33 females, 22 males).	• Outcome determined by medical examination (tendomyotic pressure pains in shoulder and neck). • Exposure based on job category. Constrained head and arm posture for the exposed group.	Cross-sectional	35%	11%	4.2	1.6–11	Hünting *et al.*, 1981	• Adjustment for potential confounders (sex), not performed in the analysis.
Conversational terminal operators (*n* = 55 females and 54 males) compared with traditional office workers (*n* = 33 females and 22 males).	• Outcome determined by medical examination (tendomyotic pressure pains in shoulder and neck). • Exposure based on job category. Constrained head and arm posture for the exposed group.	Cross-sectional	28%	11%	3.2	1.2–8.2	Hünting *et al.*, 1981	• Adjustment for potential confounders (sex), not performed in the analysis.
Out of 115 industrial workers-scissor-makers, 93 (*n* = 90 females, 3 males) were examined and completed the work history and production data collection.	• Tension neck syndrome diagnosed using objective signs (at least two tender spots and/or palpable hardenings plus muscle tightness in neck movements) combined with subjective symptoms (constant feeling of fatigue and/	Cross-sectional	61%	28%	4.1	2.3–7.2	Kuorinka and Koskinen, 1979	• Adjustment for potential confounders (sex), not performed in the analysis.

Table 3.9 Work relatedness of tension neck syndrome (Cont.)

Study population[1]	Outcome[2] and exposure information	Study design	Findings Prevalence Exposed subjects	Findings Prevalence Comparison group	Risk ratio OR	95% CI	Authors	Comments
These subjects were compared with shop assistants (n = 133 females) in a large department store.	or stiffness in the neck plus one more subjective symptom). • Exposure determined by job analysis. A work history (including calculating number of parts handled) and a work method analysis were done. Grasping with the fingers wide open and the use of force considered important.							
Film-rolling workers (n = 127 females) compared with office workers (n = 101 females).	• Outcome based on tenderness obtained by pressure measurement for those with shoulder stiffness. • Exposure based on job. Exposure assessments were based on observation and EMG and published separately (Onishi et al., 1976). Film-rollers	Cross-sectional	59%	28%	3.8	2.1–6.6	Onishi et al., 1976	• Contradictory information about the study population in the paper. • Adjustment for potential confounders not performed in the analysis.

had a work cycle time of 2.5–5 seconds and shoulder muscle fatigue due to repetitive upper limb tasks.

Study population	Design	%	%	OR	CI	Reference	Comments
Lamp assemblers (n = 95 females) compared with office workers (n = 101 females). • Outcome based on tenderness obtained by pressure measurement for those with shoulder stiffness. • Exposure based on job. Observation of work tasks showed manipulations every 3.5–12 seconds and repetitive arm movements.	Cross-sectional	60%	28%	3.8	2.1–7.0	Onishi et al., 1976	• Contradictory information about the study population in the paper. • Adjustment for potential confounders not performed in the analysis.
Teachers and nurses of handicapped children (n = 46 females) compared with office workers (n = 101 females). • Outcome based on tenderness obtained by pressure measurement for those with shoulder stiffness. • Exposure based on job. Observation of work tasks showed teachers' work tasks varied: instructed/watched children, physically assisted children (on/off bus, taking meals, using lavatory, dressing,	Cross-sectional	37%	28%	1.5	0.7–3.2	Onishi et al., 1976	• Contradictory information about the study population in the paper. • Adjustment for potential confounders not performed in the analysis.

Table 3.9 Work relatedness of tension neck syndrome (Cont.)

Study population[1]	Outcome[2] and exposure information	Study design	Findings		Authors	Comments
			Prevalence			
			Exposed subjects	Comparison group	Risk ratio / OR	95% CI	
	etc.). Nurses physically helped patients (getting up, going to bed, etc.).					
Female data entry workers (*n* = 104) compared with female workers in varying office tasks (*n* = 57).	• Outcome assessed by health examination. • Exposure according to job categories, with no exposure measurement or assessment.	Cross-sectional	47%	28%	2.3	1.1–4.6	Kukkonen *et al.*, 1983	• Adjustment for potential confounders not performed in the analysis.
Industrial workers: 287 males compared with 287 females. Of the males, 212 were exposed to high repetition and/ or high force and 75 were not exposed to either. Of the females, 226 were exposed to high repetition	• Outcome assessed by interview and medical examination. • Gender is the exposure in this comparison.	Cross-sectional	8%	1%	5.9	2.0–17	Silverstein, 1985	• Adjustment for potential confounders (work task, plant) not performed in the analysis. • The study population figures quoted represent a 90% overall response rate from those originally selected.

and/or high force and 61 not exposed to either.

Study population	Study design					Reference	Comments
Industrial workers: 212 males exposed to high repetition and/or high force compared with 75 not exposed to high repetition or high force. • Outcome assessed by interview and medical examination. • Exposure categories based on repetition and force used by hand/wrist, classified according to exposure measurements in a sample of workers.	Cross-sectional	1.4%	1.3%	1.1	0.1–10.0	Silverstein, 1985	• Adjustment for potential confounders (plant), not performed in the analysis. • The study population figures quoted represent a 90% overall response rate from those originally selected.
Industrial workers: 226 females exposed to high repetition and/or high force compared with 61 not exposed to high repetition or high force. • Outcome assessed by interview and medical examination. • Exposure categories based on repetition and force used by hand/wrist, classified according to exposure measurements in a sample of workers.	Cross-sectional	7.5%	8.2%	0.9	0.4–2.4	Silverstein, 1985	• Adjustment for potential confounders (plant), not performed in the analysis. • The study population figures quoted represent a 90% overall response rate from those originally selected.

Table 3.9 Work relatedness of tension neck syndrome (Cont.)

Study population[1]	Outcome[2] and exposure information	Study design	Findings					Authors	Comments
				Prevalence		Risk ratio	95% CI		
				Exposed subjects	Comparison group	OR			
Female students (n = 6), aged 18–29 years.	• Exposure was repetitive shoulder flexions 0–90 degrees at a rate of 15 per minute for 60 minutes with up to 3.1 kg weight in hand.	Laboratory		All 6 subjects had pain and tenderness in trapezius muscle.	N/A	N/A	N/A	Hagberg, 1981	• Muscle tenderness was probably due to muscle soreness caused by unaccustomed exercise.

[1]Based on the information available in the article, no subject was lost from the original study population when only one set of numbers appears in the table under the 'study population' or when there is no relevant information in the 'comments' column.
[2]Unless otherwise specified all outcomes in Table 3.9 were tension neck syndrome and were identified by medical examination.

There was consistency (replicability) across these studies showing increased risk of tension neck syndrome for this occupational group. Further, there was also consistency (replicability) across studies of other occupations showing increased risk of tension neck syndrome with repetitive work and constrained head and arm posture.

3.4.4.5 Predictive performance of the association

Would the association between data terminal work and tension neck syndrome predict an increased risk of tension neck with increased duration of keyboard work in any job? There are as yet too few studies to establish predictive models, with cumulative exposure-response studies especially lacking. Similarly, evidence for or against predictive performance of the association between work and this disorder is lacking for the other occupational groups studied.

3.4.4.6 Coherence of the association observed with current knowledge

Many hypotheses and mechanisms have been proposed to explain the pathogenesis of tension neck syndrome. These are not necessarily mutually exclusive. Hypotheses that might be coherent with the possibility of an association between work and this disorder are discussed below.

WORK AND THE OVERLOAD OF TYPE I FIBRE

One hypothesis for the pathogenesis of tension neck syndrome is that prolonged static contractions of the trapezius muscle during work or in daily activity result in an overload of type I muscle fibres. Type I muscle fibres are the ones used in low static contractions. When biopsies from patients with chronic trapezius muscle pain are contrasted with those from healthy subjects, the findings show that in patients with this chronic problem, type I fibres are larger and there is a lower capillary-to-fibre area ratio for type I and for type IIA fibres (Lindman *et al.*, 1991a). These changes may provide one explanation for the rapid fatigue and pain observed in such patients. Another morphological study of female industrial workers has also shown strain on type I fibres. In 8 out of 10 patients with chronic trapezius myalgia attributed to repetitive assembly work, 'ragged red' fibres were found among type I fibres (Larsson *et al.*, 1988). Furthermore, reduced ATP and ADP levels were found among these myalgic patients as compared with those of healthy referents. Lower ATP levels in type I and type II fibres among patients with chronic trapezius myalgia were also reported by Lindman *et al.* (1991b).

WORK, REDUCED BLOOD FLOW AND PAIN SENSITIZATION

Reduced local blood flow in the trapezius muscle was found to be correlated with myalgia and ragged red fibres in 17 patients with chronic myalgia thought to be associated with static load during repetitive assembly work (Larsson *et al.*,

1990). This disturbance in muscular microcirculation may lead to sensitization of pain receptors (nociceptors) in the muscle, and pain even at rest. This pain receptor sensitization means that low-threshold stimuli can activate the pain system or even that spontaneous activity can occur in nociceptive nerves (Henriksson and Bengtsson, 1991).

WORK AND ENERGY DEPLETION

Fatiguing muscular activity can result in severe depletion of ATP in a small proportion of the muscle fibres (Foster *et al.*, 1986). Energy depletion may result in activity-related pain (Layzer and Rowland, 1971). Muscular damage or energy depletion in occupational situations has been shown by findings of increased serum creatine kinase during work (Hagberg *et al.*, 1982; Mairiaux *et al.*, 1986); an increase in serum creative kinase is believed to indicate muscle damage.

WORK AND MUSCLE FATIGUABILITY

Besides causing pain, dysfunctional energy metabolism may also impair muscle function. In industrial workers with myofascial shoulder pain (tension neck syndrome) and unable to work, pathologic rapid muscle fatiguability was found by electromyography (Hagberg and Kvarnström, 1984). In patients with work related shoulder-neck myofascial pain, low elevation strength and a failure to recruit high-amplitude motor units were observed (Hagberg *et al.*, 1988).

WORK AND ECCENTRIC CONTRACTIONS

Edwards (1988) described 'hypotheses of peripheral and central mechanisms underlying occupational muscle pain and injury'. As an example of peripheral mechanisms, eccentric contractions may lead to muscle fibre Z-disc rupture (Fridén *et al.*, 1981). The rupture occurs between the small contractile units (sarcomeres). This type of muscle injury is a common finding in muscle soreness and is reversible if the muscle is allowed rest and recovery. In occupational situations, this type of mechanical damage may initiate a self-perpetuating degenerative process in muscle cells, so-called calcium-induced damage, as suggested in nutritional myopathies and muscular dystrophy (Jackson *et al.*, 1985). Edwards (1988) also suggests that the primary cause of the initial mechanical damage might be sought in altered central motor control, resulting in imbalance between harmonious motor cuff recruitment and relaxation of muscles not directly involved in the activity.

WORK AND GAMMA MOTOR NEURONS

Another hypothesis which may further explain the pathophysiological mechanism for work related shoulder-neck pain involves the gamma motor neurons (Johansson and Sojka, 1991). Local muscle ischemia or metabolites due to isometric or dynamic strain (as can occur in some work situations) stimulate muscle afferents, and may in turn activate gamma-motorneurons projecting to both homonymous and heteronymous muscles. The gamma-motorneurons influence the stretch sensitivity and discharges of secondary and primary spindle afferents. Increased activity in the primary muscle spindle afferents may cause muscle stiffness, leading to further production of metabolites in both homo- and heteronymous muscles, more stiffness, and repetition of the cycle.

WORK AND RECRUITING PATTERNS

Low static contraction during work may result in a recruiting pattern or motor programme, where only type I muscle fibres are used. This may lead to selective motor unit fatigue and damage (Hagberg, 1988). Correct motor-programme learning is important to reduce unwanted muscle strain (Forssberg *et al.*, 1991). An important preventive measure could be work tasks involving different force levels, and a variety of different motor programmes, so that energy depletion and fatigue of individual muscle fibres do not occur.

COHERENCE: OVERALL

According to the various pathophysiological mechanisms presented, an association between work and tension neck syndrome is coherent.

A NOTE ON GENDER DIFFERENCES

Shoulder-neck muscular pain is more common among females than males, both in the general population and among industrial workers (Hagberg and Wegman, 1987). Whether this skewed distribution is due to genetic difference between males and females or gender difference in exposure, both at work and at home, is not clear. For example, one possibility is that females may have an increased risk of work related myofascial pain in the trapezius muscle since females have more type I muscle fibres in the trapezius muscle than males (Lindman *et al.*, 1991a). Some of the myofascial pain may originate in type I muscle fibres.

3.4.5 Conclusion

Tension neck syndrome is an epidemiologic term for myalgias in the shoulder-neck region (also synonymous with shoulder-neck myofascial syndrome). It is defined by symptoms of pain (complaints) in the shoulder-neck region with simultaneous findings of tenderness over the shoulder-neck muscles. Reported tension neck syndrome is more common among females than among males; however, whether this difference is due to genetics or to gender difference in exposure both at work and at home is unknown.

High prevalences, 14–61 per cent, were found in the exposed group (i.e. workers exposed to static contraction of the shoulder-neck muscles, and/or use of force and repetition). The prevalences found among the referents were also substantial (1–28 per cent), although not as high as their respective exposed group in each individual study. The odds ratio ranged from two to seven. There is consistency for repetitive work and constrained head and arm posture across studies showing increased risk of tension neck syndrome among data terminal workers, typists and other professions. Predictive performance of the association has not been examined yet; cumulative exposure-response studies are lacking.

The pathophysiological mechanism is not clear, but may involve an imbalance in the nervous system, local ischemia of muscle and nociceptor sensitization. The possibility of an association between work related activities and tension nec syndrome is coherent with the various hypotheses proposed for the pathophysiological mechanisms.

3.5 Evidence of the association between work and selected joint disorders (bone and cartilage): osteoarthrosis

3.5.1 Brief description of the disorder

3.5.1.1 General background

The major joints of the body, despite differences in size and configuration, share a common structure in that the bearing surfaces of the joint are covered with a layer of articular cartilage. This white shiny tissue comprises about 50 per cent by dry weight collagen fibres in a fibrous network, proteoglycans, a complex molecule responsible for water retention of the cartilage and other specialized cells. The cartilage is spread upon a layer of smooth hard subchondral bone. These two components, in conjunction with synovial fluid secreted by the lining of the joint capsule, produce a structure capable of withstanding large loads with a very low coefficient of friction (Figure 3.8).

3.5.1.2 Osteoarthrosis

Osteoarthrosis (or osteoarthritis) can be considered a degenerative phenomenon of the joints (Figure 3.8). The articular cartilage and the subchondral bone are central to the process. An increase in cartilage water content is a very early change. The cartilage decreases in thickness despite an increase in the synthesis of the proteoglycan and collagen components. The process also includes changes in the subchondral bone with cyst formation and sclerosis. Where and how this process begins is still unclear. Osteoarthrosis is commonly considered either *primary* (idiopathic), i.e. developing in the absence of predisposing factors, or *secondary*, as a result of injury or other specific trauma to the joint.

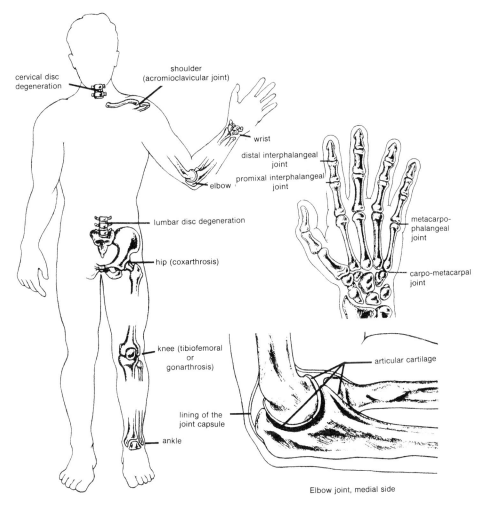

Figure 3.8 Joints of the body which can be affected by osteoarthrosis.

3.5.2 Literature reviewed

There is a large body of literature stretching back to the 1950s, which describes the prevalence of osteoarthrosis of different joints for various occupations. Peyron (1986) reviewed the main epidemiologic-etiologic evidence implying that mechanical forces were factors in the development of osteoarthrosis. He concluded that most of the epidemiologic studies pointed toward a relationship between usage and the development of osteoarthrosis. He did, however, note that a considerable number of studies demonstrated no difference with usage. A sample of the most-quoted studies on occupational osteoarthrosis is given in Table 3.10.

Table 3.10 Findings on the work relatedness of osteoarthrosis from a sample of the most-quoted studies[1] in the literature (Refer to Box 3.1 for help in interpreting the tables.)

Study population[2]	Outcome[3] and exposure information	Findings	Authors
Retired ballet dancers (n = 15 men, 28 women).	• Outcome assessed by medical history and X-rays (lateral and anteroposterior) of the hip, pelvis, knee and ankle joints (narrowing of joint space = sign of osteoarthrosis). • Exposure defined as the high level of physical activity of ballet dancers, with their joints subject to heavy weight-bearing and repeated trauma.	• *For the hip* (coxarthrosis): 13% prevalence in the dancers (statistically significantly increased when compared with prevalence data obtained from other studies of various age groups in the cities of Malmö and Copenhagen, $p < 0.001$). Cases had dancing careers of above-average duration. • *For the knee* (tibiofemoral arthrosis or gonarthrosis): 9% prevalence among the dancers (greater prevalence since there would not be any cases expected based on available data on series of individuals with similar age distribution). • *For the ankle* (bilateral ankle arthrosis – not preceded by fracture): 2% prevalence among dancers (ankle arthrosis prevalence in general population is unknown, however the condition is extremely rare). • *1st metatarsophalangeal joint (arthrosis)*: 54% prevalence among dancers (prevalence in general population unknown).	Andersson *et al.*, 1989
332 male shipyard workers compared with 179 white-collar workers from same shipyard and 173 male teachers from	• Outcome assessed by searching all records sources for X-ray referrals and reports for the knee. The films were	• The prevalence for gonarthrosis was 3.9% in exposed and 1.4% in comparison groups (difference	Lindberg and Montgomery, 1987

same city (closely matched for age).

classified with regard to osteoarthrosis of the knee joint (gonarthrosis), using Ahlbäck's criteria for gonarthrosis.

• Exposure was defined as heavy work for more than 30 years (exposed subjects were retired or close to retirement).

21 male manual workers, age matched with a comparison group of 49 subjects (from 70 subjects originally) with less manual experience.

statistically significant, $p < .05$).

• *Hand*: For the metacarpophalangeal joint, more of the exposed subjects had mild to moderate arthropathy: 15·0 (bilateral) increased $p < 0.025$.

Williams *et al.*, 1987

• Exposure was defined as 30 years or more of manual labour requiring the vigorous use of hands for manual labourers, whereas the comparison group had spent fewer than 20 years in manual labour.

• Outcome was determined by clinical examination and routine X-rays (hands/feet/cervical, dorsal and lumbar spine) for evidence of osteoarthrosis (graded 0–4).

117 male and 228 female cotton workers, 45 years and older, comprised the exposed group and were compared with age/sex matched community sample of 117 males and 228 females (who had never worked in a cotton mill). The predominant occupation of males from the community sample was coal mining. The exposed group represented an 80% to 85% response rate, depending on sex and age, from the original sample. To these subjects were added ex-cotton mill workers found in the community sample used for comparison groups.

Men only:
• *Distal interphalangeal joints (grade 3–4)*: 12% prevalence in cotton workers and 3% in comparison group (difference significant $p < 0.01$).
• *Proximal interphalangeal joints (grade 2–4)*: 20% prevalence in cotton workers and 9% in comparison group.
• *Metacarpophalangeal joints (grade 2–4)*: 15% prevalence in cotton workers and 14% in comparison group.
• *Carpometacarpal joints (grade 2–4)*: 25% prevalence in cotton workers and 12% in comparison group.
• *Wrists (grade 2–4)*: 5% prevalence in cotton workers and 12% in comparison group.

For female cotton workers, patterns of osteoarthrosis were similar to those of controls.

Lawrence, 1961

Table 3.10 Sample of studies on osteoarthrosis (Cont.)

Study population[2]	Outcome[3] and exposure information	Findings	Authors
94 male underground miners of 40–50 years of age compared with 94 male workers (47 office and 47 manual) from the same mine who had never gone underground (final study population after losses = 84 miners, 87 non-miners – 42 office and 45 manual workers).	• Outcome determined by routine general clinical examination and X-ray examination. • Exposure discussed as: 'coal mining entails much heavy lifting and laborious work, under awkward conditions beneath a low roof which tends to keep the spine in a flexed position and on an uneven and often steeply inclined floor which makes the foothold insecure, so that the frequency of exceptional stress and injury to the lumbar spine and leg joints in this occupation must be considerable, but the importance of these and other possibly causative factors must be assessed by suitably designed investigations'.	• *Severe lumbar disk degeneration*: prevalence of 43% in miners compared with 18% in manual workers ($.05 < p < .02$). There is no difference between 18% in manual workers and the prevalence of 7% in office workers ($p < .2$). • *Severe cervical disk degeneration*: prevalence of 18% in miners, 18% in manual workers and 19% in office workers. No difference between groups. • *Severe osteoarthritis of the knee*: prevalence of 6% in miners, 2% and 0% in manual and office workers respectively. No statistically significant difference between groups. • *Severe osteoarthritis of the hands*: 4% prevalence in miners, 2% and 0% in manual and office workers respectively. No statistically significant difference between groups.	Kellgren and Lawrence, 1952

Labourers (n = 332 males) from a Malmö shipyard compared with: (1) internal comparison groups – 179 white-collar workers from the same shipyard and 173 male teachers from the same city (matched for age); and (2) 438 men randomly selected from the general population (matched for age).	• For the 1122 men in the study, outcome was determined by searching all records for X-ray examinations. When X-ray reports were found, films were reviewed for hip coxarthrosis. • Exposure defined as heavy labour for shipyard workers.	• *Hip*: 3.3 per cent prevalence of coxarthrosis in labourers, 3.1 per cent in internal comparison groups, 1.6 per cent in general population. No difference between groups.	Lindberg and Danielsson, 1984

[1]Studies quoted here were all of cross-sectional design.

[2]Based on the information available in the article, no subject was lost from the original study population when only one set of numbers appears in the table under the 'study population'.

[3]Unless otherwise specified, all outcomes in Table 3.10 were determined using X-rays, with a standardized grading system (narrowing of joint space = sign of osteoarthrosis), usually Kellgren and Lawrence (1957).

There have been a large number of review articles and case series. Patterns of usage commonly cited include, for example, metacarpophalangeal (MCP) arthropathy in farmers and labourers (Williams *et al.*, 1987), work with pneumatic drills for more than two years and severe osteoarthrosis of the elbow and wrist (Copeman, 1940) and degenerative changes in the distal inter-phalangeal (DIP) and metacarpophalangeal joints of the hand in pianists (Bard *et al.*, 1984).

In contrast to the limited number of studies on other musculoskeletal disorders, there have been many studies on the relationship between work and osteoarthrosis. Some studies have also attempted to assess work exposure to musculoskeletal loads other than by using surrogate variables, such as occupational groupings. Since these latter studies are considered to provide better measurement of work exposures, we will concentrate on these studies and will not further describe the majority of studies where occupation has been used as a surrogate measure for exposure.

3.5.3 Results of epidemiologic studies

Epidemiologic studies of osteoarthrosis have in common a reasonably well accepted case definition; the work of Kellgren and Lawrence (1957) defined a 0–4 point scale for defining the severity of osteoarthrotic changes (i.e. narrowing of joint space) observed on X-ray films. Collins's (1950) or Ahlbäck's (1968) criteria have also been used on occasion. Tables 3.11 to 3.13 summarize the few studies of osteoarthrosis where work exposures have been estimated independently of a generalized occupational grouping.

3.5.4 Results and causality assessment

3.5.4.1 Strength of the association

The strength of the association between work exposure and osteoarthrosis of the hip and acromioclavicular joint is moderate when exposure is estimated independently rather than using occupation as a surrogate measure of the risk (Tables 3.11 to 3.13). The size of the effect shows risk ratios in the range of 2–4. Higher relative risks were shown by Stenlund *et al.* (1992) who demonstrated relative risks of 7–10 for cumulative load lifted.

3.5.4.2 Specificity of the association

See Box 3.2

3.5.4.3 Temporal association

For osteoarthrosis, as for many of these musculoskeletal disorders, temporality, i.e. whether work or work exposure preceded the onset of the disorder, is difficult to establish. However, there is evidence that work exposure may accelerate the degenerative process (Vingård *et al.*, 1991a,b; Kohatsu and Schurman, 1991; Croft *et al.*, 1992; Stenlund *et al.*, 1992).

3.5.4.4 Consistency of the association

The consistency (replicability) of the findings in the literature is only moderate. Most reviews cite studies supporting the occupational- or sport-relatedness of osteoarthrosis along with others that do not. However, if one concentrates on studies where work exposure was estimated independent of occupation, then a much greater degree of consistency appears; static and dynamic loads and forces in body joint(s) are consistently shown across studies of various occupations as risk factors for osteoarthrosis of the joint(s) under study.

3.5.4.5 Predictive performance of the association

The ability to predict an unknown fact based on a hypothesis drawn from an observed association (in this case an association between work and osteoarthrosis) awaits further study. There is some evidence of a cumulative exposure-response relationship between osteoarthrosis and loads lifted and vibration exposure. Indeed, recent investigations by Stenlund *et al.* (1992) suggest the presence of a cumulative exposure-response relationship.

3.5.4.6 Coherence of the association observed with current knowledge

There is coherence between the variables identified as risk factors for the development of arthrosis in recent studies where exposure has been estimated and those found in animal and *in vivo* studies. Studies on animal models have shown that continuous joint compression produces osteoarthrotic changes within joints, within relatively short time periods (Salter and Field, 1960). Continuous immobilization has also been found to produce such changes (Langenskjöld *et al.*, 1979). Dynamically varying forces, such as those that occur in running, have not been found to produce such changes in animal models (Videman, 1982a,b). High impulsive (shock) loads have been identified as a risk factor (Radin *et al.*, 1972). Thus work related exposure could lead to the development of osteoarthrosis; this is coherent with our current knowledge.

3.5.5 Conclusion

There is some evidence of an association between work and osteoarthrosis. The strength of the association is moderate. Consistency of this association does exist in studies measuring work exposure and this association is also coherent

Table 3.11 Findings from selected studies on the work relatedness of osteoarthrosis of the hip joint (Refer to Box 3.1 for help in interpreting the tables; studies were selected because exposure had been estimated independently of generalized occupational grouping.)

Study population[1]	Outcome and exposure information	Study design	Findings				Authors
			Risk factor	RR[2]	95% CI		
Of 253 male hip prosthesis recipients (as a result of idiopathic osteoarthrosis), 233 participated in the whole study. They were compared with 302 male participants from 392 men who had been randomly selected from the general population.	Questionnaire and interview on occupational exposure from the start of their career to the year of diagnosis for cases; interview for comparison group. Exposure was categorized into 3 levels (low – medium – high). Exposures collected were: static and dynamic forces, static loads only, dynamic loads only, number of heavy lifts (> 40 kg) and number of jumps between different levels.	Case control	High compared with low: Static loads only Dynamic loads only Static and dynamic loads Number of heavy lifts Number of jumps	2.9 2.2 2.4 2.4 1.5	1.7–5.0 1.3–3.7 1.4–4.0 1.5–3.8 0.9–2.5	Results adjusted for age, BMI, smoking and sports activity up to 29 years of age.	Vingård *et al.* 1991a,b

Croft *et al.*, 1992

From a population of patients of 5 rural general practices (1231 men aged 60–76), 890 men (72 per cent) responded, among whom 289 had been involved in farming and 123 had done office work only. Of this group of 412, osteoarthrosis was ultimately assessed in 179 farmers and 71 office workers.

- Hip arthrosis determined by X-rays.
- Personal interview with structured questionnaire for occupational history.

Cross-sectional

- Prevalence

Age	Farmers	Office
60–65	13.8	2.7
66–70	13.0	0
71–76	23.6	5.0

- OR (farmers vs office workers): 7.8, 95% CI 1.8–33.8.
- Neither type of farming/agriculture (dairy, stock, sheep, etc.) nor farming activities (hand or machine milking, lifting, driving tractor, etc.) could explain increase found. Heavy lifting was postulated as the likely explanation.

[1]Based on the information available in the article, no subject was lost from the original study population when only one set of numbers appears in the table under the 'study population'.
[2]Incidence density ratios (relative risks) were estimated using the odds ratio.

Table 3.12 *Findings from selected studies on work relatedness of osteoarthrosis of the knee joint (Refer to Box 3.1 for help in interpreting the tables; studies were selected because exposure had been estimated independently of generalized occupational grouping.)*

Study population[1]	Outcome and exposure information	Study design	Findings			Authors
			Risk factor	Odds ratio	95% CI	
Of 138 cases, 68 per cent responded to a questionnaire. Of these respondents, those over 55 years of age and of educational attainment of high school graduation or higher were eligible as case subjects and matched (sex and age) to community sample comparison groups to give 46 pairs of case comparison groups (28 pairs were females).	• Cases were individuals with severe osteoarthrosis of the knee and treated with knee arthroplasty from 1977 to 1988. • Exposure to work (categorized into 4 levels of physical activity at work), sports, body mass index (BMI) and previous injury, all obtained by mail questionnaire.	Case control	Cases more likely to: • have performed moderate to heavy work, for subjects of: age 20–29 years age 30–39 years age 40–49 years (The above information is based on the text, as the table in the article may contain an error.) • practise sports • have a high BMI (subjects of age 40) • have previous injury	 2.3 3.4 3.0 No difference found 5.3 4.6	 0.9–6.1 0.9–10.8 0.9–11.4 1.6–23.0 1.5–16.5	Kohatsu and Schurman, 1991

[1]Based on the information available in the article, no subject was lost from the original study population when only one set of numbers appears in the table under the 'study population'.

Table 3.13 Findings from selected studies on work relatedness of osteoarthrosis of the upper limb (Refer to Box 3.1 for help in interpreting the tables: studies were selected because exposure had been estimated independently of generalized occupational grouping.)

Study population[1]	Outcome and exposure information	Study design	Findings	Authors
Out of an original study population of 16 winders, 39 burlers and 20 spinners (all females), 16 winders, 29 burlers and 19 spinners, with at least 20 years employment in a mill, participated.	• Outcome was degenerative joint disease scored on a 0–4 scale from plain X–ray. • Using direct observation and standard-time motion analysis, exposure was defined as tasks being highly repetitive, stereotyped and complex: Winders – bimanual with power grasp and much wrist movement. Spinners – right-handed precision grip (omitting digits 4 and 5). Burlers – right-handed precision grip.	Cross-sectional	*Hands:* Degenerative joint disease score for the: • *Distal interphalangeal joints* [1,2] Right index finger Winders[3] / Spinners / Burlers 0.5 ± 0.8 / 0.9 ± 0.9 / 1.2 ± 1.0 Right middle finger Winders[3] / Spinners / Burlers 0.4 ± 0.7 / 1.0 ± 0.8 / 1.0 ± 1.1 • *Metacarpophalangeal joint*[1,2] Winders/Spinners/Burlers Right 0.4 / 0.3 / 0.2 Left[4] 0.2 / 0.1 / 0.3 Notes: (1) Differences in range of motion, misalignment and joint size also reported. (2) Other joints showed no difference between exposures (tasks). (3) Winders have less disease ($p < .05$ – Kruskal Wallis). (4) Left side has less disease ($p < .05$ – Kruskal Wallis).	Hadler *et al.*, 1978

Table 3.13 Work relatedness of osteoarthrosis of the upper limb (Cont.)

Study population[1]	Outcome and exposure information	Study design	Findings — Risk factor	Odds ratio	95% CI	Authors
Out of union files, a randomly selected study group of 75 bricklayers, 75 rock blasters and 110 foremen (engineers, not manual workers): 54 brick-layers, 35 rock blasters and 98 foremen participated.	• Outcome assessed by 5-grade severity of osteoarthrosis according to Collins (1950) using anteroposterior X-ray views of the right and left acromioclavicular joints. • Exposure measured as cumulative load lifted during working life.	Cross-sectional	*Acromioclavicular joint of the left side* Load lifted: < 710　　　tonnes 710–25 999　tonnes > 26 000　　tonnes Vibration exposure: < 9001　　hours 9001–225 200 hours > 225 200　hours Results for the acromioclavicular joint of the right side showed similar increases of osteoarthrosis in each of the risk factor categories listed for the left side; however, the difference was not always as great.	1 7.3 10.3 1 2.2 3.1	– 2.5–21.3 3.1–34.5 – 1.0–4.7 1.4–7.0	Stenlund et al., 1992

[1]Based on the information available in the article, no subject was lost from the original study population when only one set of numbers appears in the table under the 'study population'.

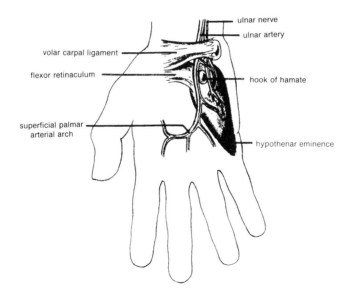

Figure 3.9 Structures involved in the hypothenar hammer syndrome.

with current knowledge. There is also some evidence of a cumulative exposure-response relationship between osteoarthrosis and loads lifted and vibration exposure.

3.6 Evidence of the association between work and selected vascular disorders: hypothenar hammer syndrome

3.6.1 Brief description of the disorder

Hypothenar hammer syndrome (Figure 3.9) consists of symptoms and signs of digital ischemia caused by thrombosis and/or aneurysm of the ulnar artery and/or the superficial palmar arch (Butsch and Janes, 1963; Kleinert and Volianitis, 1965). The most frequently reported symptoms are cold sensitivity in the affected hand, numbness, paresthesia, or colour changes without cold exposure (Stroud and Thompson, 1985). Diagnosis is based on symptoms and a manual test of the radial and ulnar artery blood supply to the hand, called Allen's test (Nilsson *et al.*, 1989).

3.6.2 Literature reviewed

There were several reports of case series and articles on surgical and diagnostic procedures. Only two cross-sectional studies were available showing an exposure-response relationship. In one of the cross-sectional studies a cumulative exposure-response relationship was also shown (Nilsson *et al.*, 1989).

3.6.3 Results of epidemiologic studies

In a study of 127 male vehicle maintenance workers, 11 (8.7 per cent) had evidence of ulnar artery occlusion in one or both hands (Little and Ferguson, 1972). Out of the 127 workers, 79 gave a history of habitually using their hands as hammers and the 11 men with ulnar artery occlusion were all considered to be part of this group of 'hammerers'. In the hammering group, therefore, the prevalence of ulnar artery occlusion was 14 per cent (Table 3.14).

In a study of plate workers and office workers, the prevalence of a pathologic Allen's test of the right ulnar artery was 37 per cent among plate workers and 20 per cent among office workers (Nilsson *et al.*, 1989). For the left ulnar artery, the prevalence was 30 per cent and 22 per cent for the plate workers and office workers respectively (Table 3.14). There was also a cumulative exposure-response relationship observed with a logistic regression model using age and vibration-years as exposure; each vibration-year gave an odds ratio of 1.06 (95 per cent confidence interval 1.01–1.11) (Nilsson *et al.*, 1989).

3.6.4 Results and causality assessment

3.6.4.1 Strength of the association

Both studies showed an association between exposure and hypothenar hammer syndrome although confidence intervals were wide for the risk evaluations, implying that the strength of the association was weak.

3.6.4.2 Specificity of the association

See Box 3.2

3.6.4.3 Temporal association

In the studies of hypothenar hammer syndrome, it was not clearly demonstrated that exposure preceded the onset of symptoms. The latency between onset of exposure and development of hypothenar hammer syndrome is unknown.

3.6.4.4 Consistency of the association

There is consistency (replicability) across the two cross-sectional studies of different occupations and reported case series, showing an association of hypothenar hammer syndrome with using the hand as a tool.

3.6.4.5 Predictive performance of the association

There is not yet any definite predictive model. However, existing studies predict that hypothenar hammer syndrome will develop if the hand is used as a hammer.

3.6.4.6 Coherence of the association observed with current knowledge

The pathogenesis of hypothenar hammer syndrome shows coherence with a possible association with work exposure. The ulnar artery is most vulnerable to injury immediately distal to the palmar carpal ligament (Figure 3.9). At this location, the hook of the hamate bone serves as an anvil against which the relatively exposed distal ulnar artery and the superficial palmar arch could be hammered upon (Conn *et al.*, 1970). Hypothenar hammer syndrome has also been described as a result of using the hand in sports such as karate (Conn *et al.*, 1970). Exposure to hand-arm vibration, such as is found in many work situations, may also be a causative factor for thrombosis in the ulnar artery or the superficial palmar arch (and thus increase the likelihood of developing hypothenar hammer syndrome). There is no scientific evidence, however, since exposure to hand-arm vibration is usually confounded by exposure in which the hand is used as a hammer.

3.6.5 Conclusion

Hypothenar hammer syndrome may be caused by using the hand as a hammer. The two cross-sectional studies available for review have demonstrated an increased risk in exposed workers.

3.7 Evidence of the association between work and bursa disorders: knee bursitis

3.7.1 Brief description of the disorder

A bursa is a closed sac with a synovia-lined membrane containing fluid. Bursae are usually located in anatomical areas with potential for exposure to mechanical friction. They are found in the knee (many bursae, including the prepatellar and superficial infrapatellar bursae – see Figure 3.10), hip (the iliopsoas bursa), elbow (the olecranon bursa), shoulder, ankle and wrist. The main function of the bursa is to facilitate the motion of tendons and muscles over bony prominences.

Bursitis is inflammation of a bursa. Bursitis may be caused by local friction and trauma, by general inflammatory diseases such as rheumatoid arthritis, and by bacteria (septic bursitis). Bursitis is most commonly found at the knee (prepatellar or superficial infrapatellar bursa) or elbow (olecranon bursa). Shoulder bursitis (subacromial bursitis) may develop in the process of degeneration of the rotator cuff tendons (see section 3.2 on tendinitis).

Table 3.14 Findings from selected studies on the work relatedness of hypothenar hammer syndrome (Refer to Box 3.1 for help in interpreting the tables and Table 3.1 to find out how the studies were selected.)

Study population[1]	Outcome[2] and exposure information	Study design	Findings				Authors	Comments
			Prevalence		Odds ratios	95% CI		
			Exposed subjects	Comparison group				
Male employees from government vehicle maintenance workshops who use their hands as hammers (n = 79) compared with male employees from government vehicle maintenance workshops who do not use their hands as hammers (n = 48).	● Outcome determined by Allen's test. When this suggested ulnar arterial block, an examination with a Doppler flow detector was performed. ● Exposure assessed by interview: habitual hand hammering defined as using the hand to hammer more than once a day.	Cross-sectional	14%	0%	16.3	2.7–100	Little and Ferguson, 1972	● Adjustment for potential confounders not performed in the analysis.

Study population	Comments		Right ulnar artery 20% 2.8		1.3–6.2	Nilsson *et al.*, 1989
Out of 112 male plate workers in a company, 89 were included in the study population and were compared with male office workers at the same plant (*n* = 61 were randomly admitted to the study population from 500 office workers).	• Outcome considered as a pathological Allen's test of reflow time through the ulnar artery. • Exposure based on job. For each job, vibration exposure was assessed by questionnaire, observation, vibration level, and duration measurements. Plate workers were exposed to vibration mainly during grinding but also used their hands as hammers. • Adjustment for age in the analysis.	Cross-sectional	37%			
			Left ulnar artery 22% 1.8		0.84–4.0	
			30%			

[1] Based on the information available in the article, no subject was lost from the original study population when only one set of numbers appears in the table under the 'study population' or when there is no relevant information in the 'comments' column.

[2] Unless otherwise specified, all outcomes in Table 3.14 were hypothenar hammer syndrome and were identified by medical examination.

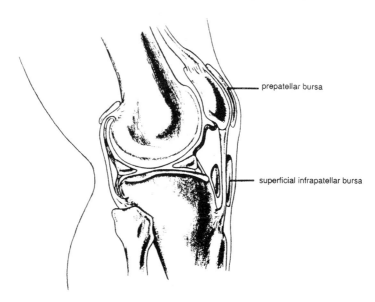

Figure 3.10 Some bursae of the knee joint.

3.7.2 Literature reviewed

Most articles published deal with the clinical appearance and treatment of bursitis. Studies have only recently dealt with the work relatedness of bursitis and, to date, the only studies published on work relatedness have looked at knee bursitis. There are no epidemiologic studies on work and shoulder bursitis or work and olecranon bursitis. The epidemiology of bursitis in the general population is unknown. Even the incidence or prevalence of bursitis in patients with general inflammatory diseases is not fully known.

3.7.3 Results of epidemiologic studies

In the USA, carpet layers comprised less than 0.06 per cent of the workforce in 1979 and yet they submitted 6.2 per cent of all compensation claims for traumatic knee inflammation (Thun *et al.*, 1987). In a cross-sectional study, the prevalence of bursitis was assessed among 112 carpet and floor layers, 42 tile and terrazzo setters, and 243 millwrights and bricklayers (Thun *et al.*, 1987). For carpet and floor layers, the prevalence of bursitis was 20 per cent compared with 6 per cent among millwrights and bricklayers. This corresponded to a prevalence ratio of 3.2 with a 90 per cent confidence interval of 1.9–5.4. The tile and terrazzo setters had a prevalence rate of 11.2 per cent which, compared with millwrights and bricklayers, gave a prevalence ratio of 1.8 (90 per cent CI 0.8–3.9). The exposure factor proposed as a determinant for the bursitis was the amount of kneeling required on the job. A kneeling score (six categories, from 'never' to 'always') was obtained by questionnaire. Carpet and floor layers scored 5.5, tile and terrazzo setters scored 4.9, and millwrights and bricklayers scored 2.9. Kivimäki *et al.* (1992) found a 19 per cent prevalence of prepatellar bursitis among 168 carpet layers, compared with 2 per cent among 146 painters when workers were examined by a physician. This corresponded to an odds ratio of 11.2 with a 95 per cent confidence interval of 3.4–38. In an observational study of a sub-sample of the layers and painters, they found that carpet and floor layers were in a kneeling position, using one or both knees, approximately 42 per cent of the observation time, whereas for painters kneeling was rare.

Other types of bursitis are thought to be due to physical exposure, and therefore could possibly result from work situations where these physical exposures are found. Vigorous hip flexion and extension have been pointed out as possible causes of iliopsoas bursitis (O'Conor, 1933; Toohey *et al.*, 1990). However, this proposed association between physical exposure and bursitis is based on case descriptions, since no epidemiologic studies are available. Septic bursitis (caused mainly by the *Staphylococcus aureus* bacteria) was found to be most frequently located in the olecranon and prepatellar bursae in a case review (Raddatz *et al.*, 1987). Skin breakage, trauma and/or occupational risk factors were associated with these infections (Raddatz *et al.*, 1987). In a review of 47 episodes of septic bursitis, half of the cases were related to occupational or recreational trauma (Pien *et al.*, 1991).

3.7.4 Results and causality assessment

3.7.4.1 Strength of the association

The strength of the studies showing the work relatedness of bursitis for carpet and floor layers is fairly high. Attributable fractions (exposed) have been estimated for both the US study (Thun *et al.*, 1987) and the Finnish study (Kivimäki *et al.*, 1992); respectively, these are 0.69 (95 per cent CI 0.47–0.81) and 0.91 (95 per cent CI 0.70–0.97). More information on attributable fractions may be found in appendix III.

3.7.4.2 Specificity of the association

See Box 3.2

3.7.4.3 Temporal association

Only two epidemiologic studies have examined the work relatedness of bursitis. Both studies were cross-sectional in design and the criterion of temporality, which means that exposure preceded the onset of the disorder, was not demonstrated in these studies.

3.7.4.4 Consistency of the association

There is consistency (replicability) across studies showing increased risk of bursitis with kneeling work for carpet and floor layers, and tile and terrazzo setters. The work relatedness of bursitis is implied by the two cross-sectional studies from the USA and Finland, and also in case reports. Friction and mechanical compression (caused by kneeling, for instance) are exposure factors attributed to bursitis.

3.7.4.5 Predictive performance of the association

Whether observed associations can be used as predictors for the association has not yet been tested for bursitis. No clear cumulative exposure-response relationship has yet been shown.

3.7.4.6 Coherence of the association observed with current knowledge

The pathogenesis of bursitis is not completely understood. Friction and mechanical compression due to kneeling may cause mechanical trauma, and may lead to local inflammation of the synovial membrane of the bursa, resulting in thickening of the membrane. Thickening of the prepatellar or superficial infrapatellar bursa, on ultrasonography examination, was found in 49 per cent of 96 carpet and floor layers, compared with only 7 per cent of the 72 house painters examined (Kivimäki, 1992). This corresponded to an odds ratio of 12.9 with a 95 per cent confidence interval of 4.8–35. Fluid collection (i.e. a sign of bursitis) in the prepatellar or superficial infrapatellar bursa was also noted for 10 carpet and floor layers, and this ultrasonic finding was associated with knee pain in kneeling postures (Kivimäki, 1992). Thus work related exposures could lead to the development of knee bursitis; this is coherent with our current knowledge.

3.7.5 Conclusion

Bursitis is located mainly in the knee (prepatellar or superficial infrapatellar bursa) or elbow (olecranon bursa). Only two cross-sectional studies were identified as having examined the work relatedness of bursitis, and these looked at knee bursitis. The strength of the association in both studies was high and kneeling was consistently found to be the risk factor associated with bursitis. The temporal association between work and bursitis was not examined, and the predictive performance of this association was not demonstrated in either study. Although the pathogenesis of bursitis is not clearly understood, that postulated is coherent with an association between work related factors and the development of bursitis. Carpet and floor layers with exposure to kneeling work were at greater risk of developing patellar bursitis than workers in occupations requiring little kneeling.

Table 3.15 Stages of occupational cervicobrachial disorder (OCD)* (according to the Committee on OCD of the Japanese Association of Industrial Health, 1973)

Stage	Criteria
1	Subjective symptoms such as general fatigue, increased irritability.
2	Muscular induration and/or tenderness in the neck, shoulder or arm region is added to stage 1.
3	The addition of clinical signs to stage 2: • exacerbation or extension of muscular induration or tenderness • positive signs on neurological tests • sensory impairment • decreased muscle strength • tenderness on percussion over the spinal processes • tenderness over the paravertebral area • tenderness along the nerve branches • tremor of the fingers or eyelids • stiffness or restriction of movements of the neck, shoulders, hands or fingers • poor peripheral circulation • severe subjective complaints
4	Many of the signs of stage 3 are strongly exacerbated, and specific syndromes are found: • organic disturbances (tenosynovitis, tendinitis, peritendinitis, etc.) • symptoms resembling cervicobrachial syndrome in orthopedics • autonomic imbalance (Raynaud's phenomenon, disturbances in equilibrium, cardiac neurosis, etc.) • mental symptoms (emotional instability, insomnia, depressive states, etc.)
5	Obvious disturbances in work and daily life.

*This term is used in Japan and is equivalent to RSI, CTD, OOS or what we have called 'work related musculoskeletal disorders' (WMSDs) here.

3.8 Evidence of the association between work and unspecified musculoskeletal symptoms or multiple-tissue disorders: CTD, RSI, OOS, OCD

3.8.1 Brief description of the symptoms and disorders

Symptoms of recurring or persistent pain, numbness, aching, burning, stiffness in specific body regions or across the upper limb with concomitant headaches, often with unspecified or not easily categorized findings on clinical examination, have been lumped into broad categories termed occupational cervicobrachial disorder (OCD), cumulative trauma disorder (CTD), occupational overuse syndrome (OOS), repetitive motion disorder and repetitive strain injury (RSI). In some cases, various specific disorders have been combined under these more general headings, by anatomical location (hand-wrist disorders) to increase the statistical power of epidemiologic studies. Criteria have been suggested for these catch-all disorders. The Committee on Occupational Cervicobrachial Syndromes of the Japanese Association of Industrial Health (1973) defined these disorders as follows (Keikenwan Shōkōgun Iinkaï, 1973):

> OCD is functional and/or organic disturbances induced by neuromuscular fatigue due to work in a fixed position or with repetitive movement of the upper extremities, though mental or environmental factors can play some role in forming the features of OCD. Some orthopedic diseases such as tenosynovitis, arthritis or scalenus syndrome may explain some or part of the disorders but new criteria with regard to the whole features of OCD should be developed.

Five stages were identified (Table 3.15). Using this approach, Japanese investigators have reported prevalence of OCD among a variety of occupations between 1960–85, ranging from 2 per cent to 28 per cent, with stage III–V prevalence of 16 per cent among telephone operators and assembly line workers and 13 per cent among film-rolling operators (Ohara *et al.*, 1982).

There have been similar attempts at identifying stages in repetitive strain injury (Browne *et al.*, 1984). In the United States, the Bureau of Labor Statistics (BLS) reported that 80 per cent of the increase in occupational diseases between 1989 and 1990 was due to disorders associated with cumulative trauma (Bureau of Labor Statistics, 1992). Unfortunately, the definition includes low back pain and noise-induced hearing loss. In general, reporting of these disorders has increased dramatically in most countries in the past five years.

3.8.2 Literature reviewed

The majority of studies have been cross-sectional questionnaire studies with job title or self-reported risk factors as surrogates for exposure. Cohort studies have tended to use sick leave due to musculoskeletal disorders and monitored changes in exposure globally. Case definitions have differed across investigations and across countries. The Nordic Questionnaire (Kuorinka *et al.*, 1987) has been used by a number of investigators to standardize questions about symptoms. This questionnaire asks about symptoms by general body region for the previous 12 months and the previous seven days. The majority of studies consider age and gender at least potential modifiers and many use time employed as a surrogate for duration of exposure. The potential misclassification bias from poorly delineated health effects or global estimates of exposure would tend to underestimate effects. A few studies address the potential impact of job satisfaction and social support on the expression of 'disorders' via reporting symptoms.

3.8.3 Results of epidemiologic studies

The results of the epidemiologic studies reviewed are presented in Table 3.16; some of these are discussed below. The prevalence of these 'unspecified disorders' varies considerably among manufacturing and service industry workers. The prevalence of symptom complexes range from 2 per cent among cashiers whose work has been ergonomically redesigned (Ohara *et al.*, 1982, previously 17 per cent), to 81 per cent among machine operators (Tola *et al.*, 1988, lifetime prevalence) and 98 per cent among sewing machine operators (Vihma *et al.*, 1982). Females tend to report a higher prevalence of neck and upper-limb symptoms than males (3:1), but rarely are they doing the same job. There is an inconsistent relationship with age, with younger workers often reporting a higher prevalence of disorders than older workers. This may reflect secondary selection ('healthy worker' effect) or generational differences in sympton reporting. Work related risk factors most consistently identified include: (1) repetitive (monotonous) hand work; (2) high physical load; (3) static load; (4) vibrating tools with hand and elbow pain; and (5) increasing intensity or duration of exposure (Table 3.16). There were no consistent findings with specific postures for hand/wrist disorders but overhead work appeared to be associated with neck/shoulder disorders. Blue-collar workers have a higher prevalence of disorders than white-collar or management workers. There is some suggestion that stress symptoms may both predict and result from chronic musculoskeletal disorders (Leino, 1989). Job satisfaction may be associated with the expression of these symptoms but studies to date have conflicting results.

3.8.4 Results and causality assessment

3.8.4.1 Strength of the association

The strength of the association with work related risk factors varied with the populations studied, precision of exposure measures and extent of controlling for confounding or effect modification. General population-based studies that asked questions about workload tended to find more modest associations than those looking at working populations. Kivi (1984) used physician-reported occupational disorders in which 'repetitive tasks' were mentioned in the report to calculate relative risks of rheumatic disorders associated with repetitive tasks in Finnish industries between 1975 and 1979. Sixty-three per cent of the occupational disorders were located in the elbow-forearm region. The highest relative risks were for butchers (RR 11.5), workers in the food industry (RR 8.8) and packers (RR 7.9) when compared with the mean incidence for all occupations. Relative risks were higher for females than males, although this may be the result of differing exposures.

In the Mini-Finland population-based study of chronic neck pain in adults (based on history and clinical examination), Mäkelä *et al.* (1991) reported 10 per cent and 14 per cent prevalence in men and women, respectively. Lifetime prevalence of symptoms was 71 per cent and period prevalence was 41 per cent (for the month preceding examination). Chronic neck pain syndrome was lowest among professionals and those never employed, and increased with age. Age- and sex-adjusted prevalence was associated with a history of injury to neck,

Table 3.16 Findings from selected studies on the work relatedness of unspecified or multiple-tissue musculoskeletal disorders: repetitive strain injury (RSI), cumulative trauma disorders (CTD), occupational cervicobrachial disorders (OCD), neck and shoulder symptoms, etc. (Refer to Box 3.1 for help in interpreting the tables and Table 3.1 to find out how the studies were selected.)

Study population[1]	Outcome and exposure information	Study design	Findings	Authors	Comments
Keyboard users: 45 cases of RSI were compared with 125 non-cases.	• Outcome based on symptoms survey and medical information in records. • Exposure based on number of hours keying per day.	Case control	• OR for RSI 12.0, 95% CI 5.0–30.0 for using keyboard > 6 hrs/day.	Oxenburgh, 1984	• No difference in work stations.
Out of 152 original participants, 136 investment casting workers participated through to full follow-up.	• Standardized interview and non-invasive physical examination (PE) used for outcomes (with set criteria, for interview and PE, for determining positive CTD). • Job analysis and video/EMG used to estimate forcefulness and repetitiveness of jobs.	Cohort: 3-year follow-up	• 35% of all job transfers due to CTD. • Compared with low force/low rep, the odds ratios were 9.8, 95% CI 3.2–30.2 for force and 8.5, 95% CI 2.7–26.4 for repetitiveness.	Silverstein et al., 1987	• No difference in job satisfaction. • Analysis controlled for age, years on the specific job and sex.
800 female sewers (representing an 80% response rate from the original population) compared with a similar sub-group of women from the Canadian Health Survey.	• Disability assessed by interview. • Exposure based on job category (plus sewers' records for work history of piece work).	Cross-sectional	• Severe disability risk ratio for musculo-skeletal disorders for sewers compared with sub-group: OR 6.9 with 95% CI 3.1–15.1. Effect increases with years doing piecework	Brisson et al., 1989	• Analysis controlled for age, smoking and employment status. • 5-year difference in the administration of the questionnaire between garment workers and the sub-

Description	Methods	Study type	Results	Reference	Comments
Out of an original population of 194 riveters and 194 manual workers, 147 and 125 responded respectively. After excluding subjects not meeting selection criteria, 101 aircraft riveters were compared with 76 manual workers as the comparison group not exposed to vibration.	• Outcome was assessed by questionnaire (using sensorineural stages and parts of the Nordic questionnaire). • Exposure was measured by accelerometer (measure of vibration); time-work studies allowed estimate of daily exposure time.	Cross-sectional	• 22% vibration-induced white fingers (VWF) in riveters *vs* 11% in comparison groups. • For wrist pain/stiffness in previous 12 months: OR 5.9 with 95% CI 2.2–18.1, increase with years of riveting. Walk-through indicated equivalent levels of repetitive work for riveters and manual workers.	Burdorf and Monster, 1991	and duration of employment. • Inadequate analysis of other potential work risk factors.
Out of 214 active garment workers, 205 were physically examined and sufficient information was obtained by questionnaire on 179 workers (162 women and 17 men). Men were excluded and the 162 women were compared with 73 female full-time workers in a hospital (excluding typists). This latter figure of 73 represented 34% of all the full-time and part-time hospital workers.	• Outcome measured with self-reported questionnaire (persistent pain = pain that lasted for most of the day, for one month or more within the last year). • Exposure assessed by questionnaire for work history and video analysis.	Cross-sectional	• Persistent wrist pain: prevalence odds ratio 3.9, 95% CI 1.4–10.9 for garment workers compared with hospital workers. Highest ratios for stitchers and finishers compared with hospital workers.	Punnett and Robins, 1985 / Punnett *et al*, 1985	group.

Table 3.16 Work relatedness of unspecified or multiple-tissue musculoskeletal disorders (Cont.)

Study population[1]	Outcome and exposure information	Study design	Findings	Authors	Comments
148 plastic assemblers and 76 former assemblers (who had left in the 4 years preceding study) were compared with 60 community referents (randomly sampled). Females only.	• Nordic questionnaire was used to obtain musculo-skeletal complaints for outcome and to obtain task info for exposure (e.g. work pace = items/hour).	Cross-sectional	• Current shoulder pain (in previous 7 days) OR 3.4 with 95% CI 1.6–7.1, and hand pain OR 2.8 with 95% CI 1.2–6.4 greater in assemblers (current and former) than in comparison group. Shoulder pain increased with job duration and pace. • No statistically significant difference between current and former assemblers for frequency of pain.	Ohlsson *et al.*, 1989	• Interaction with age differed by body region.
Out of 352 eligible active or retired manual and office workers in shipyards, 327 were available for the full follow-up (7% lost to follow-up).	• Outcome assessed by questionnaire during health check-ups or by mail. • Exposure based on job category.	Cohort: 3-year follow-up	• Neck/shoulder symptoms persisted after retirement and increased for active manual workers but not office workers: RR 2.6, 95% CI 1.0–7.2 for neck; RR 3.3, 95% CI 1.4–8.4 for shoulder.	Berg *et al.*, 1988	
22 180 Swedish employees undergoing routine	Outcome and exposure assessed by work and health	Cross-sectional	• Prevalence of neck pain requiring	Linton, 1990	• Smoking and age were modifiers. ORs

Subjects	Methods	Study type	Results	Comments	Reference
screening examination at their health care units (approximately 60% were males).	questionnaires done with routine health screening.		treatment: 14% men, 23% women. • Neck pain associated with heavy lifting (OR varied from 1.2–2.1 depending on age), monotonous work (OR varied from 1.8–3.5 depending on age), uncomfortable posture (OR varied from 1.3–2.8 depending on age) and vibration (OR varied from 0.8–2.3 depending on age). • Odds ratios increased with age and with poor psychosocial work environment: heavy lifting + psycho = OR 2.7, 95% CI 2.0–3.6; monotonous + psycho = OR 3.6; 95% CI 2.8–4.6; posture + psycho = OR 3.5, 95% CI 2.7–4.5.	increased with age for physical factors but not psychosocial factors.	
Machine operators (n = 852), carpenters (n = 696) and office workers (n = 674). Men only. These figures represent a response rate of 74% for machine	• Outcome obtained by mail questionnaire on subjective neck/shoulder symptoms in previous 12 months. • Exposure obtained by questionnaire: recall of loading factors at work.	Cross-sectional	• Machine vs office workers: OR 1.7 with 95% CI 1.5–20. • Carpenters vs office workers: OR 1.4 with 95% CI 1.1–1.6. • Twisted, bent posture	• Compared with what would be expected for women doing routine office work, men doing office work were found to have fewer loading factors.	Tola *et al.*, 1988

Table 3.16 *Work relatedness of unspecified or multiple-tissue musculoskeletal disorders (Cont.)*

Study population[1]	Outcome and exposure information	Study design	Findings	Authors	Comments
operators, 67% for carpenters and 67% for office workers from the original study population.			very often *vs* rarely: OR 1.8 with 95% CI 1.5–2.2. • Air drafts, job satisfaction and age also associated with symptoms.		• Injury, obesity and parity also predictors.
A population sample of 8000 Finns were invited to participate, and 7217 did (*n* = 3322 men, 3895 women).	• Outcome established by screening examination. If findings were suggestive, further interviews and a standardized examination by physician for diagnosis of chronic neck syndrome were done. • Exposure measured using the Nordic classification of occupations and self-reported features of physical stress (2 levels: yes and no) and mental stress (3 levels: none, mild and severe).	Cross-sectional	• Prevalence of 10% men, 14% women for chronic neck syndrome. • OR physical stress: 1.4, 95% CI 1.3–1.4; mental stress, per each additional stress level: 1.3, 95% CI 1.2–1.4.	Mäkelä *et al.*, 1991	
408 full-time sewing operators (8 hours per day) compared with 210 part-time operators (5 hours per day).	• Outcome defined as sick leave of more than 3 days for musculoskeletal disorders, with a doctor's certificate (number of days	Cohort: 1978-81	• Full-time workers compared with part-time had OR 1.26 with 95% CI 0.8–2.0 for neck and upper-limb	Waersted and Westgaard, 1991	• Controlled for many potential confounders, survival analysis most useful.

					Comments
	per worker-year). • Exposure based on hours worked per day (8 hrs vs 5).			disorders sick leave. For a 2-year follow-up, part-time work was shown to postpone the occurrence of musculoskeletal disorders by approximately half a year.	• No difference in number of meals prepared, job content or demographics.
Huang *et al.*, 1988	At lunch centre A, of the 24 full-time female workers, all answered the questionnaire and 20 underwent the physical examination. At centre B, of the 26 female full-time workers, 20 did both.	• Outcome established using the OCD Japanese questionnaire and physical examination. • Exposure established using video analysis and NIOSH methods in Work Practice Guide on manual lifting (1981).	Cross-sectional	• 50% wrist pain at centre A (35% CTS on physical exam) vs 30% (5% CTS) at B. Centre A had more manual repetitive dish-washing work and heavy lifting compared with the automated systems at Centre B.	
Leino, 1989	Sample of 902 metal factory workers (blue- and white-collar). Of these, 63% of the men and 74% of the women took part in all re-examinations during follow-up.	• Data gathered in 1973, 1978, 1983. • Outcome obtained from questionnaire and physical exam. • Exposure assessed with occupational group and questionnaire.	Cohort: 10-year follow-up	Musculoskeletal morbidity (scores based on aches and pains and on range of motion) was greater for blue-collar workers than managers, increased with increasing stress symptom scores (SSS) and age. SSS decreased over time in the cohort but decreased less for those with chronic disease.	• Stratified by sex.

Table 3.16 Work relatedness of unspecified or multiple-tissue musculoskeletal disorders (Cont.)

Study population[1]	Outcome and exposure information	Study design	Findings	Authors	Comments
2814 metal industry workers (2432 men, 382 women). These figures represent a 96% participation rate from the original study population.	• Outcome assessed with the Nordic Ministry Questionnaire on neck and upper-extremity symptoms and work activities. • Exposure: each job was categorized by the plant physician, physiotherapist and safety managers on the basis of work load, static load, repetitiveness and awkward neck/arm positions, then grouped according to overall physical work stress level (3 levels).	Cross-sectional	Sick leave prevalence for neck and upper-limb symptoms increased with each level of physical stress (0.0%, 3.3%, 7.6% for women, 1.7%, 3.5%, 3.8% for men).	Dimberg *et al.*, 1989a,b	• Smoking and age were significant contributors to OCD sick leave.
96 female electronics assembly workers, of whom 69 participated in all examinations during follow-up.	• Medical examination for OCD grouped by 4 levels of severity of symptoms. • Exposure determined by physiological and ergonomic evaluation: eg interview (work history) measures for voluntary isometric contractions, VIRA technique for evaluation of working postures and movements.	Cohort: 2-year follow-up	OCD symptoms worsened with previous heavy job and previous sick leave, and high output (after one year). Improvement of symptoms predicted by reallocation to more varied work, by physical activity in leisure time and by high productivity (after two years). Staying	Jonsson *et al.*, 1988	• A cross-sectional analysis had also been done and postural variables in the cross-sectional analysis were stronger than in the longitudinal analysis. • Low muscle strength does not predict OCD.

Subjects	Methods	Study design	Results	Reference	Comments
Out of an original population of 27 male and 35 female orchard workers doing work with pears and apples, 48 orchard workers (20 males, 23 females) completed the 3 examinations.	• Questionnaire for OCD symptoms. • Exposure based on observational/goniometric measurement of 1 female subject during a whole day's work for each different task (pear bagging, pear thinning, apple bagging). The pear work (bagging and thinning) required more overhead work than the apple work.	Cross-sectional	healthy predicted by work without shoulder elevation and job satisfaction. Shoulder flexion during pear bagging and thinning (110–140° flexion) associated with OCD symptoms more than bagging apples (30° flexion). For example, the prevalence of symptoms in the shoulder for pear work (both bagging and thinning) was 80% *vs* 40% for apple work in males, and 80% *vs* 60% respectively in females.	Sakakibara *et al.*, 1987	• Each task required similar hours per day, and similar number of work days.
81 female film-rolling workers studied before (1975) and after improvement of working conditions (1977): group A (*n* = 38) worked separately, group B (*n* = 43) worked on conveyor belts.	• Outcome determined by physical exam using a Japanese OCD questionnaire. • Exposure was assessed with time study and EMG.	Intervention	Decrease in total work time, frequent breaks, job rotation and improved work station design associated with decrease in OCD over 2 years: 44% of all workers had reduced severity (asymptomatic increased from 27% to 47% in group A and 6% to 38% in group B).	Itani *et al.*, 1979	• Multivariate analysis not done. • Numbers quoted on the study population represent those who were examined, both pre- and post-intervention, from 45 group A and 53 group B workers, available pre-intervention.

Table 3.16 Work relatedness of unspecified or multiple-tissue musculoskeletal disorders (Cont.)

Study population[1]	Outcome and exposure information	Study design	Findings	Authors	Comments
705 supermarket workers (who yielded 513 usable questionnaires).	• Outcome determined with a questionnaire on frequency of painful symptoms and discomfort. • Exposure based on observational activity analysis and a walk-through survey.	Cross-sectional	Prevalence of hand/wrist and upper-limb disorders not significantly greater in checkout vs other types of supermarket workers (7–9/10000 person-hours at work vs 5.8-4.9/10000 person-hours at work overall); higher prevalence of back, leg, foot symptoms due to standing.	Ryan, 1989	• Part-time and transient workforce.
Sick leave and turnover for 170 person-labour years during 1967–74 were compared with those for the 244 person-labour years during 1975–82, i.e. before and after ergonomic improvements respectively.	• Outcome assessed by questionnaire, sick leave statistics and records. • Exposures based on task analysis.	Cohort: 7-year follow-up	Decrease in musculoskeletal sick leave (4.7% in 1967–74 to 1.6% in 1975-82, difference statistically significant) and turnover temporally related to reduced postural load and work organization.	Westgaard and Aarås, 1985	• Multiple confounders not included in analysis. • Average employment 30 per year.
All Danish slaughterhouse workers compared with general Danish working population.	• Outcome defined as disability pension rates (medium and high levels of awards).	Cohort: 4-year follow-up (retrospective)	Overall standardized disability ratio (SDR) lower than expected, but musculoskeletal SDR greater for female slaughterhouse workers (SDR = 196).	Hansen and Jeune, 1982	• There could be a 'healthy worker' effect of healthy workers moving into highly physically demanding jobs such as slaughterhouse work

Study population	Methods	Study design	Author	Comments
104 heavy letter carriers and 92 lighter letter carriers compared with 76 meter readers and 127 postal clerks. Fewer than 5% of study subjects did not participate in the interview.	• Outcome determined by telephone interview for ascertaining problems and pain with joints (answers used to build significant work related disability score). • Exposure considered as carrying while walking, a maximum of 35 lbs (heavy letter carriers) and a maximum of 25 lbs (lighter letter carriers), walking not carrying any weight (meter readers) and neither walking nor carrying weight (postal clerks).	Cross-sectional	Increased letter bag weight associated with shoulder symptoms 23% for heavy vs 13% for lighter letter carriers, 7% for meter readers and 5% for postal clerks. Differences were significant between all groups ($p <.05$).	Wells *et al.*, 1983	• Adjusted for age, years on job, obesity and previous work. and 'weaker' ones leaving the industry. Thus number of pensions observed for industry are lower.

[1]Based on the information available in the article, no subject was lost from the original study population when only one set of numbers appears in the table under the 'study population' or when there is no relevant information in the 'comments' column.

shoulder or back, and with self-reported physical stress (lifting heavy loads, twisted posture, continuous standing, vibration, repetitive movements and machine-paced work) (odds ratio for the exposure to physical stress vs none 1.4, 95 per cent CI 1.3–1.4) and mental stress at work (monotony, hurried pace, worried about making mistakes) (odds ratio for each increase in mental stress level, i.e. none, mild, and severe, 1.3, 95 per cent CI 1.2–1.4). Various individual co-determinants were also identified. Odds ratios for hand and wrist disorders varied from two to eight for repetitiveness, and there was an odds ratio of 6 (95 per cent CI of 2–18) for workers with 10 years of exposure to vibrating tools compared with manual workers (Burdorf and Monster, 1991). Odds ratios for 'RSI' in data-entry operators keying more than six hours per day were as high as 12 (Oxenburgh, 1984).

3.8.4.2 Temporal association

Most cross-sectional studies used some variation of 'onset of symptoms since on current job' to suggest a temporal relationship, i.e. that the work related factors preceded the symptoms. In a cohort study of musculoskeletal symptoms among manual workers and office workers in the shipyards (n = 352), Berg *et al.* (1988) found that prevalence of shoulder problems increased more among active manual workers between 1982 and 1985 (from 39 per cent to 57 per cent shoulder symptoms) than among office workers (29 per cent at both times). Additionally, symptoms persisted at least three years after retirement for manual workers (30 per cent had worse shoulder symptoms compared with 12 per cent for office workers). Westgaard and Aarås (1985) reported decreasing musculo-skeletal sick leave over a seven-year period, the most dramatic decrease resulting from reduced exposure due to ergonomic and work organization improvements. On the other hand, Jonsson *et al.* (1988) reported increasing severity of OCD among electronics assemblers who were not reassigned to more varying tasks. Severity of symptoms was reduced among data entry operators (Oxenburgh *et al.*, 1985) when ergonomic and work organization changes were introduced.

3.8.4.3 Consistency of the association

There was strong consistency in the reporting of associations among symptoms, symptom-sign syndromes and work related factors (Table 3.16). These factors are highly repetitive work (repetitiveness not always defined) (Itani *et al.*, 1979; Ohara *et al.*, 1982; Silverstein *et al.*, 1987; Dimberg *et al.*, 1989a,b; Ohlsson *et al.*, 1989), static or constrained neck and shoulder postures (Sakakibara *et al.*, 1987; Jonsson *et al.*, 1988), and high physical loads or forces (Ohara *et al.*, 1982). Duration and intensity of exposure are important contributors (Punnett and Robins, 1985; Ohlsson *et al.*, 1989; Waersted and Westgaard, 1991). Self-reports of working in twisted or bent postures (Mäkelä *et al.*, 1991) and indicators of job satisfaction also appear to be consistent predictors of neck and shoulder symptoms (Tola *et al.*, 1988). Other than associating the symptoms/signs found with work related factors, competing hypotheses have been suggested, but not well defended. These have to do with artifactual increases in reporting due to increased publicity, as well as secondary rewards for 'sufferers' and health-care providers and consultants (Hadler, 1990).

3.8.4.4 Predictive performance of the association

An increase in exposure duration increased the prevalence of symptoms (Jonsson *et al.*, 1988). Waersted and Westgaard (1991) reported that sewing machine operators who worked five hours per day had fewer neck/shoulder disorder sick leave days than those working eight hours per day. However, shorter duration of exposure merely postponed the onset of symptoms by about six months, instead of preventing them entirely. Oxenburgh (1984) made similar observations about hours of keying and RSI symptoms. There is some evidence that a reduction in exposure results in a reduction in the severity of adverse health effects (Oxenburgh *et al.*, 1985; Jonsson *et al.*, 1988) or the prevalence of such effects (Westgaard and Aarås, 1985). Ohara *et al.* (1976) suggest that some of the difficulty in identifying causal relationships is related to the exchange of one risk factor for another. Manual cash register systems were replaced by electric (much lower force) systems that resulted in increased repetitiveness (keys per hour). Ohara *et al.* (1982) reported a number of studies where reduction in duration of exposure (from 480 minutes to 300 minutes) and reduction in repetitiveness of tasks resulted in lesser severity and prevalence of stages III–V OCD (Table 3.15). Itani *et al.* (1979) had similar results with female film-rolling workers, although a number of potential effect modifiers were not included in the analysis.

In a two-year follow-up study of 96 female electronics assembly workers Jonsson *et al.* (1988) reported an increase in severe cervicobrachial disorders from 11 per cent to 24 per cent in one year. When 40 per cent were transferred to more varied jobs, symptoms improved, whereas those who remained at more stereotypical jobs deteriorated further. Predictors for deterioration were high initial productivity, and for 'staying healthy' were working without elevated shoulders and satisfaction with work tasks. Satisfaction with work colleagues was negatively associated with remaining healthy.

3.8.4.5 Coherence of the association observed with current knowledge

Where specific diagnoses of musculoskeletal disorders have been combined into broad categories, the mechanisms for each disorder have been described in the section relevant to the specific tissue. The majority of studies of 'unspecified tissue' disorders, based partly on pain as an outcome, may be primarily of muscle/tendon origin. Pain perception or reporting varies by gender (Buckelew *et al.*, 1990), culture (Bates, 1987), and health belief models. Most people relate their pain to a physical cause (James *et al.*, 1991). Psychological events may be contributors to, or effects of chronic pain (Gamsa and Vikis-Freigbergs, 1991; Ursin *et al.*, 1988).

A number of studies have used EMG estimates of muscle load or fatigue that appear to correlate well with upper-limb symptoms. Edwards (1988) proposed that occupational muscle pain may be due to the contradiction between motor control of the activity in a specific posture and the need for rhythmic movement to accomplish a repetitive task; thus pain is exacerbated by poor 'learned recruitment patterns' of muscle fibres below the conscious level. Psychosocial job stress may manifest itself in muscle tension, which in turn is correlated with symptoms from the back, neck and shoulders (Theorell *et al.*, 1991).

Kilbom and Persson (1987) explored individual working technique factors that may be important in the development of work related musculoskeletal disorders of the neck, shoulder and arm. Based on a video technique (VIRA) to analyse postures and movements, there was considerable variability in shoulder posture and the percentage of time spent in different postures performing similar jobs. Among the 96 subjects in the study, there were variations in working techniques from task to task and from one individual to the next. The disorders were associated with increased number of times (repetitive) and increased percent of time (duration) in elevated shoulder postures and neck flexion. Perceived stress, previous sick leave unrelated to these disorders, individual productivity and previous work in physically demanding jobs were also significant indicators. Low muscle strength (as suggested by the work of Bjelle *et al.*, 1981) was only weakly associated. Wiker *et al.* (1990) found similar negative results experimentally with fatigue, discomfort and strength. They also found that 'apparently awkward postures' did not necessarily provoke significant discomfort or fatigue. In an electromyographic study of trapezius muscle activity during complex and simple VDU tasks among 18 subjects, Waersted *et al.* (1991) identified a subset of subjects who consistently generated higher muscle tension in complex tests, even though there was no observable difference in posture. They attributed this elevation to increased mental effort in doing the tasks. Thus, mental workload may alter (increase) the actual 'dose' for the same exposure, depending on individual capacity.

Unspecified tissue disorders are thus assumed to be muscular in origin and the mechanisms described are coherent with this assumption. Further, the possibility of an association between these disorders and work is coherent with these mechanisms.

A NOTE ON 'PERSISTENT AND RECURRING PAIN' OF UNSPECIFIED TISSUE DISORDERS

Refractory cervicobrachial pain can be considered as chronic work related musculoskeletal pain without evident tissue damage (Cohen *et al.*, 1992). The disturbance in microcirculation may lead to that pain receptors (nociceptors) in the muscle are sensitized causing pain at rest. In the sensitized pain receptor a lower than normal threshold stimuli can activate the pain system or even cause spontaneous activity in nociceptive nerves (Henriksson *et al.*, 1991). Metabolites that sensitize nociceptors are bradykinin and prostaglandins. These metabolites are reased in mechanical trauma or following ischemia. Furthermore impairment or loss of local modulation of low mechano afferents in the spinal cord may result in that mechano-receptor input is being perceived as musculoskeletal pain (Roberts, 1986; Cohen *et al.*, 1992). Although a number of studies on unspecified tissue disorders use persistent or recurring pain as part of the case definition, the chronic pain literature associated with low back pain patients does not seem as applicable to WMSDs. For example, depression, hypochondriasis and hysteria have not been associated with WMSDs. This may be because most of these studies involve workers who are still working and in milder stages of disability than those in the chronic back pain studies.

3.8.5 Conclusion

There is ample and consistent evidence that a variety of localized musculoskeletal symptoms are associated with work related risk factors of repetition, physical load, certain prolonged postures and vibration. These symptoms and their severity (e.g. sick leave, disability) increase with the intensity and duration of the work exposure and may decrease when exposure is reduced. The expression of these symptoms appears to be somewhat associated with complex work organization, psychosocial and personal mediating factors.

3.9 Summary: work relatedness of musculoskeletal disorders of the neck and limbs

In contrast to 'occupational' diseases, where there is a direct cause and effect relationship between hazard and disease (e.g. asbestos and asbestosis, lead and lead poisoning), the World Health Organization expert committee described 'work related' diseases as multifactorial, where the work environment and the performance of work contribute significantly, but as two of a number of factors, to the causation of disease:

> . . . they may be partially caused by adverse working conditions; they may be aggravated, accelerated or exacerbated by work place exposures; and they may impair working capacity. It is important to remember that personal characteristics, other environmental and socio-cultural factors usually play a role as risk factors for these diseases.
>
> (*Identification and Control of Work related Diseases*, WHO Technical Report Series 714, 1985, Geneva)

There has been increasingly widespread recognition of the multifactorial nature of work related musculoskeletal disorders, with varying levels of attention to individual, psychosocial and physical factors that may contribute to the development or prevention (salutary buffering effect) of these disorders. These factors have not been studied simultaneously with equal rigour in any scientific investigations. There are a variety of types of studies that can be used to evaluate the work relatedness of various disorders of the musculoskeletal system: clinical

Table 3.17 Summary table: is work associated with the development of musculoskeletal disorders?

Examples of musculoskeletal disorders*	Criteria for assessing the possibility of an association with work in working populations**					Conclusions
	Strength of the association	Temporal association	Consistency of the association	Predictive performance of the association	Coherence of the association with current knowledge	
Shoulder tendinitis (tendon-related disorder)	High for specific exposures.	Not demonstrated in all studies.	Replicability shown.	Not tested yet.	Association is coherent.	● Good evidence available: work exposures are associated with the development of shoulder tendinitis.
Lateral epicondylitis (tendon-related disorder)	Weak for specific exposures in the cross-sectional studies. High for specific exposure in the cohort study.	Not demonstrated.	Replicability not shown.	Not tested yet.	Pathogenesis still obscure. Predominant hypothesis would accommodate the possibility that work can trigger the pathogenesis.	● Contradictory evidence: more studies needed to be able to assess, one way or the other, the association between work and lateral epicondylitis.
Hand-wrist tendinitis (tendon-related disorder)	High for specific exposures.	Not demonstrated in all studies.	Replicability shown.	Not tested yet, some evidence of an exposure-response association.	Association is coherent.	● Good evidence available: work exposures are associated with the development of hand-wrist tendinitis.

Carpal tunnel syndrome (nerve-related disorder)	High for specific exposures.	Some evidence exists.	Evidence of replicability and survivability.	Some suggestions but nothing conclusive.	Association is coherent.	• Good evidence available: work exposures are associated with the development of carpal tunnel syndrome.
Thoracic outlet syndrome (nerve-related disorder)	Weak for specific exposures.	Aspect not studied yet.	Some evidence of replicability.	Not tested yet.	Association is coherent.	• Not clear cut, however some evidence is available that work exposures are associated with the development of thoracic outlet syndrome.
Radiculopathy (nerve-related disorder)						• In the 3 cross-sectional studies, no evidence of an association between work and radiculopathy was found. However the power of these studies was low and it is not impossible that this association exists.

Table 3.17 Summary Table (Cont.)

Examples of musculoskeletal disorders*	Criteria for assessing the possibility of an association with work in working populations**					Conclusions
	Strength of the association	Temporal association	Consistency of the association	Predictive performance of the association	Coherence of the association with current knowledge	
Tension neck syndrome (muscle-related disorder)	Moderate for specific exposures.	Not demonstrated in all studies.	Replicability shown.	Not tested yet.	Association is coherent.	• Evidence is available that work exposures are associated with the development of tension neck syndrome.
Osteoarthrosis (joint-related disorder)	Moderate for specific exposures.	Not demonstrated, but evidence that work exposures accelerate degenerative process.	Replicability shown.	Not tested yet.	Association is coherent.	• Evidence is available that work exposures are associated with the development of osteoarthrosis.
Hypothenar hammer syndrome (vascular disorder)	Very few studies available. Risk evaluations range from 1.8 to 16.3, but confidence intervals wide in all cases. Overall strength weak.	Not clearly demonstrated in the few studies available.	Replicability shown.	Yes	Association is coherent.	• Not many studies available, but those that are indicate that using the hand as a hammer is associated with the development of the hypothenar hammer syndrome.

Knee bursitis (bursa disorder)	High for specific exposures in the few studies available.	Not tested yet.	Replicability shown.	Not tested yet.	Pathogenesis not clearly understood; however, mechanisms postulated are coherent with possibility that work may trigger the pathogenesis.	● Evidence available that work exposures are associated with the development of knee bursitis.
RSI/OOS/CTD/ OCD (unspecified or multiple-tissue disorders)	Strength varied depending on population studied, precision of exposure measures and extent of controlling variables.	Yes, in some studies.	Strong replicability shown. Survivability also shown.	Some evidence available.	Unspecified disorders are assumed to be muscular in origin and the possibility of an association between work and these disorders is thought to be coherent.	● Ample and consistent evidence that a variety of localized musculoskeletal symptoms are associated with work related risk factors.

*These examples are organized by tissue type.
**Specificity of an association' is usually one of the criteria used for assessing the association in question. The information provided in Box 3.2 explains why it is not considered one of the necessary criteria for this summary table.

case series, epidemiologic studies and laboratory studies. There is no 'perfect' population study of occupational exposures and health outcomes. The strengths of population studies were evaluated here by looking at how four major components were handled in the study: selection bias, information bias, confounding and power of the study design to detect significant differences. The strength of population studies was one of the selection criteria that helped determine which studies would contribute evidence to answer the question at hand, i.e. whether there is an association between work and musculoskeletal disorders of the neck and limbs.

As our understanding of the complex nature of any disease process increases, we are continually debating the 'evidence' for causality (Weed, 1986). This debate will probably continue for a long time. For the purposes of this chapter, the framework proposed by Susser (1991) has been modified to examine the relationship between work and musculoskeletal disorders.

1. Do the results of the studies show an association between disease and exposure? If the answer is yes:
   (a) How strong is the association?
   (b) How specific is the association?
2. Do they show a temporal relationship? (Did the exposure occur before the effect?) The cause must at least be present by the time the effect occurs.
3. Is there consistency in the association?
   (a) Is the association replicated in more than one study and in other circumstances?
   (b) Does the association survive rigorous tests of alternative hypotheses?
4. Can a change in disease be predicted by a change in exposure (predictive performance)?
5. To what extent does the association fit with pre-existing theoretical, factual, biological and statistical theory and knowledge (coherence of evidence)?

Examples of specific tendon-related, nerve, muscle, joint (arthrosis), vascular and bursa disorders have been used to illustrate evidence of work relatedness. Further, there have been a number of studies of upper-limb disorders as a whole (repetitive strain injury or RSI, occupational cervico-brachial disorders or OCD, cumulative trauma disorders or CTD), although in the absence of anatomic location or structure, it can be difficult to assess them. However, since some of these studies provided important information on work related risk factors and upper-limb musculoskeletal symptoms, they were included here.

This chapter focused on the work relatedness aspects of musculoskeletal disorders. The role of individual susceptibilities as possible contributors to musculoskeletal disorders is examined in chapter 9. As seen in that chapter, some personal attributes may contribute to the development of work related musculoskeletal disorders; however, this does not diminish the importance of work exposure contributions. Depending on intensity, frequency and duration of workplace exposures, individual/personal factors may play a more or less important role. In many of the situations discussed, where there have been high work exposures, individual/personal factors are likely to be less important (although they should still be controlled for in epidemiologic studies). Based on examples of musculoskeletal disorders, a summary of the evidence found for the work relatedness of musculoskeletal disorders is given in Table 3.17. Overall there is a strong body of evidence demonstrating work relatedness of many specific and unspecific musculoskeletal disorders.

4

Identification, measurement and evaluation of risk

Evidence for the work relatedness of musculoskeletal disorders was assessed in chapter 3. It was seen that many of the disorders examined were associated with work (see summary, section 3.9). Chapter 3, however, examined the association with 'work' in general, but did not describe work exposures and thus the risk factors found in the workplace and associated with WMSDs in any great detail. Chapter 4 focuses specifically on WMSD risk factors in the workplace.

4.1 Workplace risk factors and the development of WMSDs

WMSD risk factors can either directly or indirectly influence the onset and/or course of WMSDs. Risk factors may be directly linked to the physiological processes of these disorders, they may trigger the processes or they may create appropriate conditions making the onset of WMSDs possible.

In this chapter, we use 'risk factor' as a general term for those factors at work which have an association to WMSDs. One possible approach to categorizing some WMSD risk factors is shown in Box 4.1. These risk factors are not necessarily the direct causes of WMSDs, although they may directly generate the pathological and physiological responses that produce WMSDs. These risk factors are mainly statistically or empirically defined factors associated with the occurrence of WMSDs. In most cases, the association has been first observed empirically and then confirmed with epidemiological studies.

Risk factors are not independent of one another. They may be associated with more general factors, which they influence, or they may act as surrogates for

---

Box 4.1 WMSD Generic risk factor groupings

- Fit, reach and see (subsection 4.2.2)
- Cold, vibration and local mechanical stresses (subsection 4.2.3)
- Postures (subsection 4.2.4)
- Musculoskeletal load (subsection 4.2.5)
- Static load (subsection 4.2.6)
- Task invariability (subsection 4.2.7)
- Cognitive demands (subsection 4.2.8)
- Organizational and psychosocial work characteristics (subsection 4.2.9)

Generic risk factors are grouped here in an operational way that is useful in explaining the work relatedness of WMSDs, has biological plausibility and has a strong connection with the workplace environment (see Figure 2.2 for the generic model of prevention).

---

each other. Because of this, it seems that they must be interrelated. For example, the 'fit, reach and see' variables (see Box 4.1) influence the posture of body parts and thus affect body tissues and generate adverse physiological reactions. The work organization factors may influence how the work has been organized for correct 'fit, reach and see'. The risk factors include elements and issues from different scientific domains. Some are physical and mechanical, like vibration and awkward postures. Some are physiological, like muscular overexertion, and some are psychosocial. A risk factor can have a direct effect through various mechanisms or it can influence other risk factors. For example, organizational factors such as workload and rest breaks may control the risk factors of frequency and intensity; yet organizational factors may also have a direct impact on the musculoskeletal system through stress mechanisms.

4.2 Identification of WMSD risk factors: what are they?

4.2.1 Introduction

4.2.1.1 Integrating various sources of evidence in order to identify risk factors

The type of evidence used to examine workplace WMSD risk factors in this section ranges from epidemiological, to biomechanical to physiological. This variety is necessary since chapter 3 demonstrated strong evidence of the work relatedness of a number of tissue disorders in various types of tasks, but did not provide sufficient details on the critical aspects of work resulting in the disorders. Only the most general principles could be inferred; high force, high repetition, awkward posture, overhead work, static load, etc. were described as being problematic. In addition, variables/risk factors used in epidemiological studies are often too vaguely or inconsistently defined to be used in evaluating specific situations. To compensate for this lack of information, literature on biomechanics and work physiology was examined. While often unable to directly link workplace variables to WMSDs, this literature can be used to provide specific guidance through known injury mechanisms. The injury mechanisms for WMSDs are conceptualized as a combination of influences that overwhelm a tissue's ability to adapt even though the tissue's physiological function is maintained (as opposed to acute events leading to anatomical or functional disruption) (Pitner, 1990).

4.2.1.2 Characterization of risk factors

Repetition, extreme reach, intra-carpal tunnel pressure or a pinch force are, in themselves, unlikely to be harmful if of infrequent occurrence or of moderate magnitude. We know that some motion is necessary for well-being. In order to characterize the exposure to most of the risk factors of interest, in terms of their injury potential, four main pieces of information may be required (all four are required for most of the 'physical' factors, but not for the 'organizational' factors):

- location of the anatomical structure exposed to the risk factor, e.g. the elbow,
- magnitude or intensity of the risk factor, e.g. for musculoskeletal loading this could be 'high force',
- time variation of the risk factor, e.g. for posture this could be 'repetitiveness' and
- duration of the risk factor.

Some of these aspects of exposure are discussed in greater detail below.

Magnitude or intensity of the risk factor. Some measure of the size of the risk factor is required, either in absolute terms or relative to an individual's capabilities. For example, for posture it may be the joint angle, for musculo-skeletal load it may be the weight of the object being lifted, and for psychosocial work characteristics it may be the perception of an increased workload.

Time variation of the risk factor. Biological tissue response to risk factors is highly time dependent in view of the time needed for recovery. Therefore, the variation of the risk factor over time affects risk. For example, at the 'macro' level it may be in terms of:

- variation in tasks over an hour, day or week, or
- the duty cycle, or holding time to cycle, or ratio of loaded to unloaded time of the task (Rodgers, 1987), or
- the insertion of micropauses, ten seconds every ten minutes (Sundelin and Hagberg, 1989), or
- on a larger time scale, the distribution of rest breaks, or three- vs five-day work weeks.

Duration of exposure to risk factors. The length of time a person is exposed to a particular task (i.e. risk factors) is also critical, as the latency of WMSDs may be on the order of a few days to several decades (e.g. Kivi, 1984; Castorina *et al.*, 1990; Hagberg et al., 1990).

Using the approaches described in this introduction (i.e. various sources of evidence to identify risk factors and ways of characterizing them), the rest of section 4.2 explores the mechanisms and relationships between the development of WMSDs and workplace risk factors, using the factor groupings presented in Box 4.1.

4.2.2 Fit, reach and see

Although the dimensions of the workplace do not cause musculoskeletal disorders *per se*, they can force individuals to adopt postures, experience loads and exhibit behaviours that can cause or aggravate musculoskeletal disorders. A common example is the lack of knee clearance, particularly for individuals working on conveyors and heavy machinery designed for standing; lack of clearance typically causes workers to sit back from the work surface and lean forward or twist to get close to it. This creates neck and shoulder flexion and a long arm/hand reach to access the work surface. Another example may be found in the influence of visual demands; precise work may require the worker to reduce the viewing distance, perhaps by leaning forward with a bent neck. Again, the wearing of bifocals in data entry has been associated with awkward postures of the head and neck and neck disorders (Martin and Dain, 1988). In the upper limbs, Ulin *et al.* (1990) found that, for drilling screws into a vertical surface using an in-line tool, subjects chose work heights that gave postures of the wrist in significant ulnar deviation. The authors speculated that these postures reduced the required neck flexion and moved the screws closer to eye height, i.e. closer to the optimal point for vision.

4.2.3 Cold, vibration and local mechanical stresses

This category includes stresses generated by contact with the external environment such as work surfaces and tools, along with shock and impact loading of the musculoskeletal system, and exposure to whole body vibration, hand/arm vibration and hot or cold. High noise levels, while important in industrial hearing loss, have been suggested as having other effects on worker behaviour, including poorer ability to communicate and static muscle loading (Kjellberg, 1990; Kjellberg *et al.*, 1991).

4.2.3.1 Local mechanical stresses

Whenever there is contact between the body and external objects, mechanical stresses on tissues should be considered. Local stresses can cause injury to both the skin and underlying structures, most commonly nerves, bursae and blood vessels. Common areas of consideration include the hand, wrist, elbow, shoulder and knee.

The hand is well developed for prehensile activities such as grasping and pinching; the normal contact areas are covered with conformable fat pads, covered with skin of such a structure as to maximize friction, grip conformity and sensory feedback. These areas are found on the palmar surface and include the thenar and hypothenar eminences and the palmar surfaces of all the digits. Loads, especially point loads applied elsewhere, can cause injury. The most commonly assaulted areas are the sides of the fingers, where the digital nerves and blood supply can be found. Pressure here from scissors can lead to digital neuritis; the thumb is also vulnerable, for example 'bowler's thumb' (Dobyns *et al.*, 1972). Other vulnerable sites can be found in the palm, between the two fat-covered eminences and distal to the distal wrist crease (Feldman *et al.*, 1983). The palm contains a branch of the ulnar nerve that is vulnerable to local pressure – a common problem with bicycle riders (handlebar palsy) and users of tools whose handles press into the palm (Finelli, 1975). The wrist area is vulnerable to pressure from supporting the weight of the hand (e.g. Sauter *et al.*, 1987) and also to pushing over the carpal tunnel area, especially with the wrist extended.

Local mechanical stresses can also be set up at the elbow and forearm; these are commonly used to support part of the body weight upon work surfaces. The ulnar nerve runs across the inside of the elbow (Figure 3.6, p. 62) and can be compressed relatively easily by leaning upon that side of the elbow, especially with elbow flexion – thus the origin of telegraphist's elbow (when telephone operators worked at large 'patch panels') (Feindel and Stratford, 1958; Aguayo, 1975). Carrying heavy hard loads such as lumber on the shoulder can induce acromioclavicular joint synovitis (Neviaser, 1980).

4.2.3.2 Shock, impact and impulsive loading

Large impact loads, such as occur when the hand is used as a hammer, have the potential to cause vascular disorders, e.g. 'hypothenar hammer syndrome' (Stroud and Thompson, 1985). It has been investigated in handball players, where a small hard ball is struck, often with a bare hand (Buckhout and Warner, 1980), and in baseball catchers (Lowrey *et al.*, 1976).

Eccentric contractions are recognized as having a high potential for muscular damage (Edwards, 1988; Armstrong *et al.*, 1991a). Indeed, rapid eccentric (lengthening) contractions of muscle, often seen when stopping objects, such as in resisting the kickback or torque of nut drivers and screwdrivers and during tosses, flicks and jerks, are suggested as being stressful to the musculoskeletal system (Armstrong *et al.*, 1993). Many authors have suggested that eccentric contractions are more likely to create injury to muscle tissue (Edwards, 1988; Armstrong *et al.*, 1991a).

4.2.3.3 Vibration

Vibration, whether whole body or applied to the hands, has many effects on the human. Some of these effects have been associated with the development of WMSDs.

Most power hand tools expose the operator to hand-arm vibration. The hand-arm vibration syndrome consists of three components: a vascular effect, a musculoskeletal effect and a neurological effect.

VASCULAR EFFECT OF HAND-ARM VIBRATION

Hand-arm vibration is widely recognized as leading to Raynaud's disease (also known as 'vibration white fingers' or 'white fingers syndrome'), a vascular insufficiency of the hand and fingers (Gemne *et al.*, 1987).

MUSCULOSKELETAL EFFECT OF HAND-ARM VIBRATION

Vibration may impair muscle strength and cause osteoarthrosis. The disorders linked to vibration are primarily osteoarthrosis of the joints of the upper limbs: for example, osteoarthrosis of the wrist and elbow amongst pneumatic drill operators (Copeman, 1940) and osteoarthrosis of the acromioclavicular joint (Stenlund *et al.*, 1992). Study of this area is complicated as vibration shock and manipulation of tools usually occur together.

The development of this musculoskeletal effect is thought to be related to the fact that vibration interferes with prehension. Vibration may alter sensation of the hand mechanoreceptors, leading to over-gripping to maintain control of the object (Lundström and Johansson, 1986). In addition, through the tonic vibration reflex, extra muscle activation from the forearm muscles may be elicited, leading to higher muscle loads (Radwin *et al.*, 1987). Both these effects increase the forces on the tissues of the upper limbs and may increase musculoskeletal loads.

NEUROLOGICAL EFFECT OF HAND-ARM VIBRATION

The neurological effect consists of both carpal tunnel syndrome (CTS) and neurological symptoms and disorders in the hand without a simultaneous entrapment disorder. Nilsson *et al.* (1990) found that stainless steel platers operating tools such as grinders and chipping hammers had a CTS prevalence of 14 per cent compared with 1.7 per cent among office workers. This corresponded to an odds ratio of 11. In another study, the odds ratio was 5.3 when vibration-exposed US industrial workers were compared with non-vibration-exposed workers (Silverstein *et al.*, 1987). Nathan *et al.* (1988) found an odds ratio of 4 for carpal tunnel syndrome (defined as slowing of nerve conduction velocity) when grinders were compared with administrative and clerical workers. In a case-control study, high-level vibration exposure (more than 10 hours a week) gave an odds ratio of 14 (Hagberg *et al.*, 1992). Cannon *et al.* (1981) found an odds ratio of 7 for CTS with the use of vibration hand tools. Having a vibration exposure for more than 20 years gave an odds ratio of 4.8 (Wieslander *et al.*, 1989).

MUSCULOSKELETAL EFFECT OF WHOLE BODY VIBRATION

Professional drivers are exposed to whole body vibration. Twisted postures in combination with whole body vibration may increase the load on the neck-shoulder as well as the low back (Wikström, 1993).

4.2.3.4 Cold

Cold, like vibration, has two main avenues for acting as a risk factor for chronic musculoskeletal disorders; directly, by its effect on tissue, and indirectly, from the possible problems caused by personal protective equipment used to alleviate its effects. These latter effects will be discussed in greater detail in subsection 4.2.5.

A recent study by Chiang *et al.* (1990) suggested that work in the cold was a risk factor for carpal tunnel syndrome. However, as all the workers in the cold environment wore gloves, it is not possible to identify the relative contribution of temperature and the use of gloves. The reported skin temperatures of 26–28° C are well above those used in laboratory experiments and it may be that the effect of the gloves was more important than that of the cold. In addition, the workers exposed to cold were also considered to exert higher forces.

Low temperatures have been found to have an effect upon prehension; Lundström and Johansson (1986) demonstrated that cold can increase fingertip force; Vincent and Tipton (1988) found reductions in maximum grip strength of the order of 13–18 per cent following hand and forearm immersion in cold water with skin temperature reductions of 22–23° C (resulting in skin temperatures of about 11–12° C). Many factors could explain these findings. For example, Vanggaard (1975) showed that at skin temperatures between 8 and 10° C, motor fibres of the ulnar nerve cease to conduct. Clarke (1961) found that at hand temperatures of about 13–16° C, work with the hands suffers a decrease in strength and co-ordination, can induce pain and forms the lower bound for unaffected performance.

It is also possible that cold can increase the activation of muscles. Hammarskjöld *et al.* (1992) showed that EMG activation in carpenters increased after hand cooling, as did perceived exertion and time to complete nailing tasks. Sundelin and Hagberg (1992a) found that a cool draft on the shoulders increased the activation of the trapezius muscles slightly, but significantly. This was speculated to be due to the adoption of a hunched protective posture. Studies of cold environments, such as cold storage areas, have found a large number of reports of discomfort due to cold in the neck/shoulder region (Nielsen, 1986).

4.2.4 Awkward postures

Postures with potential to cause musculoskeletal disorders generally have three characteristics, all of which may be present simultaneously:

1. extreme postures very close to the end range of motion of the joint, where loads are supported by passive tissues or that require muscle activity to maintain the posture ('E' in Table 4.1);
2. non-extreme postures allowing gravity to act about a joint to create or increase joint moments and thus create or increase loads on muscles and other tissues ('G' in Table 4.1);
3. non-extreme postures that change the musculoskeletal geometry and may put larger stresses on tendons, muscles or other supporting tissues and/or reduce the tolerance of the tissue. ('M' in Table 4.1).

Table 4.1 shows how these three types of effect are present at a number of joints in the upper limbs.

Readers are directed to *Joint Motion: Methods of Measuring and Recording* (American Academy of Orthopaedic Surgeons, 1965) for information about the various postures described in this chapter.

4.2.4.1 Postures very close to the end range of motion of the joint ('extreme postures')

An 'extreme' posture, in this sense, indicates that the end of joint range of motion has been obtained. On the other hand, for certain body parts such as in the wrist or the knee, the functional range of positions of a joint may not be obvious from the knowledge of its range of motion; for example, the knee joint has a range from straight (0 degrees) to fully flexed (about 135 degrees). It would not be correct to assume that the optimal position would be in the mid-range, around 75 degrees, and to claim that positions close to 0 degrees are extreme; the knee joint is fully functional during running and walking, very close (5 degrees) to the straight position (0 degrees) with no apparent increased risk of musculo-skeletal injury.

An extreme posture by itself may, for example, stress joint components or occlude blood flow, such as when the neck is extended to look upwards (Sakakibara *et al.*, 1987). Postures at the extreme of joint range of motion may reduce but not occlude blood flow, as in the lower leg in a squatting position. Extreme postures may also require high muscle forces to hold the limb in position, even in the absence of gravitational or other external loads; for example, to maintain the forearm in full pronation, as in keyboard work, requires high muscle activity (Zipp *et al.*, 1983). The maintenance of static loads for extended periods of time is suggested as leading to WMSD. This issue is discussed in greater detail in subsection 4.2.6. Loading the elbow joint in full

Table 4.1 Links between posture and WMSDs

Joint	Links	Postural characteristic*	References
Wrist: • Flexion/extension • Ulnar/radial deviation	Intra-carpal tunnel hydrostatic pressure	M	Szabo and Chidgey, 1989; Armstrong *et al.*, 1991b; Rempel *et al*, 1992
	Tendon contact stress and deviation of tendon paths	M	Tichauer, 1966; Armstrong and Chaffin, 1978; Keir and Wells, 1992
	Median nerve compression by tendons	M	Smith *et al*, 1987
	Extrinsic muscle activity to maintain posture	G,M	Rose, 1991; Wells *et al*, 1992;
	Pronated forearm posture	E,M	Markison, 1990.
Elbow: • Flexion/extension • Pronation/supination	Effect of dual action and flexion/extension with pronation/supination	M	Tichauer, 1966
	Muscle activation to maintain given posture	E,M	Zipp *et al.*, 1983.
	Extreme postures	E	Rose, 1992.
Shoulder: • Abduction/adduction • Flexion/extension	Joint moment	G	Westgaard and Aarås, 1984; Wiker *et al.*, 1990
	Muscle activation	G	Hagberg, 1981; Jonsson, 1982;
	Hand position	G,M	Wiker *et al.*, 1990.
	Intramuscular pressure	M,G	Järvholm *et al.*, 1988.
	Impingement	M	Beyer and Wright, 1951; Hagberg, 1981
	Tissue perfusion when limb above heart	M	Holling and Verel, 1957

Table 4.1 Links between posture and WMSDs (Cont.)

Joint	Links	Postural characteristic*	References
Neck:			
● Flexion/extension	Muscle activation	G	Schüldt *et al.*, 1987
	Endurance	G	Chaffin, 1973.
	Loads on passive tissue	E	Harms-Ringdahl and Ekholm, 1986
	Occlusion of blood vessels	E	Sakakibara *et al.*, 1987

*E = Extreme posture (close to the end of joint range)

 G = Joint loads caused by posture leading to gravitational moments

 M = Joint loads caused by posture leading to changes in musculoskeletal geometry of joint

extension produced pain and discomfort in both the ligamentous structures of the elbow joint and the inactive, but maximally lengthened passive tissues of the elbow flexor muscles (Rose, 1992). Basmajian (1961), Van Wely (1970) and Cain (1973) have also identified ligamentous load and discomfort at extremes of joint motion. Poole (1993) reported a series of cases of 'seamstress's finger', which he attributed to continued hyperextension of the distal finger joints leading to osteoarthritis.

4.2.4.2 Non-extreme postures allowing gravity to act about a joint

Our definitions of posture and 'neutrality' are based on the so-called anatomical position. The condition resulting in minimal loading of the musculoskeletal system should be the criterion used for this aspect of posture, i.e. to define neutrality. In the upright standing position, if one deviates from a posture with the arm hanging at the side, by abducting the shoulder, for instance, one places a load upon the abductor muscles of the shoulder. Any position allowing gravity or other external loads to create moments of force about the articulation of interest will create static loads (in the case of gravity) or varying loads. Thus, leaning forward creates moments about the lumbar spine that must be balanced by ligament muscle and disk tissue forces. Holding the arm straight ahead at shoulder height requires shoulder moments of approximately 10 per cent of a person's maximal strength (Takala and Viikari-Juntura, 1991). It is thought that maintaining static loads for extended periods of time may lead to WMSDs. This issue is discussed in greater detail in subsection 4.2.6.

4.2.4.3 Non-extreme postures changing the configuration or function of the musculoskeletal system

Postures that significantly change the musculoskeletal geometry may put larger stresses on tendons, muscles or other supporting tissues and/or reduce the tolerance of the tissue. It has been hypothesized (Armstrong and Chaffin, 1978) and experimentally demonstrated (Smith *et al.*, 1977), that placing the wrist in flexion causes the extrinsic tendons of the finger flexors to press against the median nerve, even if this flexion is non-extreme. Another example of such posture would be in the shoulder when reaching overhead; this can trap the supraspinatus muscle tendon between the humeral head and the acromion (Figure 3.3, p. 56) (Hagberg, 1984). Both of these situations can cause injury to the trapped structure. Flexed wrist postures are yet other examples; they may reduce the area of the carpal tunnel thus potentially increasing the pressure in the tunnel with a concomitant increase in the risk of carpal tunnel syndrome (Skie *et al.*, 1990; Armstrong *et al.*, 1991b).

Postures that do not allow muscles to function effectively can also be detrimental; this can be seen in attempts to grasp a small-diameter object with the wrist in flexion. In this posture, only small forces with large efforts are possible (Miller and Wells, 1988). This latter effect is usually restricted to muscles crossing many joints. It is also stressful to flex the elbow and keep the forearm pronated, since a major elbow flexor, biceps brachii, is a prime supinator of the forearm. These examples show that some postures require greater exertion to produce a given force.

Some postures may compromise the physiological function of the limb; holding the arms at or above shoulder level reduces muscle blood perfusion rates and thereby reduces work capacity. Holling and Verel (1957) also noted that these kinds of interference with function occurred even if the arms were passively supported in an elevated posture, indicating that this is independent of the muscle activity to maintain the elevated posture.

Certain postures may reduce the tissue tolerance. For example, for flexion of the lumbar spine, it has been indicated experimentally by Adams *et al.* (1980), using modelling approaches developed by Shirazi-Adl (1989), that a combination of flexion and twist/lateral bend reduces the tissue tolerance and may lead to rupture of the fibres of the intervertebral disk.

4.2.4.4 Postural changes (joint kinematics)

Thus far in subsection 4.2.4, we have looked at postures from a 'freeze-frame' perspective; however, postural changes occur when one is performing a task. Position, velocity and acceleration, and their angular counterparts – angular velocity and angular acceleration – can be derived from position and displacement. These are termed the kinematic aspects of posture. Recently, a number of investigators have included these kinematic variables as WMSD risk factors. In an examination of industrial jobs, Marras and Schoenmarklin (1993) found that the variables of wrist flexion, extension angular velocity and wrist flexion extension angular acceleration discriminated between jobs with a high versus a low injury risk of carpal tunnel syndrome. The authors suggested that this result was due to high accelerations requiring high forces in tendons. In another study, Marras *et al.* (1993), examining low-back injury in about 400 industrial workers, found that they were able to discriminate between jobs with a high risk versus a low risk of back injury on the basis of five risk factors. The factors included the maximum moment of force at the lumbar spine, maximum lateral velocity, average twisting velocity, frequency of repetition of the activity and maximum sagittal trunk angle.

These initial results suggest a means of combining postural effects with repetitiveness. However, as the low-back data indicate, force and posture are also important. At the wrist, gripping forces, gravity and tool use also are likely to be important contributors to WMSDs. In fact, Marras and Schoenmarklin (1991) have suggested that the acceleration criterion may not be applicable to workers using hand tools. In addition, caution must be exercised in generalizing these data. For example, there are a number of reports of VDT operators developing WMSDs in the hands and forearms. These workers have very little movement of their wrists and thus probably low accelerations. It may well be that different injury mechanisms exist when there are frequent movements and when there is little movement. This issue will be addressed further in section 4.4.

In addition to the above evidence, some findings in the clinical literature would appear to lend weight to the effect of changing postures as a risk factor for some WMSDs. Szabo and Chidgey (1989) showed that some of their patients exhibited what they termed stress carpal tunnel pressures. For this condition, they showed that repetitive flexion and extension of the wrist created elevated pressures in the carpal tunnel compared with normal subjects, and that these pressures took longer to dissipate than in normal subjects. The observed repetitive passive flexion extension appeared to 'pump up' the carpal tunnel pressure. This presentation of carpal tunnel syndrome would be similar to that of compartment syndromes (e.g. Reneman, 1975). Szabo and Chidgey (1989) also noted that active motion of the wrist and fingers may have an additional effect over and above that of the passive motions tested. Evidence of postural changes being related to WMSDs is most frequent in the wrist; however, disorders in the shoulder area have also been associated with postural changes (Hagberg, 1981).

4.2.4.5 *Posture as a WMSD risk factor: evidence in the literature*

Posture has been explicitly and implicitly associated in the literature with the development of WMSDs. Tables 4.2 to 4.5 present epidemiological and experimental laboratory studies. It is in the shoulder region and the wrist that the link between posture in workplace activities and acute effects, such as fatigue, and long-term outcomes, such as WMSDs, is best supported by both epidemiological and laboratory experimentation.

Postural risk factors at the shoulder have received the most attention in the literature, with various combinations of flexion/extension and abduction being shown to be associated with neck and shoulder WMSDs and pain in muscles and tendons. There are a number of plausible injury mechanisms by which posture could lead to shoulder problems. These were reviewed in Table 4.1. For the shoulder, because of the significant weight of the arm and the common presence of external weights in the hand, the effect of gravity is to produce large joint moments at the shoulder. This is likely to be a major pathway by which WMSDs at the shoulder are produced.

At the neck and elbow, there are few studies on which to base conclusions. The neck and shoulder share a number of muscles, the trapezius for instance, and neck pain has been found to be associated with shoulder posture. The studies reviewed concentrated upon neck flexion. It is generally agreed, however, that lateral bending and twisted postures are not recommended.

A large number of studies have been conducted attempting to relate posture to WMSDs of the forearm, wrist and hand. There are many possible mechanisms by which posture could lead to these disorders. Table 4.1 reviews these briefly. Given the small weight of the hand (compared with the loads produced by gripping, etc.) it is likely that the wrist moment produced by gravity is not the most important consideration, in contrast to the shoulder. The other mechanisms (increased carpal tunnel pressure, tendon contact stresses, forearm muscle load and wrist acceleration) will be more or less important in different work environments. This makes posture alone more difficult to relate to WMSDs in these instances than at the shoulder. However, wrist flexion has emerged from many studies as a robust risk factor.

4.2.4.6 Summary of postural risk factors

Posture is easily observable and gives valuable clues about other risk factors for WMSDs: force (through the development of gravity-induced loads), repetition (a change in posture may be used to define repetition) and lack of appropriate breaks.

A large proportion of the studies reviewed in chapter 3 used postural measures to describe the musculoskeletal exposures of workers. Three mechanisms were presented to explain how posture could lead to WMSDs: (1) extreme postures; (2) non-extreme postures creating loads; and (3) non-extreme postures changing the mechanical link system of body segments. In addition, it was shown that postural changes can also be seen as a risk factor. It can be seen that postural measures cannot effectively deal with non-gravitational forces, such as extra weight in the hands. An overview of the literature indicates that posture has shown its greatest power to predict WMSD potential in the shoulder joint, where common changes in posture, such as abduction or flexion, can create significant joint moments. However, methods of posture recording are not sensitive enough to detect some other important postural changes, such as

Table 4.2   Postural risk factors reported in the literature for the shoulder*

Risk factor:	Results: Outcome and details	References
More than 60° abduction or flexion for more than 1 hour/day	Acute shoulder and neck pain	Bjelle *et al.*, 1981
Less than 15° median upper arm flexion and 10° abduction for continuous work with low loads	Increased sick leave due to musculoskeletal problems	Aarås *et al*, 1988
Abduction greater than 30°	Rapid fatigue at greater abduction angles	Chaffin, 1973
Abduction greater than 45°	Rapid fatigue at 90°	Herberts *et al*, 1980
Abduction greater than 100°	Hyperabduction syndrome with compression of blood vessels	Beyer and Wright, 1951
Shoulder forward flexion of 30° Abduction greater than 30°	Impairment of blood flow in the supraspinatus muscle	Järvholm *et al.*, 1988; Järvholm *et al.*, 1990
Hands no greater than 35° above shoulder level	Onset of local muscle fatigue	Wiker *et al.*, 1989
Upper arm flexion or abduction of 90°	Electromyographic signs of local muscle fatigue in less than one minute	Hagberg, 1981
Hands at or above shoulder height	Tendinitis and other shoulder disorders	Bjelle *et al.*, 1979; Herberts *et al.*, 1981; Herberts *et al.*, 1984
Repetitive shoulder flexion	Acute fatigue	Hagberg, 1981
Repetitive shoulder abduction or flexion	Neck/shoulder symptoms negatively related to movement rate	Kilbom *et al.*, 1986
Postures invoking static shoulder loads	Tendinitis and other shoulder disorders	Luopajärvi *et al.*, 1979
Arm elevation	Pain	Sakakibara *et al.*, 1987
Shoulder elevation	Neck/shoulder symptoms	Jonsson *et al.*, 1988
Shoulder elevation and upper arm abduction	Neck/shoulder symptoms	Kilbom *et al.*, 1986
Abduction and forward flexion invoking static shoulder loads	Shoulder pain and sick leave due to musculo-skeletal problems	Aarås and Westgaard 1987; Aarås *et al.*, 1987
Overhead reaching and lifting	Pain	Bateman, 1983

*See chapter 3 for more details on some of these studies

Table 4.3   Postural risk factors reported in the literature for the neck*

Risk factor	Results: Outcome and details	References
Static flexion	No pain in the neck or EMG changes at 15° flexion for 6 hours. At 30° flexion, it took 300 mins for severe pain to occur. At 60° flexion, the corresponding time was 120 mins.	Chaffin, 1973
Flexion	Head inclination more than 56°: pain and tenderness in medical examination in 2/3 of the cases.	Hünting *et al.*, 1981
Dynamic flexion	Median flexion of between 19 and 39° resulted in low sick leave due to musculoskeletal problems.	Aarås *et al.*, 1988
Maximum static flexion	Rapid development of pain at end of range of motion.	Harms-Ringdahl and Ekholm, 1986

*See chapter 3 for more details on some of these studies

Table 4.4   Postural risk factors reported in the literature for the elbow/forearm*

Risk factor:	Results: Outcomes and details	References
Pronation	Steeply increasing activity of pronator teres and pronator quadratus over 60° pronation	Zipp *et al.*, 1983

*See chapter 3 for more details on some of these studies

shoulder elevation (elevation of the scapula), which may be important in sedentary tasks such as VDT data entry. Wrist flexion has also emerged as a WMSD risk factor.

4.2.5 Musculoskeletal and mechanical load

Musculoskeletal load can be thought of as the mechanical loading on the tissues of the musculoskeletal system. In this context, mechanical loading includes tension (e.g. tension in the biceps), pressure (e.g. intramuscular pressure in the supraspinatus muscle, or pressure in the carpal tunnel), friction (e.g. friction of a tendon in its sheath) and irritation (e.g. irritation of a nerve).

The literature on WMSDs has not clearly defined the specific characteristics of musculoskeletal load, due in part to a combination of conceptual differences between investigators, and technical and logistic difficulties in acquiring the

*Table 4.5   Postural risk factors reported in the literature for the hand and wrist**

Risk factor:	Results: Outcome and details	References
Wrist flexion	CTS. Exposure of 20–40 hours/week	de Krom et al., 1990
Wrist flexion	Increased median nerve stresses (pressure)	Smith et al., 1977
Wrist flexion	Increased finger flexor muscle activation for grasping	Moore et al., 1991
Wrist flexion	Median nerve compression by flexor tendons	Armstrong and Chaffin, 1978; Moore et al., 1991
Wrist flexion	Median nerve compression by flexor tendons	Keir and Wells, 1992
Wrist extension	CTS. Exposure of 20–40 hours/week	de Krom et al., 1990
Wrist extension	Increased intra-carpal tunnel pressure for extreme extension of 90°	Gelberman et al., 1981
Wrist extension	Increased median nerve stresses for extension of 45–90°	Smith et al., 1977
Wrist ulnar deviation	Exposure response effect found: if deviation greater than 20°, increased pain and pathological findings	Hünting et al., 1981
Deviated wrist positions	Workers with carpal tunnel syndrome used these postures more often	Armstrong and Chaffin, 1979
Hand manipulations	More than 1500–2000 manipulations per hour lead to tenosynovitis	Hammer, 1934
Wrist motion	1276 flexion extension motions lead to fatigue	Bishu et al., 1990
Wrist motion	Higher wrist accelerations and velocities in high-risk wrist WMSD jobs	Marras and Schoenmarklin, 1993

*See chapter 3 for more details on some of these studies.

information. This leads to difficulties in combining studies, as experienced by Winkel and Westgaard (1992b). In the following subsections, we will review the findings and issues surrounding each of the descriptors of musculoskeletal loading: force (subsection 4.2.5.1), repetition (subsection 4.2.5.2) and duration of the musculoskeletal load application (subsection 4.2.5.3), in addition to examining factors modifying the musculoskeletal load requirements of a task (subsection 4.2.5.4).

4.2.5.1 Force (magnitude or intensity of the musculoskeletal load)

APPROACHES TO DEFINING FORCE

Although an apparently simple concept, the forcefulness of tasks has been defined in a number of different ways. Force, as used in the literature, can be categorized in two main ways: first, according to whether the force is defined externally as a load or internally as a force on a body structure (the two quantities are not necessarily strongly related); and second, the force magnitude may be determined absolutely in terms of newtons or pounds or as a proportion of an individual's capacity, e.g. maximal voluntary contraction (MVC). Each approach may give different relationships with WMSDs and all combinations can be found in the literature. It is important to distinguish between the weight of the object manipulated and the force required to manipulate it. This is similar to the situation found in manual materials handling, where the guidelines available and in the widest use (the Snook tables, 1978; Snook and Ciriello, 1991; and the NIOSH guidelines, 1981) all explicitly reject consideration of the object weight or the force required alone as a useful predictor of injury. To define the riskiness of each it is necessary to consider the position of the load with respect to the body, and thus acknowledge that this relationship influences the challenge (energy and force requirements) of the task. For the upper limbs, the grasp used is important in defining force, as it greatly changes the muscular requirements necessary to apply a given force to an object (Chao *et al.*, 1976), as does the coefficient of friction (Buchholz *et al.*, 1988a).

Studies in the literature usually consider or use only one of the above-mentioned approaches; for instance, most studies taking an electromyographic approach utilize a relative measure, e.g. a percentage of a maximal voluntary contraction. Whereas many studies of hand and wrist WMSDs use grip force, Silverstein *et al.* (1986) used kg as a measure of hand load.

POSSIBLE MECHANISMS LINKING FORCE TO THE DEVELOPMENT OF WMSDs

Force and forceful exertions have long been associated with the development of WMSDs. Force may potentially create damage through a number of mechanisms. Obviously, very high forces may create immediate rupture of tendons and ligaments or damage to muscle tissue if the relevant tissue tolerance is exceeded. For chronic situations, the temporal variation of the force becomes of critical importance in assessing the injury potential to the tissues of the musculoskeletal system. However, very few field studies relating workplace exposures to WMSDs characterize this important aspect of force. While this may be adequate to establish force as a risk factor, without temporal variation it is difficult to define optimal levels of force to reduce the potential for WMSDs. On the other hand, most of the studies inducing fatigue by repetitive activity in laboratory settings do report the time course of the force application.

Since the tissues of the musculoskeletal system all have different physiological responses to loads, muscle, tendon and nerve responses will be discussed separately. In addition, we will restrict ourselves to a consideration of dynamic or varying loads, with static loads being discussed separately in the next section.

MUSCLE RESPONSE TO DYNAMICALLY VARYING FORCES

The link between dynamic forces on muscle and WMSDs is currently subject to extensive investigation and has been for many decades. Muscle is a complex structure and it is beyond the scope of this text to review the vast literature pertaining to muscle function and pathophysiology under dynamically varying forces. The reader is referred to subsection 3.4.4 or recent reviews such as that of Armstrong *et al.* (1991a), which attempt to relate muscle use to injury.

If fatigue, not muscle disorders, is the outcome of interest, then there is a large body of literature that can be used to predict the force and temporal variation (including rest pauses) required to achieve fatigue (e.g. Rohmert, 1973; Byström, 1991; Dul *et al.*, 1991; Rose, 1991). The difficulty of relating these short-term laboratory exposures to the development of WMSDs is discussed in section 4.4.7.

TENDON RESPONSE TO DYNAMICALLY VARYING FORCES

There are four main injury mechanisms for injury or degradation of tendon: (1) permanent deformation as a result of mechanical microfailure; (2) fraying of the tendon due to mechanical wear; (3) shearing of the synovium; and (4) compromise of the nutrition to the structure (Goldstein, 1981; Smutz *et al.*, 1992). Tendon, like other biological materials, exhibits viscoelastic behaviour when loaded continuously or intermittently. Under a constant load, the tendon will elongate immediately (elastic response) and continue to creep or elongate slowly (viscous response) (Goldstein, 1981; Goldstein *et al.*, 1987; Smutz *et al.*, 1992). It has been hypothesized that, under dynamically varying forces, creep could occur and cumulative strain could develop in the tendon. If this cumulative strain exceeds some unknown amount, then an inflammatory response may develop (Goldstein *et al.*, 1987). Abrahams (1967) observed permanent deformation in horse tendon when strained beyond 2–3 per cent. Smutz *et al.* (1992) tested human tendons under loads typical of typing, but did not observe strains that they believed could lead to WMSDs of the tendon. More recently, Fisher *et al.* (1993) have extended this work to predict optimal sequences of force production to both minimize WMSD potential (to tendon) and maximize productivity. However, their model predictions remain to be validated. The work relatedness of tendon disorders is strong, and a plausible mechanism for dynamic loads to cause WMSDs to tendons exists, but there is little direct evidence that this particular hypothesis explains work related tendon disorders.

NERVE RESPONSE TO DYNAMICALLY VARYING LOCAL STRESSES

Nerve tissue may be irritated by local mechanical stresses – for example by overlying tendons in the carpal tunnel, or by stretching that may impair blood circulation. Contact stresses can be developed in nerves as the result of impingement of overlying structures. This has been best described in the wrist, where the flexor tendons can trap the median nerve against the flexor retinaculum if the muscles are activated with a flexed wrist (e.g. Smith *et al.*, 1977; Keir and Wells, 1992). If motion occurs under this condition, such as wrist movement while gripping an object in the hand, the nerve will be subject to shear and possible irritation.

It has been demonstrated that nerves slide within the limbs to accommodate motion of the joint (McLellan and Swash, 1976). While the stretching of nerves in the order of 6 per cent has been shown to produce changes in nerve function (Wall *et al.*, 1992), it is not known whether these conditions exist *in-vivo*. It has been hypothesized that stretching nerves may produce a range of symptoms of a generalized nature in the limb concerned (Quintner and Elvey, 1990). This hypothesis, however, remains untested.

The response of nerve tissue to static load or, more properly, constant hydrostatic pressure, has been explored by a number of authors. Lundborg *et al.* (1982), for example, showed that at some critical pressure between 30 and 60 mm Hg, the microcirculation of the nerve is severely reduced and nerve conduction is impaired. The response of nerve tissue to dynamically varying pressure is not well established.

There is an increasing body of information showing that the combination of finger movements, wrist movements and gripping force affects the pressure in the carpal tunnel (Szabo and Chidgey, 1989; Armstrong *et al.*, 1991b; Rempel *et al.*, 1992). For example, Armstrong *et al.* (1991b) showed in a pilot study that peak pressures over 30 mm Hg could be generated in a simulated bottle-transfer operation.

BRIEF OVERVIEW OF LITERATURE FINDINGS ON FORCE

Irrespective of the possible mechanisms by which force could be linked to WMSD development, studies have found relationships between force and WMSDs. Some are briefly listed here, and the reader is directed to chapter 3 for further details on the individual studies. In these studies, force may be estimated by questionnaire, biomechanical models, in terms of weight lifted, or by electromyographic activity. In the shoulder and neck, the following authors found high force to be a risk factor for WMSDs: Wells et al. (1983); Herberts *et al.* (1984); Silverstein (1985); Berg *et al.* (1988); Linton (1990). In the forearms and hand, the following authors found high force to be a risk factor for WMSDs: Little and Ferguson (1972); Silverstein (1985); Silverstein *et al.* (1987); Nathan *et al.* (1988); Kurppa *et al.* (1991); Viikari-Juntura *et al.* (1991). At the hip, knee and acromioclavicular joint, high forces have been associated with musculoskeletal disorders by Kohatsu and Schurman (1991), Vingård *et al.* (1991b) and Stenlund *et al.* (1992).

4.2.5.2 Repetition (time variation of the musculoskeletal load)

While the time variation of any risk factor is important, it is absolutely critical for musculoskeletal load. The term repetitiveness has been used to describe this time variation, but it is used in a number of ill-defined and sometimes confusing ways. When used as a qualitative descriptor, the term repetition has connotations of the same work elements being repeated many times. Kivi (1984) used cycle time to judge repetitiveness and he called work with a cycle time of less than one minute monotonous. In our discussion, we have defined this concept of 'monotony' as invariability – a topic that is detailed further in subsection 4.2.7.

The time variation may have a wide range of time scales. Short-term aspects of variation, up to a day, will be treated by defining a number of relatively arbitrary subdivisions. This variation may be at the 'micro' level in terms of 'gaps' (or number of times muscle activity drops to zero per minute). At the 'macro' level, it may be in terms of the duty cycle, holding time to cycle, or ratio of loaded to unloaded time of the task (Rodgers, 1987), or in terms of 'repetitiveness', which is commonly used to define the short-term variation in force or effort. The insertion of micropauses, ten seconds every ten minutes, may define longer-term variations (Sundelin and Hagberg, 1989). Still-larger time scales describe the distribution of rest breaks, or three- vs five-day work weeks.

Quantitative descriptions of repetitiveness are usually defined as a frequency of actions or work activities. These range from output-based measures of parts/hour, knife cuts/hour, keystrokes/hour to movement-based approaches, e.g. cycle time and the number of tasks per cycle (Punnett and Keyserling, 1987), number of movements (Kivi, 1984), number of efforts or exertions (Stetson *et al.*, 1991), cycle times of 30 seconds or less or repeating subcycles occupying more than 50 per cent of the basic cycle (Silverstein *et al.*, 1986).

Figure 4.1 shows different types of temporal variability that are either of interest physiologically or are used in descriptions of movement or tasks. The figure uses posture or force as examples, but any WMSD risk factor can vary with time.

Reporting of repetitiveness of a task can be defined as the cyclical use of the same tissues, either as a repeated motion or a repeated muscular effort without movement. At the very least: (1) the body area muscle or joint involved; (2) the cycle time; and (3) the number of movements per unit time, should be reported. Other information over and above this should be added where possible. If a static component exists, this should be reported separately, as should the similarity or invariability of the exposure. With an increase in our ability to collect quantitative information on joint angles and forces exerted, etc., there is a need to analyse or process these data. Moore and Wells (1992) and Radwin and Lin (1993) have presented approaches to describing the time variation of the risk factors in quantitative terms.

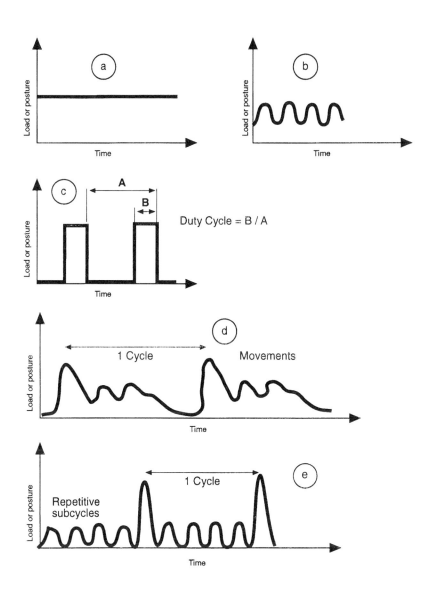

Figure 4.1 Types of temporal variability that are either of interest physiologically or used in describing tasks: (a) illustrates static load or posture; (b) shows a static component with a dynamic varying load; (c) shows an on/off pattern quantified by the duty cycle; (d) shows two cycles of a task with a number of movements or efforts per cycle; and (e) shows two cycles of a task with a repeating subcycle.

LITERATURE FINDINGS ON 'REPETITION' AS A RISK FACTOR

Cox (1985) reviewed the literature on the stressful aspects of repetitive work and its effects on workers. Repetitive work has been associated with physical problems such as musculoskeletal problems (Salvendy and Smith, 1981). Silverstein (1985) and Sjøgaard *et al.* (1987) found associations between repetitiveness (over 120 cycles per hour) of the task and WMSDs. Hagberg (1981) demonstrated that acute shoulder tendinitis could be developed with repetitive shoulder flexions for one hour. Other studies have also found a positive association: Onishi *et al.* (1976); Kuorinka and Koskinen (1979); Luopajärvi *et al.* (1979); Sällström and Schmidt (1984); Amano *et al.* (1988), Ohlsson *et al.* (1989).

A number of other studies have attempted to assess the effect of increasing the repetitiveness of tasks. Odenrick *et al.* (1988) observed an increase in trapezius muscle activity as a group of workers assembling chainsaws stepped up their work pace. Ohlsson *et al.* (1989) demonstrated an increase in shoulder pain with a faster work pace, but no such relationship was found in other body regions. The effect of work pace on rating of perceived discomfort was shown to be quite strong by Ulin *et al.* (1990): increasing the speed of driving screws from 8/min to 12/min increased discomfort from 'somewhat hard' to 'very hard', or from 4 to almost 7 on a 10-point Borg rating scale. Westgaard and Bjørklund (1987), however, found no change in muscle activation of the neck and shoulder musculature as a motion was increased from 88 to 176 movements per minute. To illustrate the difficulty with these types of studies, Arndt (1987) examined wrist flexor electrical activity during postal sorting. As a result of requesting workers to either speed up or slow down, it was found that the muscle activity increased or decreased respectively: however, it was found that the subjects' measured work pace did not change as expected.

In the arm, repetitiveness is almost always used as a risk factor and was so identified by the following studies: Cannon *et al.* (1981); Silverstein *et al.* (1987); Nathan *et al.* (1988); Wieslander *et al.* (1989); Chiang *et al.* (1990); Barnhart *et al.* (1991). Despite some inconsistencies in certain studies, overall repetitiveness or frequency of movements is seen as a WMSD risk factor. (The reader is directed to chapter 3 for further details on some of the individual studies mentioned here.)

4.2.5.3 Duration of the musculoskeletal load

The time worked per day can be considered either as contributing to temporal variability, i.e. to repetition (subsection 4.2.5.2), or as affecting the total duration of exposure; we will consider it here under total duration of exposure. WMSD latency can vary from a few days to a decade (e.g. Kivi, 1984; Castorina *et al.*, 1990). The duration of exposure (independent of age and ageing) is therefore of great interest. The following studies found that work time per day had an effect for particular tasks:

- for VDT tasks: Knave *et al.* (1985); Rossignol *et al.* (1987); Aronsson *et al.* (1988); Grieco *et al.* (1989), Jeyaratnam *et al.* (1989), Kamwendo *et al.* (1991);
- for garment workers: Waersted and Westgaard (1991).

The following studies found that exposure duration had an effect (WMSDs) for:

- piecework: Brisson *et al.* (1989);
- assembly work: Hägg *et al.* (1990);
- vibration exposure: Hagberg *et al.* (1990);
- sewing: Hviid-Andersen and Gaardboe-Poulsen (1990) and Waersted and Westgaard (1991);
- other occupations: Westgaard and Aarås (1984) and Kamwendo *et al.* (1991).

4.2.5.4 *Factors modifying the musculoskeletal load requirements of a task*

There are a number of factors that modify the force requirements of a task or increase the risk; these may make the task either easier or more difficult for a worker. They include such factors as grip type, wrist posture, grip size, gloves and work behaviour. Wrist posture has already been considered under posture, but has an additional effect in that changing the wrist posture changes the effort required.

GRIP TYPE

The hand may interact with tools and the environment in a multitude of ways. Different grips are needed to conform to the many shapes and sizes of objects handled, and to apply force to and manipulate objects. Of concern is the fact that some grip types require high tendon and muscle loads to apply a given external force. If applying force is the goal, a pinch grip requires approximately five times higher tendon and joint loads than a power grip (Chao *et al.*, 1976). This would suggest that external forces applied by the hand can be misleading if the grip type is not specified. This is also reflected in the maximal forces that can be exerted by the hand in different grips.

WRIST POSTURE

Neutral wrist postures have been reported to permit maximal power grip (Anderson, 1965; Kraft and Detels, 1972; Eastman Kodak Company, 1986). This is opposed to the teachings of hand surgeons and clinicians, who find the best and most efficient posture for manipulation to be in extension and slight ulnar deviation (Bunnell, 1942; Norkin and Levangie, 1992). O'Driscoll *et al.* (1992) showed that subjects felt most comfortable and also exerted a maximal grip effort in ulnarly deviated (7° ± 2) and extended wrist positions (35° ± 2), while Miller and Wells (1988) found increased strengths at 20° extended for both pinching and grasping. Moore *et al.* (1991) observed increased forearm flexor EMG as subjects completed a task in neutral and flexed postures compared with extended postures. The effect of wrist posture on other modes of hand manipulation is unclear; however, Wells *et al.* (1990a) have suggested that the optimal postures of the fingers and wrist need to be considered together. They suggest that a reciprocal arrangement exists between finger and wrist postures; the more flexed the fingers (as in grasping), the more extended the wrist and, conversely, the straighter the fingers (as in VDT operation), the straighter the wrist. Force has been found to be a WMSD risk factor and wrist postures that increase the effort to exert a given external force could be considered to increase the risk of fatigue, if not of WMSDs. How we can balance the competing effects of extended postures that might increase tendon loads against straight postures with increased muscular effort is not clear.

GRIP SIZE

Several authors (Ayoub and Lo Presti, 1971; Replogle, 1983; Amis, 1987; Buchholz *et al.* 1988b) have suggested that an optimal grip size could be found that would optimize exertions. Ayoub and Lo Presti (1971), as part of an EMG study, discovered that the overall gripping force that could be exerted on a cylindrical handle was maximal at a diameter of 3.7 cm; they had investigated handles from 3.1 to 6.3 cm in diameter. Results from Amis (1987) showed a trend different from those presented by Ayoub and Lo Presti (1971), in that he found no significant difference between the gripping strength on cylinders 3.7 and 3.1 cm in diameter. Replogle (1983) reported that the maximum torque is achieved when the diameter of the cylinder is approximately 5 cm. Torque increases as the diameter of the handle is squared, up to the point where the fingers and palm just touch without overlapping. This 'grip-span diameter' is approximately 2.5 cm. It was also stated that the grip span and maximum torque diameters do not vary greatly between males and females. Conversely, Grant *et al.* (1992) found that handle diameters 1 cm less than the user's inside grip diameter reduced efforts and recommended that grip sizes be matched to the user's hand. They are supported by Fransson and Winkel (1991) who observed differences in forces exerted depending on hand size. Again, we must consider that grip sizes that increase the effort to exert a given external force could increase the risk of fatigue, if not of WMSDs.

GLOVES

Gloves are commonly worn to protect the user from potentially damaging environmental conditions, cold, heat, corrosive or irritating conditions, and to prevent cuts and lacerations. It has been suggested that wearing gloves increases the musculoskeletal load requirements of a given task. A number of authors have noted degradation in the maximum hand grip force when gloves are worn. Hertzberg (1955) found decrements of maximum force of 14.2 to 28.1 per cent, with the largest decrement at the largest sized grasps. Vincent and Tipton (1988) noted significant reductions of 16 per cent with 'a universally sized three-fingered neoprene glove'. Cochran *et al.* (1986) used a range of brands and styles of gloves and showed differing degrees of grip force reduction, ranging from 7.3 per cent for a cotton glove to 16.8 per cent for a combination of leather over cotton gloves.

WORK BEHAVIOUR

It is logical to assume that the methods of carrying out work tasks can affect the risk of acquiring WMSDs. For instance, if the employee adopts non-optimal 'techniques' when using tools, lifting objects or manipulating products, then it is likely that poor postures, excessive repetitions or excessive tool gripping forces could result. These non-optimal 'habits' appear to be prevalent in many work settings (Stammerjohn *et al.*, 1981; Armstrong *et al.*, 1982; Grandjean *et al.*, 1983; Silverstein, 1985). There are various reasons for the use of non-optimal work practices, including poorly designed workstations, poor work task methods, inadequate employee training, employee inexperience, employee preference for the improper method, poor tool design, psychological stress and work pressure. For example, the work of Parenmark *et al.* (1988) suggests that electromyographic feedback of shoulder muscle activity may be effective in correcting techniques leading to high shoulder muscle loads.

It has been recognized in meat processing that employees carrying out the same tasks with the same tools often do the job quite differently (OSHA, 1991). Employees differ (often substantially) in the number of knife strokes, the posture of the body parts, the force used to grip the knife, the force used to move through the cutting of the meat and the motion patterns used to complete a task. Similar observations have been made by Kilbom *et al.* (1986) with respect to shoulder girdle movements by assembly workers.

It has been shown that when lifting objects people apply more force than is strictly necessary to maintain control of the object (e.g. Westling and Johansson, 1984). Presumably they do so to maintain a safety margin. It is also likely that individuals wearing gloves may be uncertain of their control and this may result in higher grip forces. This idea was taken further by Patkin and Gormley (1991), who suggested that excess effort due to excessively forceful grip, co-contraction or inappropriately timed and graded co-contraction is responsible for high loads on various tissues. However, they provided little evidence for their suggestions.

Behaviour at work can be compared with learning a skilled sport such as tennis or golf, where 'the proper technique' is learned to become proficient. We should note that in sport settings many 'correct' techniques are advocated, rather than a single correct technique. Similarly, in ergonomics there are a number of schools of thought concerning correct posture or technique. For example, conventional postural recommendations for VDT operators will often stress an upright trunk posture, thighs level with the ground, upper arms vertical, forearms horizontal, etc. It has been credibly argued, though, that a different posture, including a reclined trunk and other variations, may be preferable (Grandjean *et al.*, 1983; Ong *et al.*, 1988; Verbeek, 1991). For manual lifting, Authier and Lortie (1992) could not identify a 'correct' method. Despite these differences in approach, it seems clear that the behaviour of the individual employee in applying the 'techniques' of conducting the job is likely to have an effect on the risk factors involved (e.g. force) and the potential risk of acquiring WMSDs, as well as on the successful prevention of WMSDs.

4.2.6 Static load on the musculoskeletal system

Static load on the musculoskeletal system is almost universally proposed in the ergonomics literature as a risk factor for WMSDs, yet it is not easy to define in quantitative terms when a static load is present. Static load is generally agreed to be present when a limb is maintained against gravity, such as in overhead work, but can be considered to be present even with motion of the limbs if the load on tissues does not return to zero (Aarås, 1991).

Quantitative definitions of static load and static posture and the distinction between them are infrequent in the literature. The terms static effort, static posture, static exertion and isometric contraction are often used interchangeably. Table 4.6 shows the combinations of musculoskeletal load and posture that can be observed to define 'static'. Blocks A, B and C correspond to such commonly used terms as 'static effort' and 'static exertion' that may imply non-zero loads and/or postures without motion. As can be seen from the table the terms do not describe the same work activities and likely represent different risks of WMSDs owing to the different tissue responses. Static contractions may refer to musculoskeletal load with a non-zero value or a muscular effort with no movement, the latter term being more correctly called an isometric contraction. Jonsson (1988) defined a static level of force from the electromyogram using the 10th percentile of the amplitude.

It appears that three aspects are used to infer the presence of static postures: observed postural fixity, work related constraints and job design. Postures that do not return to neutral, or that continuously load one muscle group and are so maintained for time scales of 1–10 minutes, are thought to exhibit postural fixity. The importance, or risk of the posture, must be evaluated using the intensity of the effort required to hold the posture. Where the intensity of the effort is high, as in overhead work, one could tend to lower the time limit of exposure. Where the demand is low, as in slightly abducted arms, one could tend to keep the upper time limit. Where postures are constrained by equipment, there is a greater likelihood of static postures, as in pronation of the hand when using a standard keyboard or abduction of the arms when working at a table that is too high. Poor job design can also contribute to the static nature of a task when job categories include only single repetitive or monotonous tasks with little variety. Under these conditions, it is likely that static postures will prevail.

Table 4.6 The concept of 'static' and the relationship between musculoskeletal load and posture

		Musculoskeletal load	
		None	Static	Varying
Posture	Static	Static posture but tissues under minimal load, e.g. relaxed or asleep.	(A) Load (e.g. shoulder muscle activity) does not drop to zero and posture is unchanging, e.g. shoulder posture in VDT work.	(B) Load (muscle activity) does drop to zero but posture is unchanging, e.g. operating an inline power tool.
	Dynamic	Passive motion due to outside influences	(C) Load (e.g. shoulder muscle activity) does not drop to zero, however, motion of limbs occurs, e.g. overhead work.	(D) Postures and loads not described by other conditions.

It is possible that tasks combining both static load and static postures (Block A in this table) represent the greatest risk of WMSDs.

Static loads arise from our operating in a gravitational field and having to support our position against this ever-present force. The human musculo-skeletal system is well designed to support these kinds of loads. The mere presence of a static load is not then cause for concern, but instead its duration and, as will be argued in the next section, its variability (or lack thereof). As with many other WMSD risk factors, it is difficult to reconcile the changes in functional status of the tissue with potential pathophysiological mechanisms; for example, are the changes seen in muscles during short-term laboratory experiments on muscle fatigue relevant to the development of WMSDs in the workplace or are they simply functional changes in response to the experiment? We will first discuss putative mechanisms on links between low-level static contractions and WMSDs and then move on to the limited evidence on this relationship. Although static load on muscle has received the most attention in the literature, all musculoskeletal tissues are potentially at risk. Since the tissues of the musculoskeletal system all have different physiological responses to static load, muscle, tendon and nerve responses will be discussed separately.

MUSCLE RESPONSE TO STATIC LOAD

The link between static load on muscle and WMSDs is currently subject to extensive investigation and some plausible hypotheses have been advanced. Possible mechanisms for the development of WMSDs from low-level conti-nuous contractions have been investigated in the laboratory under fatiguing contractions of relatively short duration (compared with the durations of exposure experienced in the workplace). The difficulty of linking these acute laboratory exposures with the development of WMSDs is discussed in subsec-tion 4.4.7. Larsson *et al.* (1988) and Dennett and Fry (1988) have shown that in both the trapezius and first dorsal interosseous muscle of the hand in patients with chronic muscle pain there are changes in the small, slow motor units consistent with local hypoxia (i.e. low oxygen content) and have suggested that this is due to long-term low-level contractions.

Contraction of a muscle increases its intramuscular pressure; this offers an alternative means of examining the effect of static contractions. If the muscle is encased in a tight compartment, as in the leg or forearm, or is encased in bone, as in the case of the supraspinatus, the intramuscular pressure can rise considerably with only moderate load on the muscle (Järvholm *et al.*, 1988). Compartment syndromes have been observed in the leg, but also in the forearm and hand (Reneman, 1975). The impaired circulation, as a result of the high intramuscular pressure, could lead to muscle disorders and may also affect nerves passing through the compartment.

The most extensive information with application to WMSDs is available for the supraspinatus muscle. This muscle is important in abduction of the shoulder and general stability of the glenohumeral joint. It is encased on three sides in a bony channel in the scapula and on the fourth side in a ligamentous sheet. Even small amounts of shoulder abduction or flexion cause large increases in intramuscular pressure (Järvholm *et al.*, 1988; Järvholm *et al.*, 1990). It is thus likely that many tasks, ranging from overhead work even to keyboard work with some abduction or flexion, will elicit increased pressure response. This may impair blood flow and, as hypothesized by Larsson *et al.*, 1988, lead to chronic muscle damage.

As is apparent from the foregoing, most reports of muscle disorders arise in the shoulder and neck region (tension neck syndrome). Reports of muscle disorders in the rest of the upper limb are almost unknown. Ranney *et al.* (1992) have reported a significant number of presumed muscle disorders in the wrist extensors of manual workers and suggested that the lack of previous reports is due to examination procedures that simply did not allow the reporting of muscle disorders. Wells *et al.* (1992) have also reported preliminary findings of similar levels of electromyographic activity (force) in the extensors of the wrist to those suggested as potentially leading to WMSDs in the shoulder musculature.

TENDON RESPONSE TO STATIC LOAD

The tendon response to static load is similar to that with dynamic load; see subsection 4.2.5.1 on tendon response to force.

NERVE RESPONSE TO STATIC LOAD

The response of nerve tissue to static load or, more properly, constant hydrostatic pressure, has been explored by a number of authors. Rydevik and Lundborg (1977), Rydevik *et al.* (1981) and Lundborg *et al.* (1982) showed that at some critical pressure between 30 and 60 mm Hg, the microcirculation of the nerve is severely reduced and nerve conduction impaired. Chronic exposure to these pressure magnitudes has been observed in patients with carpal tunnel syndrome and is plausibly the reason for the development of the disorder. Gelberman *et al.* (1981) found resting pressures in the carpal tunnel to be on the order of 2.5 mm Hg, while CTS patients presented with pressures around 32 mm Hg. Similar data was reported by Luchetti *et al.* (1990), where CTS patients had pressures greater than 30 mm Hg. These data support static pressure as a risk factor for nerve disorders.

4.2.7 Task invariability

Invariability implies both physiological and psychological sameness. This usually occurs when the task is highly repetitive and unchanging, and when the same tissues are loaded in the same manner. The concept of variability (or invariability) can have a number of time scales; from a micro level, where the state of the tissues on a second-by-second basis is important for the maintenance of homeostasis (e.g., the EMG gaps discussed below), to an intermediate stage where rest breaks and duty cycles are considered.

The pattern of muscular activity during repetitive, stereotyped work has been found to be important in the development of muscular pain (Veiersted *et al.*, 1990). This leads naturally to the idea of what have been termed 'micropauses' or short breaks measured in seconds, at frequent times, typically every ten minutes, to break up monotonous loading (Sundelin and Hagberg, 1989). Work by Winkel and Bendix (1984) and more recent work by Veiersted *et al.* (1990) has suggested that EMG 'gaps' or periods when the muscle is not contracting (operationally defined as a period of at least 0.2 seconds when electromyographic activity is below 0.5 per cent of a maximal voluntary contraction) should be considered in the development of chronic shoulder muscle problems. These studies found that subjects without pain had more gaps. A study by Laville (1982) suggests a relationship between repetitiveness and invariability: two groups of data entry clerks entered either simple data or more complex data with more decision-making requirements. The speed of data entry (or repetitiveness) of the 'simple' group was much higher, as might be expected, but most interestingly, the postures of the 'simple' group were invariable and postural immobilisation was observed. The authors interpreted this immobilization as part of maintaining a stable reference system to permit work within precisely dimensioned and controlled motor and visual space co-ordinates.

The view that 'tasks can define posture' is supported by a preliminary study by Parsons and Thompson (1990); VDT operators showed small motions of 5–15 degrees of the neck with infrequent baseline shifts, and they hardly ever moved the neck into extension (thus never provided much recovery for the neck extensors). Shop assistants who had greater task variety demonstrated a wider range of motion and higher frequencies of movement of the neck (Parsons and Thompson, 1990). Both groups showed a similar mean head angle that fell well within the range of 5–15 degrees of flexion; this has been termed 'natural head slump' by Kroemer and Hill (1986).

In the legs and feet, monotonous loading gives rise to much discomfort but apparently few injury claims. As a result of continuous standing or sitting, the worker is more likely to suffer from pain, aching, swelling, varicose veins in the legs and feet as well as low back pain (Couture, 1986). It would appear from the review conducted by Couture (1986), the work of Buckle *et al.* (1986) and Ryan (1989) that standing for periods of time in excess of 3–4 hours per day is associated with elevated rates of foot and leg discomfort. People standing for longer than this amount of time had pain and discomfort about 5–7 times as often as those standing for less than 3–4 hours. Sitting for 3–4 hours had similar effects. Combinations of sitting, standing and walking produced the least discomfort.

4.2.8 Cognitive demands

The extent of mental effort (cognitive demands) that a task requires has been found to influence psychological stress and employee behaviour. Work load, such as quantitative and qualitative underload and overload, as defined by

Frankenhaeuser and Gardell (1976), has been related to stress. They found that jobs with quantitative overload and qualitative underload resulted in acute stress reactions, such as catecholamine excretion, and in negative effects on perceived well-being, job satisfaction and health. Quantitative overload is a significant stressor because it affects the extent of exposure and the frequency of musculoskeletal actions. Thus, cognitive demands can and may have an effect on stress and WMSD risk.

Non-specific reactions to high cognitive demands, such as increased catecholamine excretion, can increase static muscle loads more than those due to posture alone. This additional muscular activity or 'tenseness' could contribute to muscle overload. Westgaard and Björklund (1987) found that electromyographic activity of the trapezius, rhomboid and erector spinae muscles increased during tasks that had little requirement for muscle load, but were considered visually or cognitively demanding. Waersted *et al.* (1987) showed that VDT operators had 'attention-related muscle load' in the trapezius muscles. Weber *et al.* (1980) found an increase in neck muscle tension with more concentration and van Boxtel and van der Ven (1978) found similar results from intensive reading. Takala and Viikari-Juntura (1991) failed to show any effect of 'psychic load'; however, the authors acknowledged that this extra cognitive load may have been small in comparison with the musculoskeletal loads, which were relatively large, around 10 per cent of MVC.

Similar findings were reported in a study of a repetitive ball-bearing packaging task, using observations based on an integrated EMG of the trapezius and splenius capitus muscles (O'Hanlon, 1981). Four approaches to the work were used; in all approaches, the subjects removed ball bearings from a conveyor and placed them in a box. In the first set-up, they only removed the ball bearings without counting or discriminating the sizes. The other three work approaches were:

- only counting the ball bearings as they were put into boxes,
- only discriminating between two sizes, and
- both discriminating and counting.

It was found that only removing the ball bearings produced significantly lower EMG levels than simply discriminating between two sizes and performing both the discriminating and counting tasks.

In summary, cognitive demands may play a role in the development of WMSDs, either via increased muscle tension or through the more general stress reaction.

4.2.9 Organizational and psychosocial work factors

Psychosocial work factors are the worker's subjective perceptions of organizational factors, which in turn are the objective aspects of how the work is organized, supervised and carried out. Thus, organizational and psychosocial factors may be the same (e.g. career structuring in an organization), but psychosocial factors carry 'emotional' value for the worker. Smith and Sainfort (1989) have proposed a model of job design that integrates the psychological and biological aspects of work within an ergonomic framework. The theory states that poorly designed working conditions (and other workplace features such as technology) can produce a 'stress load' on the individual and an 'imbalance' in the job design. This stress load can have both physiological and psychological consequences such as biomechanical loading of muscles or joints, increased levels of catecholamine release, or adverse psychological mood states. The extent of the load is influenced by an individual's psychological 'perceptions' of the job demands, which are the product of the physical characteristics of the load, the individual's personality and past experiences and the social situation at work. The stress load is also greatly influenced by its objective physical properties independent of the perception of those properties.

If the psychological perceptions of the work are negative, this can lead to adverse psychological and physiological strain reactions. These in turn can lead to physical problems such as muscle tension or elevated catecholamine and cortisol production. In addition, they can lead to inappropriate behaviour at work, such as using poor work methods, using excessive force to accomplish a task or failing to rest when fatigued (these types of worker responses were discussed under work behaviour in subsection 4.2.5.4). Such influences may lead to health problems, including WMSDs; examples of how some psychosocial factors could lead to WMSDs are provided below.

4.2.9.1 Career considerations, job future and worker's role

Career considerations such as over-promotion, under-promotion, status incongruence and lack of job security have been linked to worker stress (Cooper and Marshall, 1976). Burke (1988) argued that different work stressors appear at different career stages. For instance, starting a career can be stressful because of ambiguity about work expectations and uncertainty concerning adequate working procedures. Job future ambiguity has also been shown to have a stress connection in various occupations such as factory and office workers (Caplan *et al.*, 1975; Sainfort, 1991). The possibility of job loss can also create stress (Cobb and Kasl, 1977). Indeed, workplaces that have limited career opportunities or where there is the potential for job loss put employees at increased risk of job stress (Cobb and Kasl, 1977; Cohen, 1983; Smith *et al.*, 1992). These same conditions may create a working 'climate' of distrust, fear and confusion that could lead employees to 'perceive' more aches and pains and report more WMSDs. Other aspects, such as role conflict and role ambiguity also have negative emotional consequences (Caplan *et al.*, 1975). In a meta-analysis of research on role ambiguity and role conflict, Jackson and Schuler (1985) found that both role ambiguity and role conflict were associated with a range of affective reactions such as tension and anxiety. Kahn and Cooper (1986) found that a lack of role clarity was the main predictor of job dissatisfaction and mental health among word processor operators. Even with the limited evidence available to date on the relationship between career, job future and worker's role as factors involved in WMSDs, it does seem that they could play a role in the development of WMSDs and that job design should evolve around recognizing the need to address the psychological needs of the employee as well as the physical demands of work (Herzberg, 1966; Maslow, 1970; Lawler, 1986; Smith and Sainfort, 1989).

4.2.9.2 Work schedules and overtime

Shiftwork has been shown to have negative mental and physical health consequences (Rutenfranz *et al.*, 1977; Monk and Tepas, 1985). In particular, night and rotating shift regimens affect worker sleeping and eating patterns, family and social life satisfaction and injury incidence (Rutenfranz *et al.*, 1977; Smith *et al.*, 1979). Overtime influences the duration of exposure to workplace stressors when the employee may be fatigued and unable to respond at peak efficiency. This could lead to risky work methods to remain on schedule, and possibly an increased risk of WMSDs. Caplan *et al.* (1975) found that unwanted overtime was a far greater problem than simply the amount of overtime. Overtime may also have an indirect effect on worker stress and health (including WMSDs). Although a link between WMSDs and the perception of work schedule and overtime is conceptually feasible, no study has yet addressed this association.

4.2.9.3 Workload and the effect of work pacing

Workload is a recognized risk factor for occupational stress and injuries (Cooper and Marshall, 1976; Smith, 1986). Caplan *et al.* (1975) found that workers in jobs with higher perceived workloads also exhibited greater physiological and psychological indicators of stress. Workload has a synergistic relationship with work pacing in that fast-paced, short-cycle, highly demanding tasks that do not provide opportunities for variety, interruption of the work cycle or rest breaks produce perceptions of heavy workload and associated stress and strain responses (Salvendy and Smith, 1981; Cooper and Smith, 1985). As workload increases, there is more work pressure and heightened performance demands, forcing employees to remain constantly at their activities. This increases the level of repetitive motions and psychological stress. It may also encourage employees to take risky shortcuts in carrying out their tasks, which may increase the risk of WMSDs (see work behaviour, subsection 4.2.5.4). Kuorinka and Koskinen (1979) showed that quantity of work correlates with musculoskeletal complaints. As Smith *et al.* (1981) demonstrated in their study of VDT operators, inappropriately established workloads can create substantial psychological stress and high levels of musculoskeletal health complaints.

Work pace is a very important workload factor. The speed or rate of work can influence worker well-being and physical health especially when the speed is controlled by a machine. According to dictionary definitions, the terms speed and pace can be used interchangeably, to describe rates of motion. However, there are distinct differences in their use in the scientific literature. Work pace refers to the method by which the rate of work is controlled. A job can be completely machine-paced where the speed of the operation and the work output is controlled by a source other than the operator (Murphy and Hurrell, 1980). Contrary to machine pacing, a self-paced task allows the operator to work at a self-determined speed without the pressure of maintaining the rate of the machine. Therefore, task speed is one aspect of work pace. It refers to the rate or frequency at which tasks are performed. The type of work pace, whether machine- or self-paced, is used to control the speed. There are a variety of pacing conditions, which present different cognitive and musculoskeletal loads (Murphy and Hurrell, 1980; Dainoff *et al.*, 1981; Salvendy, 1981).

Machine-paced work tasks are more stressful than non-paced tasks (Salvendy and Smith, 1981). Machine-paced work is even more stressful because it is often characterized by quantitative overload, high work pressure, repetitiveness and lack of control (Smith, 1985). Traditional WMSD risk factors such as the frequency of repetitive motions and the duration of exposure are tied directly to the work pace and workload requirements (work standards) established by management.

Lack of participation (Margolis *et al.*, 1974; Caplan *et al.*, 1975) and control (Coburn, 1979; Karasek, 1979) in task design can produce emotional problems and even increase the risk of cardiovascular disease. Karasek (1979) proposed that a lack of decision latitude associated with high levels of workload could lead to a range of stress reactions, job dissatisfaction, depression and higher risk of heart disease.

Studies of work pace usually involve measurements of non-specific reactions to different pacing techniques. The monitoring of catecholamine excretion rates, heart rate and self-reports of mood has indicated that machine pacing produces a greater stress response than do tasks with more operator control (Johansson and Lindström, 1975; Johansson *et al.*, 1978). A greater stress response was identified as greater adrenalin excretion, higher heart rate and more negative moods. However, stress responses will vary with different degrees of machine and self-pacing. Salvendy (1981) identified that self-paced tasks can be self-perceived as more stressful (using a stress index) if there is inadequate performance feedback; in self-paced tasks, greater mental load results from the requirement to keep track of the work output.

4.2.9.4 Social environment at work

The social environment at work can be a source of social support and thus reduce worker stress, but can also be a source of stress. Group pressure, negative social interaction and relationships with upset clients are all potential stressors.

In addition, lack of social support can make employees more 'vulnerable' to work stressors (Gore, 1986). The relationship with the supervisor can also be an important social consideration. Work environments where supervisors aggressively pursue production, are non-supportive of employees and monitor employee performance closely cause substantial psychological stress as well as musculoskeletal health complaints (Smith *et al.*, 1992). Supervisor support is a positive social factor that can decrease the effects of job stress (House and Wells, 1978).

4.2.9.5 *Technical work environment*

The sociotechnical environment influences the health of employees. Technology may have inherent characteristics that make it stressful, such as physical and mental requirements. There is some evidence to show that video display terminal (VDT) technology can be a source of physical stress, causing visual and musculoskeletal problems (Smith *et al.*, 1981; Marriott and Stuchly, 1986; Smith, 1986). One influence of technology is workers' fear of lacking adequate skills to use the technology and fear of job losses as a result of increased technological efficiency (Östberg and Nilsson, 1985; Smith *et al.*, 1987). Buchanan and Boddy (1982) showed that the introduction of new technology had both negative (e.g. decreased task variety, control and feedback) and positive effects (e.g. increased pay and opportunities for promotions).

4.2.9.6 *Organizational and psychosocial factors: summary*

Work psychosocial factors are the worker's subjective perceptions of work organizational factors. Examples of psychosocial factors were presented: career considerations, clarity regarding the worker's role, work schedule, workload and work pace, and social and technical work environments. Few studies have been conducted on the relationship between these factors and WMSDs, but of those performed, some have shown evidence of a relationship with musculoskeletal health complaints and pain. In addition, various possible examples of the mechanisms for a relationship with WMSDs were postulated in subsection 4.2.9, showing that it is plausible that an association exists between these factors and WMSDs. Further, it is also worth noting that some intervention studies on the more 'physical' risk factors have shown a lack of success when only physical risk factors were controlled; these studies have led researchers to believe that psychosocial variables are important contributors to the chain of events leading to WMSDs (Bigos *et al.*, 1991).

4.2.10 Interactions between risk factors

All risk factors interact. For example, 'repetition' is linked with the 'invariance of a task' and 'static load'. The physiological and psychological loads are also not independent of each other. They interact and may even reinforce each other or act in synergy. For instance, a poorly designed keyboard can put physiological loads on the employee's wrists and thus may produce muscle strain (Grandjean *et al.*, 1983). This may also be perceived as uncomfortable and painful, thus representing a psychological load that can lead to stress responses (Amick and Smith, 1992). The load on the individual can be influenced by the physical demand, psychological response to the demand as mediated by perception, or both. When the load becomes too great, the person displays stress responses, i.e. maladaptive emotions, behaviours and biological reactions. When these reactions occur frequently over a prolonged time period, they lead to health disorders. Such disorders reduce the available resources for dealing with subsequent loads and job demands (and thus increase the possibility of WMSDs), and a circular effect begins unless external resources are made available or the loads and demands are reduced. Figure 2.2 (p. 9) illustrates one model showing how these interactions could occur.

4.2.11 Summary: risk factors identified for WMSDs

The type of evidence used to examine workplace risk factors in this section ranges from epidemiological to biomechanical and physiological. Work risk factors were examined in turn using the categories listed in Box 4.1:

- fit, reach and see (subsection 4.2.2),
- cold, vibration and local mechanical stresses (subsection 4.2.3),
- awkward postures (subsection 4.2.4),
- musculoskeletal load (subsection 4.2.5),
- static load (subsection 4.2.6),
- task invariability (subsection 4.2.7),
- cognitive demands (subsection 4.2.8),
- organizational and psychosocial work variables (subsection 4.2.9).

Research to date has concentrated mostly on those risk factors of a more physical nature (e.g. musculoskeletal load, posture), so evidence is available to support links between these factors and WMSDs. Although fewer studies have been conducted on the relationship between cognitive, psychosocial and behavioural factors and WMSDs, plausible mechanisms for such associations do exist.

4.3 Measurement of WMSD risk factors: some concepts and examples of tools

4.3.1 Introduction

This section will be confined to some concepts associated with the measurement of WMSD risk factors and some examples of methods. WMSD risk factors span a range from purely physical attributes of the workplace to work organization and psychosocial issues (see section 4.1). Section 4.3 will concentrate on physical risk factors; it begins with an introduction to various concepts that should be considered in the measurement of WMSD risk factors and ends with examples of possible measurement methods. It should be noted that some of the terminology found in the literature is used loosely, for example, 'exposure measurement' to mean 'measurement of exposure' or 'dose' or 'stress'/'strain'. Readers are referred to the original articles for the definitions used by the authors for the terms.

4.3.2 Important background concepts for measurements: exposure, dose, acute response and long-term effect/health outcome

Exposure, dose, acute response and effect (health outcome) are concepts established in occupational health and epidemiology to describe various stages in the causal chain of occupational illness (Checkoway *et al.*, 1989). These concepts (Figure 4.2) can guide the type of measures required for quantification of musculoskeletal risk factors. One could consider exposure to be 'external' to the individual worker and unaffected by a worker's personal characteristics. Such measures would be, for example, weight lifted or line speed. Dose measures might be the electromyographic activity of a given muscle or, for example, the moment of force in abduction at the shoulder. An acute response could be a shift in the EMG spectrum to lower frequencies coupled with an increase in amplitude in a given muscle. While the main outcomes we are interested in are WMSDs (Figure 4.2), an acute response such as fatigue may be equally worthy of measure as an early indicator of stress or as a surrogate measure of strain. As we move from exposure to dose, the worker's individual characteristics have a role to play in modifying the dose that results. As we continue down the chain through acute response to effect, yet more individual factors enter into the picture.

As an example, we might consider an exposure to consist of work at a height of 1.8 m; the shoulder posture will depend on the worker's height. This might lead to a 'dose' at the shoulder, in the supraspinatus muscle for example, of a muscle force for a certain time. The dose could depend, for instance, on the work technique chosen by the person, i.e. a worker's individual effect. The acute response may be muscle fatigue, perhaps influenced by muscle size, another individual effect. The long-term health outcome depends on many factors such as tissue type, general health, recent viral infections, etc. As a result of many individual characteristics, there can be a variance in outcomes across workers. Despite the contribution of workers' individual factors, it can be demonstrated that working conditions make substantial contributions to musculoskeletal disorders: in the case of shoulder tendinitis, for instance, the workplace contribution is very high (see chapters 3 and 9). Examples of choices available

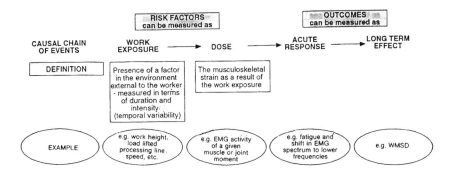

Figure 4.2 Concepts of exposure, dose and outcome.

for measuring WMSD risk factors are shown in Figure 4.3. As can be seen in this figure, as we move down the arrow we move from measures of exposures to measures of dose and outcome.

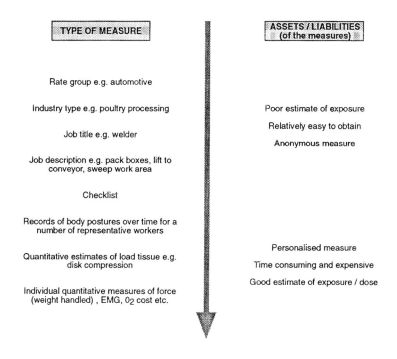

Figure 4.3 Example of options for measuring WMSD risk factors.

4.3.3 Information to be considered for selecting measures for WMSD risk factors

As mentioned previously, we are concerned here primarily with physical risk factors. However, the basic principles discussed should apply, in theory, to other risk factors.

4.3.3.1 Information to be considered for selecting measures: sampling versus continuous measures

Most measurement techniques currently in use look at work as a series of events (e.g. a lift in a poor posture, a reach above shoulder height) and may include the associated frequency of occurrence of each event. However, injuries in jobs with long cycle times may result from the time-history of the tissue loading. This means that techniques are needed to assess factors such as rest time as a proportion of the work cycle, and static, mean and peak force levels that result from a task taking minutes or hours to perform. Techniques capable of providing this kind of information include frame-by-frame analysis of posture or activity (e.g. Armstrong *et al.*, 1982), electromyography (Christensen, 1986a,b) and goniometry (e.g. Aarås and Stranden, 1988). In jobs with long cycle times, decisions have to be made concerning measures using sampling versus continuous recording.

The time course of risk factors is important for most of those categorized in Box 4.1. For sampling techniques, see for instance *Work Sampling*, Richardson and Pape (1982). The kind of inferences that can be made from sampled or continuous data are very similar with one exception: with sampling we lose information on sequences and interrelations.

4.3.3.2 Selecting measures: type of strategy to be used for data collection

For WMSD risk factors, three general types of measurement strategies appear useful for data collection on exposure measurement: self-reports by workers, observations and interviews by trained persons; and measurement by some form of instrumentation (for physical risk factors). Figure 4.3 illustrated the general concept of some of the measurement options available. Some types of measure, such as job classification, will not be considered further, since they provide only limited information on WMSD risk factors.

Measures of risk factors can be obtained by self-reports from workers. Either with spontaneous self-reports or by completing questionnaires, checklists or other methods (more on checklists below), workers may report symptoms of strain, pain, discomfort, exertion or information on their work environment (i.e. risk factors). They may also describe their own tasks in terms of time spent walking, bending or lifting in a manner similar to data collected by a trained observer. Measures of psychosocial stressors can also be obtained by self-reports. Self-reporting questionnaires and diaries are critical to the success of efforts to assess large numbers of workplaces and workers because they are more cost-effective than other approaches. They are also useful in obtaining historical data about previous exposures. However, the reliability and validity of most of the currently used questionnaires and checklists for musculoskeletal risk have not been determined. When evaluations have been made, the validity of the questionnaire has been only moderate (Kilbom *et al.*, 1984 as cited in Hagberg, 1992). Baty *et al.* (1986) used a combination of observations by trained personnel, measures of posture and self-reported postures. They found poor agreement for many self-reported postures and activities and suggested that all such questionnaires be validated against other objective measures before use. These results are similar to those reported by the MUSIC 1 group (Wiktorin *et al.*, 1993), who also noted that infrequent activities were reported better than common activities. Other problems of note with self-reports are the type of scale used – many questionnaires attempt to identify the presence or absence of a risk factor. Continuous or ordinal scales of measurement in self-reports of exposure would aid in the development of badly needed cumulative exposure-response relationships. Hagberg (1992) reported that extensive use of pictorial and other devices aids comprehension and recall considerably, and increases the validity of questionnaires. Carayon *et al.* (1987a,b) made similar observations in office environments.

A trained observer can use checklists or other methods and make notes from visual observation or video-assisted observation about exposures. The documentation and description of activities at work is basic to most measurement systems, and in many cases is the only measure used of exposure to physical risk factors. The use of portable microcomputers has helped in the description of work elements. Such descriptions can be simply in terms of predetermined activities to obtain a catalogue of actions, or can be expanded to allow the recording of postures of the back, shoulders or upper limbs, (e.g. Armstrong *et al.*, 1982; Keyserling, 1986; Fransson *et al.*, 1991; van der Beek *et al.*, 1992). With the use of checklists, a categorical decision can then be made for each factor (i.e. presence vs absence of a risk factor). Waikar *et al.* (1990) recently presented an expert system to estimate risk from the presence or absence of a number of risk factors. Criteria for devising checklists were also offered by Easterby (1967). However, with trained observers as with self-reports, the level of exposure, more than just the presence or absence of the risk factor, is important and depends a great deal on the size of the risk factor and interactions among risk factors. Size and interactions may not be visible to the naked or even video-aided eye (so even a trained observer would have difficulties).

Lastly, measures based on instrumentation can be used to directly measure physical stressors. Calculation of loads on body tissues is a more direct approach to establishing injury risk. Some data are available for certain risk factors. Relevant data on lumbar disc compression force tolerance are available. Maximum voluntary 'muscle force' (strength) data are also available, although they represent loads well below muscle injury threshold. Data on tissue injury thresholds as a result of repetitive low-level loads are scarce even though tissue damage has been shown for this type of loading (e.g. Larsson *et al.*, 1988).

4.3.4 Examples of methodologies used to measure physical WMSD risk factors

Using the risk factor categories summarized in Box 4.1, examples of various methods used for measuring WMSD risk factors are presented in this subsection.

4.3.4.1 Methods for measuring 'fit, reach and see'

Fit and reach can best be measured by comparing work space and workstation design against recommendations in the literature. Measurement of workstation characteristics can be performed using sketches and a tape measure. More recently these measures have been acquired using three-dimensional imaging or other automated systems (e.g. Das *et al.*, 1989). Many texts provide recommendations on dimensions for seated workstations, standing workstations and sit/stand workstations (e.g. Roebuck *et al.*, 1975; Eastman Kodak Company, 1983). This will usually determine whether reaches are beyond the capabilities of an individual or population, and/or whether clearance for limbs is of sufficient dimensions (e.g. knee room).

Requirements for vision vary from low to demanding; visually demanding tasks, such as inspection or interactive tasks using VDTs, require special attention as this may also dictate postural requirements. For VDTs, for example, visual requirements can be assessed using the *Guideline on Office Ergonomics* (CSA, 1989).

4.3.4.2 Methods for measuring cold, vibration and local mechanical stresses

The stresses of impacts, impulsive loads and local mechanical stressors can be identified using observational techniques. Effects of cold environments can be assessed using questionnaire approaches (Nielsen, 1986) or quantitatively using thermocouples on both the body and fingers. Hand-arm vibration evaluation requires specialized equipment and skills. For vibration, the reader is referred to texts such as Wasserman (1987) and Pelmear *et al.* (1992).

4.3.4.3 Methods for measuring posture

Measures of posture, usually the relative joint angle, the angle of a segment to gravity or the position of the limb end-point (for example, the hand relative to the shoulder) or a hand posture, can be obtained in many ways in the workplace; they may be estimated directly or from video (NIOSH, 1990), measured using commercial video-based and imaging systems, or measured directly using goniometers for static positions or electrogoniometers for dynamic recording.

Direct estimation of posture is the main way in which posture has been recorded in many workplace settings. In order to simplify the process, many recording and classification schemes have been developed. These schemes may provide:

- a compact manner of recording posture such as posture targetting (Corlett *et al.*, 1979; Baleshta and Fraser, 1986), OWAS (Karhu *et al.*, 1977), or a special notation scheme;
- a worksheet for recording the posture of many anatomical joints, for each element of the task (Armstrong *et al.*, 1982; Drury, 1987);
- a computer interface to permit semi-real-time transcription of posture from video (Keyserling, 1986; Kilbom *et al.*, 1986); or
- use of palm or lap-top computers for field recording of posture (Kerguelen, 1986; Fransson *et al.*, 1991).

Lastly, continuous measures of the posture of anatomical joints can be obtained from electrogoniometers. These may measure up to three degrees of rotational freedom of a joint simultaneously (Marras and Schoenmarklin, 1993).

The postural data collected can be processed to give, for example, percentage of time spent in various angles (Kilbom *et al.*, 1986; Drury, 1987; de Krom *et al.*, 1990; Stetson *et al.*, 1991), the number of repetitions (Armstrong *et al.*, 1982), and wrist angular velocity and acceleration (Marras and Schoenmarklin, 1993).

4.3.4.4 Methods for measuring musculoskeletal load

BACKGROUND INFORMATION ON APPROACHES USED TO MEASURE MUSCULOSKELETAL LOAD

The measurement of musculoskeletal load is best approached in terms of anatomical structure and localization. It may be assessed directly by invasive measures such as pressure in the carpal tunnel, indirectly by considering external force, repetition of angular acceleration of the wrist joint or in performance terms by endurance time in performing tasks. The approaches currently used to measure exposure to musculoskeletal loads in the workplace are examined below, broken down into biomechanical, physiological and psychophysical categories. An examination of the various methods available for measuring musculoskeletal load by anatomical region then follows.

BIOMECHANICAL METHODS FOR MEASURING MUSCULOSKELETAL LOAD

This approach assesses risk by determining the size and effects of forces acting on and produced by the body as a result of loads applied to some body region in the execution of a task. Fingers, wrists and shoulders are of particular interest in manipulative tasks, the neck and upper back in manipulative tasks with seated postures, and lumbar discs and various body joints in materials handling.

One method uses computerized biomechanical models to estimate disc compression and shear forces, and the moments of force (strength) at the elbow, shoulder, etc. Examples of such computerized biomechanical models include the 'Static Strength Prediction Program' (Chaffin and Andersson, 1984) and WATBAK, produced at the University of Waterloo (Norman and McGill, 1984).

Biomechanical methods that 'invade' the body include the use of swallowed or intrarectal pressure transducers for measuring intra-abdominal pressure (IAP) (Davis and Stubbs, 1977), needle-pressure transducers inserted directly into the disc for the measurement of intradiscal pressure (IDP) (Nachemson *et al.*, 1986) and intramuscular pressure (Järvholm *et al.*, 1988), and buckle transducers to measure achilles tendon forces while people walk, run, jump and perform other normal tasks (Komi *et al.*, 1987). IAP measurement has been used in the workplace, IDP measurement is a hospital research method only, and tendon buckle measures have been used in basic laboratory research.

PHYSIOLOGICAL METHODS FOR MEASURING MUSCULOSKELETAL LOAD

Measures of metabolic and respiratory demand, usually based on heart rate or direct measures of oxygen cost, have been used to assess manual materials handling tasks and to set musculoskeletal load limits and work rates for repetitive manual materials handling tasks. More recently, electrophysiological methods have been used in the assessment of local muscle fatigue and muscle loading. Myoelectric signals (electromyograms, or EMGs) measure the activity of contracting muscles. Researchers in Nordic countries have concentrated on developing specific guidelines for low-level 'static' loading of the muscles of the shoulder and, in particular, the trapezius muscle, based on the worker's response to the task (e.g. Jonsson, 1982; Aarås and Westgaard, 1987). Electromyograms have also been used to monitor neural activation levels of muscles (Schultz *et al.*, 1982; McGill and Norman, 1986; Reilly and Marras, 1989).

PSYCHOPHYSICAL METHODS FOR EVALUATING MUSCULOSKELETAL LOAD

Psychophysical methods utilize the plausible, implicit assumption that effort and comfort in performing a task are related to keeping tissue loads safe. They use workers' estimates of discomfort or effort using cross-modal matching (e.g. Wiker *et al.*, 1990). These methods allow detection of fatigue or discomfort from any tissue at a large number of possible sites (Corlett and Bishop, 1976), whereas most instrumentation (e.g. biomechanical and physiological methods) is specific to certain tissues at specific sites. These methods appear to have good sensitivity and to be usable in situations involving low effort (Corlett and Bishop, 1976; Kuorinka, 1981; Boussenna *et al.*, 1982). It should be noted that these methods can take the form of questionnaires (see subsection 4.3.3.2).

MUSCULOSKELETAL LOAD: METHODS FOR THE HAND/WRIST/FOREARM

Resources available in most workplaces should allow for a useful subset of these measures to be made for the hand/wrist/forearm area:

- number of hand efforts,
- estimates of hand forces applied,
- estimates of duty cycle as well as hand forces.

Number of hand efforts. The number of exertions by the hand can be defined as an indication of the repetitiveness of the hand activity. Stetson *et al.* (1991) used this approach and found it reliable and useful in distinguishing between tasks.

Estimates of hand forces applied. Without instrumentation it is possible to estimate force exertion only at a small number of critical instants of the task. There are four main approaches suggested in the literature. (1) Where the worker must squeeze a tool or operate a machine, it will be possible to measure the minimal force required to perform the task or operate the machine. (2) Force matching has been suggested by Drury (1987). In this approach, workers are asked to grasp a force transducer and match the effort they use at identified parts of their work cycle, usually when the force is at a maximum or at parts of the task when posture is poor. (3) Rodgers (1987) suggested that by using a 10-point 'Borg' scale, workers could estimate their exertion in per cent MVC. (4) Armstrong *et al.* (1982) and Silverstein *et al.* (1987) have estimated grip force from object weight. This last method poses some difficulties in that the approach does not account for hand/object friction or 'over-gripping'. Stetson *et al.* (1991) extended this approach and defined high force hand exertions by combining grip type with object weight or force. For example, an object weighing over 1 kg being held in a pinch grip would be classified as high force, as would a power grip with an object weighing over 4 kg. This classification of force was developed from the research of Silverstein *et al.* (1987) and was used to separate workers into two exposure groups. The research was not intended to determine tolerable loads and this magnitude of force is therefore not usable as an upper limit on acceptable force.

Estimates of duty cycle and hand forces. The effect of an external force on the musculoskeletal system depends not only on its magnitude but also upon the time variation, such as the duration or duty cycle of that force. Physiological experiments have demonstrated that the duty cycle or ratio of the time a force is exerted to the total time for the cycle (contraction time plus rest time) is important in the development of muscle fatigue and perhaps chronic injury (e.g. Rohmert, 1973).

MUSCULOSKELETAL LOAD: METHODS FOR THE SHOULDER REGION

The shoulder region, while being distinct in function from the 'shoulder' proper, shares many of the same muscles. This makes the two areas functionally linked. One can identify three areas studied in the literature: the neck proper (e.g. Sakakibara *et al.*, 1987), the shoulder rotator cuff, including the long head of the biceps, the supraspinatus, infraspinatus and deltoid (e.g. Järvholm *et al.*, 1988) and the neck/shoulder represented chiefly by the trapezius, but also by muscles such as the levator scapula (e.g. Jonsson, 1982). Methods to measure the musculoskeletal load in the 'neck' proper will be discussed later.

Two methodologies appear to provide the most information for measuring musculoskeletal load in the shoulder region: joint moment and electromyography. Posture relates to two of the major injury mechanisms of rotator cuff injury: (1) muscular load, especially during overhead work and (2) the changes in the configuration of the glenohumeral joint, also with overhead work and its pote itiation of biceps and supraspinatus tendinitis. Shoulder loads can best be estimated by the moment required for equilibrium. For example, Wiker *et al.* (1990) estimated shoulder moments at the glenohumeral articulation using a biomechanical model incorporating the subject's anthropometry, load weight and posture. Shoulder moments showed high correlations with indices of discomfort and fatigue. A number of biomechanical models are available to compute shoulder (and other) joint moments from video or other imaging systems (e.g. WATBAK, Norman and McGill, 1984). Such measures are usually made at selected key points in time, such as sustained postures or during peak loads or lifts in awkward postures.

While electromyography has been used to estimate shoulder load, biomechanical estimates seem preferable in most workplace settings unless specific technical expertise in using electromyography is available. If, however, electromyography expertise is available, then EMGs provide valuable complementary information to supplement moment information and allows insight into sedentary tasks where postural changes are minimal and close to 'neutral'. The muscles of most interest include the trapezius, deltoid, infraspinatus and supraspinatus (e.g. Hagberg, 1981; Jonsson, 1982). EMG recordings for all these muscles, save the latter, can be obtained via surface recordings. Electromyography has a major advantage over biomechanical models in that it allows a continuous recording to be made which can then be used to assess the overall or cumulative load.

MUSCULOSKELETAL LOAD: METHODS FOR THE NECK

Postural information appears to provide the best approach for assessing exposure to musculoskeletal load for the neck (see subsection 4.3.4.3 for details on these methods), although there are a limited number of studies using electromyography (Schüldt *et al.*, 1987).

4.3.4.5 Methods for the measurement of static load and invariability

Static load and invariability can be assessed through two main routes: posturally or electromyographically. The approach discussed for measuring posture and musculoskeletal load (subsections 4.3.4.3 and 4.3.4.4) can be applied to the evaluation of static load and invariability; in this case, however, the temporal aspects of posture or the variability of muscle contraction levels are evaluated. Static load is assessed from postures that are thought to create musculoskeletal overload: this is usually gravitational in source, but can have other sources such as extreme postures. Typically, posture is used to infer the presence of a static load; however, if loads are held in the hands, then biomechanical models that include shoulder joint moments are preferred. Typically, for lightly loaded situations, both measures are highly correlated (e.g. Wiker *et al.*, 1990). Variability/invariability is assessed as:

1. time between postural changes that reduce load to zero or close to zero (to permit recovery), as in lowering the arms during overhead work, or
2. time between postural changes that load a different muscle group, as in changing trunk posture during data entry, or
3. total time (per day) spent in a single posture creating static load.

Goniometry may have utility in sedentary situations; Aarås and Stranden (1988) used goniometers on the arms to determine static shoulder load. Similarly, using goniometry, Parsons and Thompson (1990) showed that VDT operators showed small motions of 5–15 degrees with infrequent baseline shifts that never moved the neck into extension.

Electromyographic measures have had much success in evaluations of this area; measures of static load (Jonsson, 1982) and gaps (Veiersted *et al.*, 1990) appear to be simple and related to the development of musculoskeletal problems. More recent techniques such as exposure variation assessment (EVA) may prove useful as well (Mathiassen and Winkel, 1991). While electromyography is not routinely used in industry, with improvements in technology it should become a useful, albeit specialized, method of assessment in trained hands. Marras (1990) presents a useful introduction to the use of electromyography in industry.

4.3.5 Examples of methodologies to measure combination of WMSD risk factors

Tables 4.7 and 4.8 show examples of general analysis methods which permit the collection of structured information on work factors associated with WMSDs. Some methods are more specific to WMSD risk factors, some aim at a more extensive analysis of various work factors. A few methods also propose criteria for acceptable values of work factors.

4.4 Task optimization

4.4.1 Introduction

When goals are set for the prevention of WMSDs, a crucial question will be asked: what optimal level of work factors should be used to minimize the hazard? Are there standards or other recommended values for the level of the various factors found in the workplace? In this chapter, we use 'risk factor' as a general term for those factors at work which have an association to WMSDs. While most of these factors are a normal component of the work, we use the term 'hazard' to describe a situation where a 'risk factor' increases the risk of a WMSD. Optimality is defined here as a level of risk factors which, according to present knowledge, should minimize the hazard of WMSDs.

Understanding the relationship between exposure and WMSDs is important for minimizing the hazard. The problem is that cumulative exposure-response relationships for WMSD risk factors are not well known in detail. The few studies that can be used to demonstrate exposures and responses show different kinds of relationships. In some cases the hazard (development of WMSD) increases along with the increasing risk factor. In other cases, extremes of the risk factor ('too little or too much') are associated with the hazard; the relationship is 'U' shaped. The first example in Table 4.9 demonstrates monotonically increasing risk with increasing risk factor level. The final three examples show a 'U' shaped relationship.

4.4.2 Norms and norm-like data on WMSD risk factors

Standards and norms are important means in the improvement of working conditions from the point of view of WMSDs. While there are many different standards and other data that could be considered in this context, space does not allow a comprehensive listing of the available material. We want to point out some sources that the reader may find useful for obtaining more information on this topic. One group of standards of interest for WMSD risk factors are codes of practice aimed at setting principles and procedures to reduce the number of WMSDs. These codes of practice may be strong encouragement to follow given principles in the prevention of WMSD.

There is extensive literature on ergonomic guidelines for job design. Many of these guidelines and data are of direct interest for the prevention of WMSDs. Examples are guidelines for workplace design, design of hand tools and other equipment, design of signs and recommendations on illumination (Diffrient *et al.*, 1974, 1981a,b; Eastman Kodak Company, 1983, 1986). Pheasant (1987) has prepared a review on standards and guidelines for designers. Another important source of authoritative material on WMSD-related problems are the documents prepared by the International Labour Office in Geneva. Their material ranges from general publications to international conventions and recommendations.

Examples of 'best practice' documents:

- Ergonomics Program Management, Guidelines For Meatpacking Plants (OSHA, 1991).
- Australian draft code of practice for manual handling (Australian National

Table 4.7　Sample of instruments for recording and evaluating tasks: examples of tools that provide some summary of the information

Instrument	Description	Input information	Output information	References
OWAS	Observational scheme for whole-body activity. Uses work sampling. Has been implemented on portable computers.	At every sample period a posture for the back, arms and legs is selected from a set of diagrams. An additional code is used to identify loads producing a five-digit code.	Assigns Intervention priority using an empirically derived weighting scheme.	Karhu *et al.*, 1977
ARBAN	Observational scheme for whole-body activity.	Sampled analysis of posture, force, vibration and static load is recorded as well as perceived effort. Postures are chosen from a photographed set.	From a set of weightings derived from laboratory-tested guidelines based upon EMG, endurance time and perceived effort, a time history of five risk factors and an overall risk is produced.	Holzmann, 1982; Wangenheim *et al.*, 1986
RULA (rapid upper-limb assessment)	Observational screening tool to identify high-risk tasks. Includes consideration of trunk posture. Tool includes general ergonomic checklist.	Posture of the upper limbs from diagrams as well as identification of muscle use, forces or load and frequency.	Assigns intervention priority using an empirically derived weighting scheme.	McAtamney and Corlett, 1993
Ergonomics workplace analysis	Whole-body assessment tool using observer's and workers' reports on a wide variety of stressors from posture to work organization. Whole-body assessment.	Includes workplace dimensions and characteristics, lighting, postures, lifting, job content, communication, decision making. On a five-point scale from examples.	Profile of analyst's ratings (1 to 5) as well as worker's assessments ($++$ to $-$).	Ahonen *et al.*, 1989
Hand-wrist stressors	Checklist of risk factors for upper-limb disorders.	Questions on risk factors for upper-limb WMSDs including physical stress, force, frequency and posture.	Number of 'NO' responses indicates injury risk.	Lifshitz and Armstrong, 1986

Ergonomic analysis	Worksheet approach to document upper-limb postures and forces on a continuous basis.	Posture of major upper-limb joints as a proportion of maximum range of motion and force application.	As recorded plus daily damaging wrist motions (DDWM).	Drury, 1987
Carpal tunnel syndrome risk	Expert system approach to predicting risk of carpal tunnel syndrome.	Field data on a wide range of work characteristics.	Estimate of CTS risk.	Waikar et al., 1990
Higher Productivity and Better Place to Work: Action Manual	A 46-item action checklist, which aims at finding solutions to problems. Covers a variety of issues in working conditions including WMSDs. Checklist and manual included, trainer's manual available.	Observations and interviews on each item (i.e. most recognized WMSD risk factors).	Decision algorithm: corrections needed or not, priority evaluation.	Thurman et al., 1988
Safety-Health and Working Conditions : Training Manual	An 82-item (grouped under 27 titles) checklist on working conditions aiming at analysis of improvement needs. WMSD issues partly and indirectly covered. Checklist and training manual available.	Observation and interview on each relevant item (i.e. some WMSD risk factors).	Short description of each item, improving need and priority assessed.	ILO, 1987
Les profils de postes	A checklist of items. Covers repetitive tasks well. Level of acceptability rated on scales.	Observations on most WMSD risk factors.	Profile of the workplace analysed.	Régie nationale des usines Renault, 1976

Table 4.8   Sample of instruments for recording and assessing injury risk in tasks.

Instrument	Description	Input information	Output information	References
UAW/GM – upper extremity checklist	Checklist of risk factors for upper-limb disorders.	Questions on risk factors for upper-limb WMSDs, including repetitiveness, forceful exertions, awkward posture, hand-tool usage, and local mechanical contact stress.	Score for each risk factor: from 'some' exposure to exposure for more than 1/3 of the cycle.	Keyserling *et al.*, 1993
Ergonomic work analysis grid	As part of a set of forms to help in adapting work and working conditions, an ergonomic work analysis grid is presented.	Postures adopted, object handled, weights, durations, frequencies noted for activities of the task. Postures chosen from 30 diagrams. Overall assessment by evaluator and worker from easy to very difficult.	As recorded	Interdisciplinary task force on rehabilitation, 1988
Physical demands analysis	Designed as a handicapped employment form, this form is commonly used to assess the physical demands of a task for placement of injured workers.	Checklist record of actions encountered along with weights and frequencies. Sections include strength, mobility, sensory, work environment, condition of work and personal protective devices.	As recorded	Ontario Ministry of Labour, 1988
Observational analysis of the hand and wrist	Observational technique to quantify likely injurious manual activities.	Record of number of hand exertions with and without power tools, pinch grips, high force, palm striking, involuntary wrist flexion, extension and ulnar deviation.	As recorded	Stetson *et al.*, 1991

Table 4.9 *Examples of linear and non-linear relationships between levels of various risk factors and health outcomes*

Activity	Example of outcome when *low level* of risk factor is present	Example of outcome when *high level* of risk factor is present	References
Trunk Flexion	Lower risk of back disorders	Higher risk of back disorders	Punnett *et al.*, 1991
Low-limb activity	Foot swelling and edema in sitting	Foot and leg pain after prolonged walking	Winkel, 1987
Upper-limb movement	Static muscle contraction and tension neck syndrome	All WMSDs associated with highly repetitive activity	Silverstein *et al.*, 1986; Larsson *et al.*, 1988
Bone load	Calcium loss due to immobilization or micro-gravity	Stress fractures in walking	Stewart *et al.*, 1982; Eisele and Sammarco, 1993

Occupational Health & Safety Commission, 1992), ISO 6385:1981 (ISO, 1981 on work systems).

- The American National Standards Institute (ANSI) is actively pursuing a standard project (Z-365) on control of cumulative trauma disorder.

Examples of documents related to work place issues:

- ISO, 1992a; ISO, 1992b; CSA, 1989 (office environment and VDT) ANSI B11 TR 1-1993, Ergonomic Guidelines for Design, Installation and use of Machine Tools.

Some standards or norm-like data regulate the work load (e.g. time limits for keyboard use):

- In Japan various documents and guidelines have been produced which regulate work and working conditions: Guide no. 705, Suppl. I, 20 Dec. 1975 (VDT work and cash register operations); Regulation no. 94, 19 Feb. 1975 (trigger operated tools).
- ISO 9241 (keyboard use).

Normative data can also be found in the various criteria documents (e.g. Criteria for a Recommended Standard: Occupational Exposure to Hand-Arm Vibration, NIOSH, 1989)

- Revised NIOSH Equation for the Design and Evaluation of Manual Lifting Tasks (Waters *et al.*, 1993) is an example of a normative document which has had important influence on manual handling.
- Swedish Ordinance Concerning Work Postures and Working Movements, AFS 1983:6.
- Worksafe Australia: Guidance Note for the Prevention of Occupational Overuse Syndrome in Keyboard Employment, 1989.

4.4.3 Which risk factors should be considered in task optimization?

WMSDs may be found at many possible sites of injury. Although the pathomechanics of injury at each body region are similar for nerves, tendons, etc., the anatomical differences suggest that the importance of risk factors in terms of WMSD development may vary with the body region and particular work tasks. Tables 4.10 to 4.12 present some of the possible factors that could be involved in task optimization by body region. From such a range of factors it follows that many approaches to optimization, involving different risk factors, could be used.

In the following subsections, we shall review possible approaches to assessing the optimality of a task using: (1) pain and discomfort; (2) psychophysical approaches; (3) engineering time systems; (4) local muscle fatigue; (5) maximal force; (6) posture; and (7) electromyography. These approaches were chosen

*Table 4.10   Suggested factors in assessing the optimality of a task for the hand, wrist and forearm**

Suggested factors of importance	References
External load	Silverstein *et al.*, 1986
Grip type	Schuind *et al.*, 1992
Wrist posture (flexion/extension, ulnar/radial deviation) as well as finger posture (includes angular velocity and acceleration)	de Krom *et al.*, 1990; Marras and Schoenmarklin, 1991
Forearm posture (pronation/supination) (includes angular velocity and acceleration)	Zipp *et al.*, 1983; Marras and Schoenmarklin, 1991
Blood flow	Hansford *et al.*, 1986
Vibration	Radwin *et al.*, 1987
Temperature	Chiang *et al.*, 1990
Wrist moments	Snook, 1992
Supination/pronation moment	Vanswearingen, 1983
Characteristic of muscle: (a) Muscle activation of finger and wrist flexors and extensors	Rose, 1991; Wells *et al.*, 1992
(b) Fatigue	Baidya and Stevenson, 1988
(c) Biochemical status	Byström and Sjøgaard, 1992
Intramuscular pressure	Rydholm *et al.*, 1983
Intra-carpal canal pressure	Szabo and Chidgey, 1989; Rempel *et al.*, 1992
Median ulnar nerve contact stresses either by local external stress or by compression by the flexor tendons	Smith *et al.*, 1977; Keir and Wells, 1992
Tremor	Wiker *et al.*, 1989
Self-reports of effort and discomfort	Corlett and Bishop, 1976; Snook, 1992

*Some effects changed by elbow angle

because they have attracted the most attention and may give appropriate insight into task optimality.

4.4.4 Surveying pain, discomfort and WMSDs: one possible approach to determining the optimality of a task

Pain is one of the main symptoms of most WMSDs. Self-reported symptoms of pain/discomfort can be used in three main ways: (1) as indicators of current

Table 4.11   Suggested factors in assessing the optimality of a task for the shoulder

Suggested factors of importance	References
Shoulder posture (includes hand position with respect to shoulder): abduction, flexion/extension and internal/external rotation, elevation/depression, protraction and retraction	Bjelle *et al.*, 1981; Hagberg, 1981; Aarås, 1991
Temperature	Sundelin and Hagberg, 1992a
Hand load	Wiker *et al.*, 1989
Vibration, especially low frequency	Stenlund *et al.*, 1992 Wikström, 1993
Shoulder moment	Wiker *et al.*, 1990
Task shoulder moment/maximum moment	Wiker *et al.*, 1990
Characteristics of muscle: (a) Muscle activation (trapezius, deltoid, infraspinatus supraspinatus (b) Fatigue (c) Biochemical status	Hagberg, 1981a; Aarås, 1991 Christensen, 1986a Larsson *et al.*, 1988
Intramuscular pressure, especially in supraspinatus	Järvholm *et al.*, 1988
Blood flow changes induced by arm elevation	Holling and Verel, 1957
Tremor	Wiker *et al.*, 1989
Self-reports of effort and discomfort	Corlett and Bishop, 1976

Table 4.12   Suggested factors in assessing the optimality of a task for the neck

Suggested factors of importance	References
Posture: flexion/extension, lateral/bend and rotation, head retraction/protraction and combinations	Schüldt *et al.*, 1987
Vibration	Wikström, 1993
Neck moment	Bishop *et al.*, 1983
Characteristic of muscle: (a) Neck muscle activation (b) Fatigue (c) Biochemical status	Schüldt *et al.*, 1987 – –
Self-reports of effort and discomfort	Corlett and Bishop, 1976

musculoskeletal illness; (2) as predictors of future illness; and (3) as a sign of excess strain in acute work simulations. The third use will be described under the section on psychophysical approaches.

Examining the relationship between pain and discomfort and current WMSDs is important, since high correlations would make it possible to use 'symptom surveys' instead of expensive physical examinations in epidemiological studies.

The use of discomfort for surveillance purposes is based on the belief that it is an indication of strain and a precursor to many WMSDs. Most WMSDs, at least in their later stages of development, have pain among their main symptoms. The development of discomfort (as opposed to fatigue) is also believed to be an indicator of early stages of WMSDs, especially if there have been many episodes of discomfort and it has extended after the work shift.

Recording of current pain/discomfort symptoms has been advocated by many as a practical means of determining excess strain (e.g. Corlett and Bishop, 1976) and has been related to more-objective measures of musculoskeletal load. It is typically used to localize pain/discomfort symptoms to a body region. Discomfort has been used widely in the literature to assess the optimality of a job and, in view of the body parts where discomfort is present, those parts of the workstation or task that can be improved. Follow-up studies have shown it to be of use in reducing fatigue (e.g. Corlett and Bishop, 1976).

As we noted earlier, it is not clear that fatigue and discomfort are necessarily precursors to WMSDs, nor that comfort is necessarily protective; however, pain and discomfort monitoring remains an important approach for assessing the optimality of a task.

4.4.5 Establishing psychophysical data: one possible approach to determining the optimality of a task

Psychophysical methods were discussed briefly under methods for measuring musculoskeletal load (subsection 4.3.4.4). Here, they will be used to determine optimal levels of task risk factors.

4.4.5.1 Introduction: background on psychophysical approaches

Psychophysical methods for examining human perception of sensory stimulation go back to the last century. Psychophysics is used to define the thresholds of discrimination in humans of sensory stimulation. These can be absolute thresholds of detection of stimulation, or thresholds that differentiate levels of stimulation and perception.

These techniques have proven very successful in mapping the range of stimulus intensities that can be discriminated for all of the senses. The evidence suggests that there is a wide range of threshold discrimination ability for all of the senses in the general population that varies with age, disease, method of testing, motivation of the individual being tested, nature of the instructions given by the tester and presence of the tester during testing (automated versus manual testing).

Psychophysical methods record individuals' sensory perceptions that can be effective in establishing acceptable limits for muscular exertion and discomfort. Therefore, psychophysical methods use pain and discomfort not as an indicator of current WMSDs or as a predictor of future WMSDs, but as signs/stimuli, which the worker can perceive and interpret, of excess strain in the current work situation.

4.4.5.2 Psychophysical approaches and determining the optimality of a task

Psychophysical approaches for determining the optimality of manual tasks have been popularized by the work of Snook for tasks involving pushing, pulling and lifting (Snook, 1978; Snook and Ciriello, 1991) and recently on the upper limbs (Snook, 1992). This approach has empirical support from two workplace studies demonstrating that the use of psychophysically determined acceptable loads reduced low-back injury reporting (Snook, 1987; Herrin *et al.*, 1986). Whether these studies addressed the large numbers of peak overload or back over-exertion injuries vs back disorders of a chronic onset is not clear. Herrin *et al.* (1986) reported that their measures of peak demand correlated better with injury than did average loads.

In order for the psychophysical approach to successfully prevent WMSDs and help determine task optimality, the workers in the task must be able to discern sensations associated with strain or potential injury to the tissues at risk in the task during work or simulated work. For upper-limb and even low-back WMSDs this may not be the case.

Two questions arise concerning the use of psychophysical approaches:

1. can a worker or experimental subject judge non-injurious work load/speed?
2. does an absence of pain and discomfort symptoms – i.e. comfort – imply a minimal risk of developing WMSDs?

The work of Wiker (1992) sheds some light on the factors workers may use in judging the optimality of tasks. His work would suggest that workers are sensitive to strain sensations from the cardiovascular system and shoulder and knee, but not those sensations of mechanical loading of the low back, even though mechanical loading of the low back has been well accepted as being a major factor in the development of low-back disorders.

Gamberale and colleagues (1987) urged caution in the use of psychophysically determined limits for the optimality of a task. They found that workloads chosen by individuals were dependent on previous experience. More importantly, there was no consistent relationship between the workloads found acceptable by the subjects and their physical characteristics and performance capacity as determined from a battery of static strength tests.

In considering the issue of comfort as a predictor of minimal risk, we will move from materials handling to upper-limb disorders. Two examples, using the carpal tunnel syndrome, will help illustrate peoples' difficulties in being able to discern sensations associated with strain or potential injury. One of the occupational risk factors for CTS is 'pressure over the base of the palm which compresses or irritates the median nerve' (Armstrong and Silverstein, 1987; Lundborg *et al.*, 1982). Recent work by Fransson and Kilbom (1991), who mapped the pain threshold of different parts of the hand, showed that the most sensitive areas were over the side of the knuckle on the index finger, where tools often press. The vulnerable area over the carpal tunnel at the base of the palm was not as sensitive to pressure. A worker may avoid putting pressure over the side of the knuckle and place it instead over more vulnerable areas. In addition, working in a flexed wrist posture with a pinch grip, an acknowledged combination of risk factors for the development of CTS, does not seem to produce discomfort, although its long-term effects appear to be deleterious to the median nerve.

A similar caveat on the use of psychophysical approaches for determining WMSD potential can be extrapolated from the research experience of Fellows and Freivalds (1991). In this study, an attempt was made to improve a hammer by padding the handle. After foam padding was applied to the handle, the workers preferred the padded version. However, the grip force required was higher than that used without the padding. It is possible that increased comfort was preferred in spite of a possibly greater risk of injury due to increased static hand grip force.

Tendinitis in the forearm-hand is well reported and, while it can be of acute onset (e.g. Hagberg, 1984), typically has a long latency. It is not clear that any sensations of strain emanate from the tendon to help a worker judge the optimality of a manual task. Even in Hagberg's acute onset experiment, pain was delayed.

Similarly, development of chronic problems in the shoulder/neck region, for example, tension neck syndrome, has been associated with continuous activation of low threshold motor units in the trapezius and other muscles of the shoulder (Larsson *et al.*, 1988; Sjøgaard *et al.*, 1988; Veiersted *et al.*, 1990). The situation, although potentially injurious over a time scale of months, apparently gives rise to little or no sensation of strain and requires biofeedback to allow conscious control of muscle activation. Recent reports by Byström *et al.* (1991) show that the introduction of micropauses into a hand-grip task increased endurance and reduced symptoms of fatigue, but resulted in potentially harmful changes in muscle homeostasis that could last many hours. Workers' perceptions as a guide to the optimality of a task may be 'fooled' by certain work situations that minimize muscle discomfort. Mathiassen (1993) has even suggested that people cannot determine the risk to their musculoskeletal health in some situations, since important information from active muscles never reaches conscious levels, and they may not be able to recognize a physiologically excessive work situation.

When faced with competing demands, workers may choose to use work methods believed to be potentially injurious. Where the upper limbs are concerned, Ulin *et al.* (1990) found that, in drilling screws into a vertical surface using an in-line tool, subjects chose work heights that gave postures of the wrist in significant ulnar deviation. The authors speculated that this reduced the neck flexion required and moved screws closer to eye height at a point closer to the optimal point for vision.

A recent series of publications (Snook, 1992) applied the psychophysical approach to elementary wrist movements. These papers do not discuss the relevance of psychophysically determined loads to chronic forearm and hand injury. Data from these studies indicated that the most discomfort in these wrist flexion activities was recorded by the subject at the fingertips from holding the apparatus handle. Such repetitive wrist flexion tasks have a strong potential to induce CTS. The discomfort felt was not at the site of probable injury, and thus gave rise to concerns that sensations not relevant to fatigue/pain/discomfort were driving the workers' definition of an acceptable task. The problem with using psychophysical approaches is that they do not link workers' perceptions to the injury risk (WMSD in our case) with any degree of certainty; further, as far as workers are concerned, these methods may underestimate the injury potential of upper-limb tasks. Snook himself suggested in 1985 that psychophysical approaches will likely be replaced '. . . when and if more objective methods are available'.

4.4.6 Using engineering measures: one possible approach to determining the optimality of a task

4.4.6.1 Introduction: background on motion analysis and time study

Motion analysis and time study (MTS) of people at work has been used as far back as Mosso in France, over 150 years ago, when he hooked a pneumograph onto workers to study their walking gait while doing tasks. Motion analysis is a methodological approach for evaluating actions taken during skilled activity to define their nature, frequency, duration, timing and sequencing. Such an evaluation can provide valuable information about the type of motions and their effectiveness in completing work tasks. When motion analysis is combined with time study, determinations can be made about the 'efficiency' of any one motion or set of motions in contributing to the successful completion of the task. Those motions that are not making a useful contribution can be eliminated, or the nature of specific motions or the sequencing of several motions can be revised to provide more effective completion of the task. The basic concepts for the measurement of workplace motions and their time study for efficiency improvement were proposed by Taylor (1911), and institutionalized in industrial and management engineering by Gilbreth (1911).

As defined by Barnes (1958), the primary purposes of motion analysis and time study are to: (1) find the most economical way of performing an operation; (2) standardize an operation; (3) determine a time standard (work output determination); and (4) specify the training needs of workers. All these aspects can make some contribution to ergonomic improvements to help control WMSDs. French researchers have criticized the basis of motion and time studies, which separate the 'prescribed work activity' from the 'actual work activity' (e.g. Leplat, 1989).

4.4.6.2 Motion analysis and time study and determining the optimality of a task

In theory, time and motion studies can be used to quantify optimality and provide a good tool for understanding the nature of workers' skilled activities. Predetermined time and motion measurement studies do contain a codified, generalizable database on the motor skills that workers can achieve under good conditions. They can provide valuable data on what is being done properly and what is being done poorly with regard to many WMSDs risk factors. They can provide the basis for improvements in the motions that go into work tasks.

Experience in manufacturing, meat processing and automated office tasks has demonstrated that many employers do not use motion analysis and time study to establish work standards. Often, the standards that are set are based on the economic returns sought by the company rather than on scientific methods. For this reason, motion analysis and time study is a good first step in conducting ergonomic evaluations to assess the basis of work standards and ensure that they are not the primary source of worker WMSD complaints. But, it must also be recognized that current motion analysis and time study methods only consider some WMSD risk factors. Thus, they can serve only as general information of the level of the production standards which, combined with surveillance data on WMSD's may give guidance for minimizing hazard. Indeed, some research dealing with the use of predetermined time standards has shown that such production standards did not provide adequate protection against back injury (Mital and Sanghavi, 1986). Sundelin showed that activities at an MTM standard rate could produce muscle fatigue (Sundelin and Hagberg, 1992b). Armstrong *et al.* (1982) used motion analysis techniques as the basis of a structured approach for describing and evaluating work in a poultry processing plant. They extended standard analyses and added a detailed description of posture. Using a frame-by-frame analysis, task elements were identified and the postural requirements and forces exerted were described. Those parts of the task that exhibited high risk could be addressed and the work methods after modifications could be compared.

During physically demanding lifting tasks, comparisons have been made of MTM standard work rates and their resulting demands on workers (Garg *et al.*, 1986; Mital and Sanghavi, 1986). Workers determined lifting rates by psychophysical techniques and this rate was compared to MTM standard time rates. It was found that the MTM method significantly overestimated the lifting rate for males and females of lifting tasks that varied with posture (sitting or standing) (Mital and Sanghavi, 1986). This overestimation was between +32 per cent and +192 per cent of the psychophysically determined rate. Mital and Sanghavi (1986) also determined the load to be lifted at a specific rate using psychophysical methods. These loads were used to determine the MTM standard time. It was found that the MTM method overestimated the rate by +217.5 per cent to +455 per cent of the rate compared with psychophysical methods. The use of MTM-1 standard times was evaluated in a grocery warehouse that called for heavy lifting (Garg *et al.*, 1986). It was found that 25 per cent of the subjects could not meet the performance standard set by the MTM-1 method. The limitations of pre-determined MTM methodologies, which possibly lead to these discrepancies, could be due to the inadequate consideration of: (1) the weight handled (Chaffin and Andersson, 1984; Mital and Sanghavi, 1986); (2) the force exerted (Mital and Sanghavi, 1986); (3) postural requirements; and (4) tasks that occur less than 5 per cent or 10 per cent of the workday (Chaffin and Andersson, 1984).

On the basis of the studies ci ed, motion analysis techniques may have the ability to describe a task and, with suitable extensions, identify elements of a task that may lead to WMSDs. From the limited evidence available from manual materials handling tasks, there does not appear to be agreement between work rates predicted by motion analysis and time study techniques and those predicted from psychophysical experiments. In most studies, motion analysis and time studies predicted higher work rates than did psychophysical techniques. Little is known about the relationship between work rates predicted by motion analysis and time study techniques and upper-limb WMSD potential. However, on the basis of the evidence from manual handling research, standard MTM methods may underestimate the hazard and caution is needed in using these techniques as the sole approach to optimizing tasks for minimal WMSD potential.

4.4.7 Local muscle fatigue: one possible approach to determining the optimality of a task

Many researchers believe that there is a link between muscle fatigue and the development of WMSDs. Some mechanisms explaining this possible link were discussed in subsection 3.4.4. The link has not been proven, however.

The methodology used to evaluate fatigue and endurance time has much apparent attractiveness as an approach for determining acceptable muscle force and time variations (duty cycles) in the task. Models to determine such acceptability have been proposed by many authors, including Rohmert (1973); Kogi (1982); Rodgers (1987); Bishu *et al.* (1990); Byström (1991); Dul *et al.* (1991); Rose (1992), most of them based on acute work simulations. However, Mathiassen and Winkel (1992) criticized experiments relying on endurance time (e.g. Rohmert, 1973 and Dul *et al.*, 1991) and derived measures, such as Rodgers (1987), stating that they are not valid for developing guidelines for acceptable physical workloads aimed at preventing WMSDs. They noted that the models developed on the basis of such short-term laboratory experiments are poor indicators even of short-term physiological status.

Most models predict that a mean force of between 10–20 per cent of an MVC can be maintained 'indefinitely' without fatigue (e.g. Rohmert, 1973; Byström, 1991). This may be acceptable in terms of fatigue, but may be hazardous in terms of WMSD potential. This appears to run counter to other findings on the chronic effect of much lower loads, typically 1–5 per cent MVC, in the shoulder girdle (e.g. Westgaard and Aarås, 1984). It should be emphasized that these studies refer to continuous muscle contractions (static load) applied for long periods of time and in no way imply that higher exertions on an intermittent basis have a potential effect on WMSDs. In fact, it could be hypothesized that moderate intermittent contractions may have a salutary (protective) effect. In addition, these models predict that maximal efforts are acceptable, in terms of 'no fatigue', as an everyday occurrence, with suitable rest pauses. Although data on the upper limbs is scarce, it does not seem prudent to design tasks to require jobs of a maximal nature except in exceptional circumstances. Doolittle and Kaiyala (1987), for example, have suggested an upper limit of 75 per cent maximal force on the low back during preplacement tests to reduce the chance of injury.

In summary, local muscle fatigue could be used for assessing the WMSD potential of tasks. However, there are caveats, chiefly concerned with: (1) lack of applicability to longer-term work situations; (2) overestimation of the MVC that can be maintained; and (3) suggestions that maximal efforts are acceptable. The approach also suffers from a lack of epidemiological corroboration of its effectiveness in predicting load variations associated with safe work (i.e. with no ill health). It is worth noting that the European Union has developed draft guidelines on the hand force required to operate machines, apparently based on the use of fatigue to determine task optimality (CEN, 1991).

4.4.8 Using maximal force capability: one possible approach to determining the optimality of a task

The approach of using strength capability as a guide in job design and pre-employment and preplacement selection has a long history in manual materials handling (e.g. Chaffin *et al.*, 1978; Keyserling *et al.*, 1980; Doolittle and Kaiyala, 1986). Briefly, these studies have suggested that as the demand increases relative to an individual's strength capacity, so too does the risk of injury. More specifically, they suggest that those employees who are matched to job demands according to their strength (measured either statically or dynamically) are subject to lower injury occurrence and severity. In employees whose measured strength was less than the weights handled or forces exerted during work, the reverse was true. However, as discussed in subsection 9.6.4 methodological difficulties within these studies gave rise to questions about the results.

How can this information on maximal force be used to help in reducing WMSDs of the upper limbs (i.e. to help determine task optimality)? The available evidence is inconclusive. Wiker *et al.* (1990) conclude that '. . . the utility of strength measures alone as a guide for specifying acceptable workplace postures, *when task exertions are small*, is uncertain (emphasis added).'

Strength approaches could best be used to limit maximum or peak demand to acceptable levels and cannot address low-level demands. It follows that, if measured strength is used to determine the optimality of a task, the criterion should not be set at 100 per cent of measured strength, but at a lower level. Tables 4.13 to 4.16 review some of the data on maximal strength for various articulations of the upper limbs.

4.4.9 Assessing posture: one possible approach to determining the optimality of a task

4.4.9.1 Introduction: background on posture

In addition to the epidemiological studies using posture, there are a number of other work physiology studies that give us plausible postural data upon which to base guidelines for the optimality of a task. In the literature, these data have usually been used to support a two-stage classification of posture: recommended and non-recommended.

As with other risk factors, posture cannot be considered alone; for example, the effect of external loads other than gravity has an enormous influence on the effect of posture, especially at the shoulder. It seems that as we move from the neck and shoulder to the wrist, the power of posture alone to predict increased risk and therefore determine task optimality declines. We could speculate that this is due to the decreased effect of gravitational loads and the increased effect of gripping force – which is, by definition, not considered in a purely postural approach. Three types of postural guidelines appear necessary to determine task optimality: guidelines for continuously maintained or static postures, for joint motion, and for postures in which significant force generation is present.

4.4.9.2 Optimality of a task and static postures

Static postures that lead to static musculoskeletal loads are regarded as strong risk factors for WMSDs. It is, of course, possible to have non-neutral postures without musculoskeletal load, for instance if the limb is supported. There is considerable evidence that many postures lead to WMSDs. For example, overhead work is strongly related to WMSDs in the shoulder region. What is not generally available is the range of postures that do *not* appear to lead to the development of WMSDs. As posture is not the only risk factor, this is understandable. A few authors have attempted to define optimal static posture; however, as in most other approaches, only the intensity or magnitude of the posture is considered, thereby restricting its generalizability.

Comparing a group of telephone assembly workers to a control group with similar musculoskeletal disorders allowed Aarås *et al.* (1988) and Aarås (1991) to make a number of suggestions regarding the shoulder region. They suggested that a median angle of less than 15 degrees and a median ab/adduction angle of less than 10 degrees was likely optimal when external load is low. For the neck, Aarås *et al.* (1988) found that workers with median neck flexion of 19–39 degrees had low rates of sick leave related to problems in the neck and shoulder region. They suggested that these workers had neck movement, and thus the optimal range for neck flexion was wider than that suggested for immobile postures (Chaffin, 1973). Chaffin (1973) found that neck flexions maintained at angles exceeding 30 degrees produced rapid fatigue and an angle of 15 degrees produced no discomfort over the course of a day (six hours). Again, this emphasizes the importance of the variation in a risk factor, not just its magnitude or intensity.

4.4.9.3 *Optimality of a task and motion of joints*

Movement of body segments about the joints is normal and required for healthy function. However, some types of motion appear to be contraindicated and could help determine the optimality of a task. This would include highly repetitive wrist flexion or extension in workers with early symptoms of carpal tunnel syndrome (Szabo and Chidgey, 1989), flicking, jerking or other motions with high wrist accelerations (Marras and Schoenmarklin, 1993) or rapid eccentric muscular contractions (Armstrong *et al.*, 1991b).

4.4.9.4 *Optimality of a task and postures with significant force generation*

As was noted in subsection 4.2.4.3, many postures change the way in which the musculoskeletal system operates and make some postures inadvisable for generating large forces because of the potential for WMSDs. Table 4.17 presents some of the postures which, when coupled with large force generation, would make the task non-optimal.

4.4.10 Optimality of a task and electromyography

Electromyography has been used extensively in the description of occupational tasks, especially in the shoulder region. It can be used to assess the activity of a muscle or muscle group either in absolute units of microvolts or, more commonly, in terms of relative (as compared with maximal) activation. In addition, it can be used, with appropriate processing, to indicate local fatigue of a muscle (e.g. Christensen, 1986a,b). For this interpretation see subsection 4.4.7.

A few authors have attempted to define optimal range of muscle activity. However, as in most other approaches, only the intensity or magnitude of the activation is considered, thereby restricting its generalizability. On the basis of laboratory testing, Jonsson (1982) suggested limits of 2–5 per cent, 12–14 per cent and 50–70 per cent of the maximal voluntary contraction (MVC) for the

Table 4.13   Data on maximal strength capabilities in set postures: shoulder

Joint function	Unit[2]	Strength ± (SD)[1]		Population studied and notes on the study	References
		Males	Females		
Elevation					
● both	(N)		818(236)	Bank cashiers aged 20–50	Takala and Viikari-Juntura, 1991
right			433(121)	A sling connected to a strain gauge force	
left			394(123)	transducer was placed over both acromions.	
Abduction					
● right	(N)		96(28)	Bank cashiers aged 20–50	Takala and Viikari-Juntura, 1991
left			88(29)	A sling was set on the lateral epicondyle of the elbow.	
● 95%ile	(N.m)	101	57	25 men and 22 women employed in manual jobs.	Stobbe, 1982; Chaffin and Anderson, 1984
50%ile		71	37	Measured with 90° vertical shoulder abduction;	
5%ile		43	15	elbow at 90°; hand supine relative to the head;	
				cuff placed medial to medial surface of forearm.	
Mean		68.5(15.3)	36.3(9.3)		
Range		43–105.8	13.3–57.8		
● 20°	(N.m)	56.2(13.5)	32.4(6.5)	Males aged 51–65 and	Kuhlman *et al.*, 1992
90°		51.8(12.9)	26.7(5.8)	females aged 50–65	
30°	(N.m)	73.2(10.5)		Males aged 19–30.	—
120°		46.9(9.7)		All given data are isometric. Isokinetic data available.	
Abduction					
● 95%ile	(N.m)	115	54	25 men and 22 women employed in manual jobs.	Stobbe, 1982; Chaffin and Anderson, 1984
50%ile		67	30	Measured with 90° vertical shoulder abduction;	
5%ile		35	13	0° horizontal shoulder; elbow at 90°; cuff placed	
				on inferior surface of upper arm 2 cm medial to	
Mean		68.8(22.8)	33(11.5)	elbow joint.	
Range		32.8–120.7	13.3–57.8		

Medial rotation (N.m)

• 95%ile	83	33
50%ile	52	21
5%ile	28	9
Mean	50.9(13.7)	20.7(5.9)
Range	27.6–87.5	8.5–33.3

25 men and 22 women employed in manual jobs. Medial rotation was measured with the cuff medial to the wrist and with the same joint angles as abduction.

Stobbe, 1982; Chaffin and Anderson, 1984

Lateral rotation

• −60° (N.m)	33(7.1)	15.4(3.4)
0°	29.9(5.8)	17.8(4.4)
−60° (N.m)	45.3(7.8)	
0°	39.0(5.7)	
60°	32.5(6.0)	

Males aged 51–65
Females aged 50–65
Vertical dynamometer with humerus abducted 45° and 30° horizontal flexion.
Males aged 19–30
Isokinetic data available. All given data are isometric.

Kuhlman *et al.*, 1992
—
—

• 95%ile (N.m)	51	28
50%ile	33	19
5%ile	23	13
Mean	35(7.7)	19.7(4.2)
Range	22.6–51.6	13–28.3

25 men and 22 women employed in manual jobs. Measured with 5° vertical shoulder abduction; horizontal shoulder at 0°; elbow at 90°; hand semiprone; cuff placed posterior to the wrist.

Stobbe, 1982; Chaffin and Anderson, 1984

Flexion

• 45° (N.m)	68.6(14.1)	32.4(5.9)
0°	89.9(19.8)	43.1(9.8)

Females (mean age 22.5) and males (mean age 22). Average of concentric and eccentric strengths. Isokinetic strengths are also available.

Koski and McGill, in press

Horizontal flexion

• 95%ile (N.m)	119	60
50%ile	92	40
5%ile	44	12
Mean	83.5(22.4)	39(11.5)
Range	43.1–121.4	10.5–61.1

25 men and 22 women employed in manual jobs. Measured with 90° vertical shoulder abduction; horizontal shoulder at 0°; elbow at 90°; hand prone relative to the floor; cuff placed medial to the nearest surface of the forearm.

Stobbe, 1982; Chaffin and Anderson, 1984

Table 4.13 Data on maximal strength in set postures: shoulder (Cont.)

Joint function	Unit[2]	Strength ± (SD)[1]		Population studied and notes on the study	References
		Males	Females		
Horizontal extension	(N.m)			25 men and 22 women employed in manual jobs. Extension was measured at 60° horizontal shoulder angle; hand horizontal and prone and the same joint angles and cuff placement as above.	Stobbe, 1982; Chaffin and Anderson, 1984
● 95%ile		103	57		
50%ile		67	33		
5%ile		43	19		
Mean		71.5(19.4)	34.4(11)		
Range		42.7–103.7	18.6–57.6		

[1]SD = Standard deviation
[2]N = Newtons, and N.m. = Newtons acting at a distance of 1 metre from a joint

Table 4.14 Data on maximal strength capabilities for specific grips

Grip type	Unit[2]	Strength ± (SD)[1]		Population studied and notes on study	References
		Males	Females		
● Power	(N)	466.4	241	Skilled, manual and sedentary workers; Jamar handgrip dynamometer used for measurement.	American Medical Association, 1990
● Power	(N)			30 right-handed dominant adults aged 20–40	Pryce, 1980
0°		289.7(93)		A 16% difference in grip strength with changes in flexion, extension and ulnar deviation.	
15° ulnar dev.		289.2(106.3)			
30° ulnar dev.		263 (106.9)		Pacific Scientific grip device connected to a Gould 200 lb load cell adapter was used for measurement.	
● Power	(N)	487.4(17.3)	308(11.2)	Males and females aged 18–40. Stoelting handgrip dynamometer	Imrhan, 1989
● Power	(N)			20 healthy, right-handed adults (10 men and 10 women aged 20–51);	O'Driscoll *et al.*, 1992
wrist flexion		294.3(19.6)		Jamar dynamometer and electrogoniometer were used for measurement; unconstrained motion of wrist in saggital and transverse planes was allowed.	
wrist extension		323.7(29.4)			
radial deviation		274.7(29.4)			
ulnar deviation		294.3(29.4)			
● Power	(N)	503.1(72.5)	310.5	Applicants for employment at Kaiser Steel Corporation, Steel Manufacturing Division (1128 males and 80 females); Jamar dynamometer used for measurement. Data for the major hand.	Schmidt and Toews, 1970

Table 4.14 Data on maximal strength capabilities for specific grips (Cont.)

Grip type	Unit[2]	Strength ± (SD)[1]		Population studied and notes on study	References
		Males	Females		
● Power	(N)			12 male and female students aged 20–25	Miller and Wells, 1988
70° flexion		96.2		Power grasp performed on small diameter (4.5 cm) dynamometer to simulate tool handle.	
neutral		245.3			
60° extension		210.9		There was a 3–5% variation in grip strength with changes in pronation and supination.	
20° radial dev.		210.9			
neutral		245.3			
30° ulnar dev.		181.5			
● Pinch	(N)			Same as Miller and Wells, above.	Miller and Wells, 1988
70° flexion		18.6			
neutral		44.1		Pinch performed on pinch dynamometer.	
60° extension		44.1			
20° radial dev.		39.2			
neutral		44.1			
30° ulnar dev.		39.2			
● Pinch	(N)			Female dentists aged 22–55 with elbow unsupported while sitting.	Catovic *et al.*, 1991
Thumb + fingers			77–96	Gripping device with strain gauges	
Thumb + forefinger			34–49	Ranges of pinch strength include forces at various angles formed by forearm and clavicles in the frontal positions.	
Thumb + middle finger			28–38	Data while subjects standing available.	
● Lateral pinch	(N)	92.2(3.4)	63.8(1.7)	Adults aged 18–40 Preston pinch meter	Imrhan, 1989
					American Medical Association, 1990

• Lateral pinch	(N)	73.56	48.07	Skilled, manual and sedentary workers Pinch gauge	Hazelton *et al.*, 1975
• While using a suitcase grasp Forces at: Index finger Long finger Middle finger Little finger	(N)	117.7 155.7 108.8 85.3		30 right-handed males Digital dynamometer and dynograph recorder. Strengths are the distribution of forces among 4 fingers during suitcase grasp at distal phalanx. Changes in wrist positions produce 15% variations in strength distribution.	

[1] SD = Standard deviation
[2] N = Newtons, and N.m = Newtons acting at a distance of 1 metre from a joint

Table 4.15   Data on maximal strength capabilities in set posture: elbow

Joint function	Strength ± (SD)[1] in N.m[2]		Population studied and notes on study	References
	Males	Females		
Flexion			25 men and 22 women employed in manual jobs in industry. Flexion was measured with the elbow at 90°; forearm horizontal and hand semi-prone. Cuff was placed medial to the wrist.	Stobbe, 1982; Chaffin and Anderson, 1984
● 95%ile	111	55		
50%ile	77	41		
5%ile	42	16		
Mean	77.6(18.8)	40.2(10.1)		
Range	41–115.2	15.7–55.4		
Extension			25 men and 22 women employed in manual jobs in industry. Extension was measured with the elbow at 70°; the hand semi-prone and the cuff placed medial to the wrist.	Stobbe, 1982; Chaffin and Anderson, 1984
● 95%ile	67	39		
50%ile	46	27		
5%ile	31	9		
Mean	47.1(9.8)	27.2(8.3)		
Range	29.5–68.3	8.2–39		

[1]SD = Standard deviation
[2]N.m. = Newtons acting at a distance of 1 metre from a joint

Table 4.16 Data on maximal strength capabilities in set postures: wrist and forearm

Joint function	Strength ± (SD)[1] in N.m²		Population studied and notes on study	References
	Males	Females		
• Wrist flexion	7.97(2.43)	5.62(.92)	Female lab assistants	For females no. – Nordgren, 1972; for males no. – Bäcklund and Nordgren, 1968 both as quoted in Armstrong, 1987
Wrist extension	11.39(5.8)	6.41(1.36)	Male medical students	
• Radial deviation				
Dominant	17.9(4.8)	9.7(2.3)	Normal college students, 26 female and 6 male, aged 20–28	Vanswearingen, 1983
Non-dominant	15.5(2.6)	9.4(2.1)		
• Ulnar deviation				
Dominant	14.9(4.7)	8.6(1.7)	Normal college students, 26 female and 6 male, aged 20–28	Vanswearingen, 1983
Non-dominant	13.9(4.1)	8.3(2.1)		
• Pronation				
hand position – 60°	2.75		3 naval ratings, aged 18, 19 and 27; 0° = palm facing medially elbow angle at 90°.	Salter and Darcus, 1952
0°	6.74			
+ 90°	11.77			
• Supination				
hand position – 60°	8.73		3 naval ratings, aged 18, 19 and 27; as elbow angle changes, variations of 16% occur in maximum torque.	Salter and Darcus, 1952
0°	7.06			
+ 90°	4.02			

Table 4.16 Data on maximal strength capabilities in set postures: wrist and forearm (Cont.)

| Joint function | Strength $\pm$ (SD)[1] in N.m^2 | | Population studied and notes on study | References |
	Males	Females		
● Pronation			Female lab assistants	For females no. –
Key	2.8(0.9)	1.7(0.4)	Male medical students	Nordgren, 1972;
Handle	14.1(3.4)	6.3(2.1)	Key: pronating and supinating while	For males no. –
			using a lateral pinch grip;	Bäcklund and Nordgren,
			Handle: pronating and supinating	1968 both as quoted in
			while using a hand grip on a T-bar.	Armstrong, 1987
● Supination			Female lab assistants	For females no. –
Key	2.6(0.6)	1.5(0.4)	Male medical students	Nordgren, 1972;
Handle	12.9(6.1)	5.9(1.7)	Key: pronating and supinating while	For males no. –
			using a lateral pinch grip;	Bäcklund and Nordgren,
			Handle: pronating and supinating	1968 both as quoted in
			while using a hand grip on a T-bar.	Armstrong, 1987

[1]SD = Standard deviation
[2]N.m. = Newtons acting at a distance of 1 metre from a joint

Table 4.17   Recommended and non-recommended postures for force generation according to the literature*

Joint or action	Example
Pinch grip	Not recommended for force generation in pinch grip where power grasp can be used.
Finger	Not recommended to perform flexion against resistance if local loads on middle phalanages. Not recommended to apply force hyperextended.
Wrist	Recommended to use wrist angle 0 to 35° extension and 0 to 10° ulnar deviation for gripping.
Thumb	Not recommended to perform extension/abduction of thumb especially with ulnar or radial wrist deviation.
Wrist and forearm	Not recommended to move wrist or pronate/supinate while exerting force.
Wrist and forearm	Not recommended to pronate with wrist and/or fingers flexed. Not recommended to supinate and extend wrist. Not recommended to be in wrist extension and elbow flexion. Not recommended to lift heavy objects with palm facing down.
Forearm	Not recommended to repetitively supinate and pronate.
Elbow and forearm	Recommended to supinate against resistance with flexed elbow.

*Information is not available to provide intensity or frequency of exposure

10th, 50th and 90th percentiles of the amplitude probability distribution function (APDF) of an electromyogram. The APDF technique summarizes the magnitude of the activation.

Comparing a group of telephone assembly workers presenting similar musculoskeletal disorders with a control group allowed Aarås (1991) to make a number of suggestions regarding the shoulder region. He suggested that a static (10th percentile) trapezius muscle EMG of between 1 and 2 per cent MVC coupled with a 40th percentile value of below 5 per cent MVC was likely to be optimal. Veiersted *et al.* (1990) suggested that the presence of short periods of very low activity 'gaps' may be protective with respect to WMSDs.

Electromyography possibly offers a promising approach for assessing the optimality of tasks; it addresses musculoskeletal loads on muscle due to the workplace and the effects of other risk factors such as techniques or cognitive factors. It is specific to a particular body region on an individual. The results of 10 or so individuals can be combined to assess the optimality of a task (Jonsson, 1982). However, the recording and interpretation of electromyogram results are technically complicated during dynamic tasks and many factors can alter the signal characteristics. EMG has been used to assess existing work systems and is beginning to be used in the design process.

4.4.11 Comments on the identification of optimal levels

Only physical aspects of the task were discussed in section 4.4. No quantitative data are currently available to investigate the possible optimality of other risk factors, such as psychosocial factors, for instance; research on WMSD-related psychosocial factors is relatively new.

Although some studies have shown acute and chronic effects of different intensities of work risk factors and the mitigating effects of some work organizational features such as breaks, the limited number of studies and the shortcomings of their design make it difficult to develop quantitative recommendations for task optimality. Some recommendations also appear contradictory: for example, Fisher *et al.* (1993) have recently suggested, based on theoretical work derived from a cumulative strain model in tendon by Goldstein *et al.* (1987), that short breaks may reduce cumulative strain in tendon (and thus the risk of tendinitis). Byström *et al.* (1991), on the other hand, have suggested that micropauses could lead to greater injury potential for muscle than long periods between breaks. Winkel and Westgaard (1992a) have discusssed the difficulty of defining exposure and creating guidelines for the optimality of a task for the neck and shoulder area.

Care must be taken in applying methods discussed above, especially when it comes to determining work standards. Yet they provide a first approximation to what may be appropriate to minimize hazard, which is better than relying on 'economic' determinants as is often the case where such methods are not used. There is no single approach that deals effectively with all work factors that increase hazard of WMSDs. MTM methods as well as fatigue and endurance models may underestimate risk. Psychophysical methods have a large inherent variability and may also underestimate the risk. Maximum strength values should not be used as acceptable strength criteria.

In other areas of occupational health and safety, safety margins have been introduced to reduce hazard for the most susceptible worker populations, but this has not consistently been done to protect workers from WMSDs. Just as these disorders are multifactorial in origin, the assessment of risk in the workplace will have to combine methods to adequately characterize multiple and simultaneous risk factors.

5

Health and risk factor surveillance for work related musculoskeletal disorders

5.1 Introduction

In the course of their work, consciously or unconsciously, people interact with and react to many facets of their work situation. For the purposes of this discussion, an outcome is the manifestation of such a reaction to a work environment or task by a worker. Such reactions may be positive or negative. For instance, a reaction to a given work demand may be increased or decreased productivity, improved or reduced job satisfaction, or improved or impaired musculoskeletal health (for example, developing a WMSD is an outcome). Each work situation produces many outcomes of various intensities and at various levels. Positive and negative outcomes may co-exist as reactions to the same work situations. It is extremely difficult to monitor in great detail all these outcomes simultaneously in a constantly changing dynamic work environment. In view of these limitations, surveillance strategies are designed to recognize: (1) patterns of health and disease in groups of people; and (2) the most obvious risk factor patterns in the workplace that may contribute to health/disease patterns. While there may be a variety of other important patterns, the focus of this chapter is on methods for helping early recognition of patterns of work related musculoskeletal symptoms and disorders and their risk factors in the workplace (since this book focuses on workplace contributions to the development of WMSD, and how to detect and prevent workplace problems, this chapter will discuss workplace risk factor surveillance).

Surveillance has been defined as:

> The ongoing systematic collection, analysis and interpretation of health and exposure data in the process of describing and monitoring a health event. Surveillance data are used to determine the need for occupational safety

and health action and to plan, implement and evaluate ergonomic interventions and programs.

(Adapted from Klaucke *et al.*, 1988)

Health (WMSD) and risk factor surveillance provide employers and employees with a means of systematically evaluating WMSD and workplace ergonomic risk factors by monitoring trends over time. This information can be used for planning, implementing and continually evaluating ergonomic interventions. A surveillance system is an integral part of an overall ergonomics programme. Every employer, large or small, should have a WMSD surveillance system as part of the firm's health and safety programme. In small or medium-sized firms, which may not have the in-house personnel to design and run surveillance programmes, employers may decide to handle some portion of the surveillance programme in-house and contract out the rest. Regardless of how such programmes are organized, all the information and tasks mentioned in this chapter should be considered. It is important to remember that company and individual confidentiality in surveillance data should be assured in statistical summaries of data.

5.2 Surveillance: an overview

5.2.1 Brief description of surveillance system components

5.2.1.1 Risk factor and case definitions

Surveillance systems aim for early identification of patterns of work related musculoskeletal symptoms and disorders and their risk factors. Therefore, definitions for risk factors and cases of WMSD must be decided upon for surveillance purposes. Surveillance case definitions and clinical diagnostic criteria may differ considerably. For example, the NIOSH surveillance case definition of carpal tunnel syndrome may not be appropriate for determining whether a carpal tunnel release is a necessary treatment, but may be quite useful in identifying causes or risk factors so they can be reduced (Katz *et al.*, 1991).

5.2.1.2 Data collection instruments

Surveillance data collection instruments are characterized by their 'practicality, uniformity, and frequently, their rapidity rather than their complete accuracy' (Last, 1986). In that respect there is a role for both 'passive surveillance', which relies on information collected from existing databases, and 'active surveillance', which uses specifically designed tools and information solicited from a particular group or population.

Table 5.1 Examples of tools/instruments for WMSD surveillance

Surveillance on:	Methods of surveillance	
	Passive	Active
Health (WMSDs)	• Company dispensary logs • Insurance records • Workers' compensation records • Accident reports • Transfer requests • Absentee records • Grievances	• Checklists • Questionnaires • Interviews • Physical exams
Workplace risk factors (associated with WMSDs)	Not really used for WMSD risk factors yet*	• Checklists • Questionnaires • Job analysis

*The use of surrogate measures for exposure (e.g. job title or firm's department) could be viewed as 'passive surveillance'.

Passive surveillance uses existing records and data (e.g. company dispensary logs, insurance records, workers' compensation records, accident reports and to a lesser extent, transfer requests, absentee records and grievances) to identify WMSD cases and patterns and potential problem jobs (Table 5.1). Since passive surveillance data are the most readily available, they should always be considered first in any surveillance programme.

Passive surveillance records are often useful in helping to determine the frequency with which active surveillance tools should be used and the interventions required, and in assessing the effectiveness of the current ergonomics programme. Passive surveillance is used almost exclusively to survey health outcomes (WMSDs in our case). In theory, passive surveillance could also be used to study risk factors; in practice, however, to our knowledge, no existing records have been used to compile information on the risk factors thought to be associated with WMSDs (Table 5.1). It should be noted, however, that records may already exist which could be used for passive risk factor surveillance. In order to assess the suitability of a job for the return to work of an injured worker, currently a very brief job analysis or physical demand analysis must be completed on that job in some provinces, states and countries. Such records could possibly provide some passive surveillance data on risk factors. It should also be mentioned that surrogate measures of exposures (e.g. a worker's job title or his/her work department in the firm) could be viewed as 'passive surveillance'. In summary, passive surveillance involves the use of existing records (Table 5.1).

Active surveillance involves 'actively seeking' information; there can be both health (WMSD) active surveillance and risk factor active surveillance (Table 5.1). Active health (WMSD) surveillance often obtains information before individuals would normally feel compelled to report it. Health questionnaires are widely used for this purpose. Since most musculoskeletal disorders produce some symptoms of pain or discomfort, questionnaires are useful in identifying new or incipient problems as well as for assessing the effectiveness of medical interventions and ergonomic controls. Such questionnaires can also be used to obtain employee perceptions about aggravating factors and job improvement ideas. Symptoms questionnaires should not be viewed as clinically 'diagnostic'. In addition to questionnaires, medical interviews and examinations can also be used for active health (WMSD) surveillance (see section 5.3 for further details). On the other hand, active risk factor surveillance uses checklists and job analysis (see section 5.4 for further details).

5.2.1.3 Analysis and interpretation of data

Once surveillance data are available, tools and methods will be needed to analyse and interpret these data. Surveillance is not only the observation and interpretation of data, but must also lead to the appropriate action or follow-up. Results of surveillance programmes in industry are closely linked with the development and evaluation of ergonomics programs (see subsection 5.2.3.)

5.2.1.4 Training

Finally, individuals responsible for the surveillance programme require training to ensure that both active and passive surveillance systems are administered properly. Training is also necessary to guarantee that the information and data are analysed, interpreted and used correctly. Training is discussed in chapters 8 and 9.

5.2.2 Passive and active surveillance: in brief

The surveillance system must be able to link the occurrence of WMSDs to work related risk factors. Ideally, the surveillance system should make it possible to identify workplace risk factors before symptoms develop. Surveillance data can be obtained in two ways to identify WMSD and workplace risk factors: passive surveillance and active surveillance. Table 5.2 summarizes the differences between these two methods.

5.2.2.1 Passive surveillance

Passive surveillance is used mostly for health (WMSD) surveillance since, in practice, no existing records have been used to obtain information on risk factors associated with WMSDs. Passive surveillance implies that the data is collected principally for another purpose, such as medical management, workers' compensation or a bill-paying system for medical benefits. Examples of this approach using workers' compensation data have been described by Tanaka *et al.* (1988) and Franklin *et al.* (1991). Workers' compensation data are usually

Table 5.2 Information on passive and active surveillance methods

Passive surveillance*	Active surveillance
Information source and method already exist and are usually designed for other administrative purposes	Information source and method specifically designed for surveillance
Relatively inexpensive	Modest to quite expensive
Usually requires additional coding of information for the purpose. For instance, surrogate(s) of exposure: e.g. job titles	Since tools are 'tailor-made', includes at least job title information and other data considered important by surveillance analyst. Will include data for linking of information between risk factor and WMSD data
Examples: health and safety logs, medical department logs, workers' compensation data, early retirement, medical insurance, absenteeism and transfer records, a cident reports, product quality, productivity	Examples: for WMSD surveillance (confidential questionnaires without personal identifiers, questionnaire interviews, physical examinations); for risk factor surveillance (workplace walk-throughs, job checklists, postural discomfort surveys)

*Used mostly for health surveillance since, in practice, no existing records have been used to obtain information on risk factors associated with WMSDs.

already by 'body part' using 'nature of the injury' coding (e.g. ANSI Z-16.2), which poorly describes 'diseases'. The use of ICD-9 coding in records should be already by 'body part' using 'nature of the injury' coding (e.g. ANSI Z16.2), which poorly describes 'diseases'. The use of ICD-9 coding in records should be encouraged. There are national and international efforts currently underway to standardize the coding used. Linkage to specific exposures in passive health (WMSD) surveillance data is generally inadequate. Job titles, which can usually be found in these existing records, are a poor surrogate measure for exposure; for example, 'assembler', 'labourer' or 'computer operator' give no indication of intensity, duration or frequency of specific job risk factors for WMSDs. Production data have often been used as a surrogate measure for 'frequency', but these data are difficult to link to specific risk factors and WMSDs.

5.2.2.2 Active surveillance

Active surveillance involves designing a special system to obtain the information needed to identify and control specific patterns and processes that may put employees at risk of developing WMSDs. Information is then actively sought and entered into the surveillance system. The assumption is that passive surveillance does not give the full picture. Employees with WMSDs or symptoms may not be reporting them through passive surveillance channels for a variety of reasons; further, passive surveillance may identify problems only at a

---

Box 5.1   Active health and risk factor surveillance: levels 1 and 2

Health (WMSD) surveillance:
- level 1 uses self- or group-administered anonymous symptoms questionnaires
- level 2 uses health interviews and/or physical examinations by trained health care providers

Risk factor surveillance:
- level 1 uses quick checklists
- level 2 uses in-depth job analysis

---

later and more severe stage of development. Active surveillance should be workplace-specific so that unique risk factors particular to that workplace can be more readily identified and monitored. One of the major advantages of an active surveillance system is that standard definitions of 'WMSD cases' and 'risk factors' can be decided upon, used, and then compared over time or between departments or workplaces. The focus of active surveillance should be on job-related risk factors and adverse health outcomes (including but not always restricted to WMSDs). There are two levels of active health and risk factor surveillance (Box 5.1). As a general rule of thumb it should be remembered that for both health and risk factors surveillance level 1 active surveillance is a quick appraisal of the situation, whereas level 2 is an in-depth appraisal.

5.2.3 A brief note on links between surveillance, ergonomics programmes and screening

5.2.3.1 Ergonomics programmes and surveillance

A surveillance programme is part of a firm's ergonomics programme, which in turn is part of its occupational health and safety programmes. Important goals of an ergonomics programme are to identify, prevent or reduce work related musculoskeletal disorders. The effectiveness of any ergonomics programme can be measured against this goal. Health and risk factor surveillance provide the means for this systematic evaluation of the ergonomics programme by monitoring trends over time and using this information for planning, implementing and continually evaluating ergonomic interventions.

5.2.3.2 Screening and surveillance

Screening is the application of at least one test (or examination) to individuals in order to sort out apparently well persons who are probably developing the disorder (WMSD in our case) from those who are not. Screening tests are not diagnostic; persons with suspicious screening findings are referred to their physicians for diagnosis (Last, 1988). Screening clearly has many roles. The purpose of screening tests could be: (1) to predict future risk; (2) to target individuals for early treatment; and (3) to estimate current functional capacity. Some of these roles could be pertinent to a surveillance programme.

With respect to the use of screening 'to predict future risk', there are no screening tests at present that can predict which individuals will develop WMSDs. For example, muscle force or endurance tests have been proposed to predict propensity for these disorders. Takala and Viikari-Juntura (1991) demonstrated no difference in muscle force or endurance in sedentary workers with and without neck-shoulder symptoms. Chapter 9 contains a brief overview of the literature on this topic, showing that there is no scientific evidence to substantiate the use of screening tests for determining workers at risk of developing WMSDs. Screening for this purpose is not recommended.

Although no screening tests can predict which individuals will develop WMSDs, there could be a role for screening for current functional capacity within a surveillance programme. If a surveillance system has identified workers in high-risk jobs, periodic screening tests for early detection of decreased function (e.g. nerve conduction, loss of strength) may be appropriate for medical follow-up of these individuals at the same time as their jobs are being redesigned to reduce the risk factors. In such circumstances, temporary medical removal and/or immediate job modification may be appropriate if subtle signs of increasing dysfunction or disorder are identified.

5.3 Health (WMSD) surveillance

Health (WMSD) surveillance can be done with both passive and active surveillance methods. Regardless of the method used, in each case an acceptable surveillance WMSD case, i.e. WMSD outcome, must be defined and an appropriate tool selected to measure this outcome. One way to decide which type of health surveillance (passive or active) will be used, and when, is discussed later in section 5.5.

5.3.1 WMSD outcomes: general information

Work related musculoskeletal outcomes can be classified into different categories, such as self-reported symptoms, medical findings and behavioural signs. Most of these can be used for surveillance purposes. Self-reported symptoms may include discomfort, aches and pains, experience of mental strain and fatigue. Symptoms are the workers' experience of sensations that are not believed to be 'normal'. Examples of symptoms are pain and paraesthesia, as in tingling and numbness. Medical findings such as clinical signs may be the workers' experience of dysfunction. Examples of self-experienced signs are impairment of grip strength, difficulty in buttoning clothes, or loss of range of movement. Behavioural signs of WMSD outcomes may include absence from work. Other behavioural signs are frequency of consumption of medical care and of medication, injury reports and claims.

There is no universally agreed classification for work related musculoskeletal disorders that could be used for surveillance purposes, although various classification schemes have been proposed (Waris, 1979; Kuorinka and Viikari-Juntura, 1982; Silverstein and Fine, 1984). A modified version of the Spitzer *et al.* (1987) classification scheme for spinal disorders might be used. Criteria for classification should include:

- biological plausibility (compatible with current knowledge of pathophysiology),
- exhaustive classification (can encompass all clinical cases seen in occupational health),
- mutually exclusive categories,
- reliability (given disorder classified the same by several practitioners),
- clinical usefulness (facilitate making clinical decisions as well as evaluation of care),
- simplicity (simple to use and neither calling for complex paraclinical examinations nor encouraging superfluous investigations),
- consistency (consistent with medical experience and current diagnostic settings), and
- analogy (similar sets of criteria for diagnoses are used in the classification regardless of body region).

A variety of national and international medical organizations are currently addressing this issue of classification.

5.3.2 Passive health surveillance (using existing records)

5.3.2.1 Tools (passive health surveillance)

Possible tools (existing records) that could be used in passive health surveillance have already been discussed (Table 5.1 and subsection 5.2.2).

5.3.2.2 Case definitions (passive health surveillance)

Clinical diagnostic criteria and surveillance case definitions may differ considerably. For example, the NIOSH surveillance case definition for carpal tunnel syndrome includes:

- characteristic symptoms and either physical exam findings or abnormal nerve conduction, and

- the presence of certain risk factors (irrespective of duration, level or frequency).

This definition is quite useful for identifying work related causes so they can be controlled (Katz *et al.*, 1991), but may not be appropriate for determining the efficacy of carpal tunnel surgery and ability to return to work. On the other hand, simplifying the definition of a surveillance case as 'persistent symptoms limited to a group of workers doing the same job' may be sufficient cause to look more closely at the job immediately rather than waiting until all the criteria of the NIOSH case definition, especially the more clinical components, have been met.

Figure 5.1 presents an example of a decision tree for determining WMSD surveillance case status. Using passive surveillance data where the data have been coded using ICD-9, potential WMSD cases are identified, the various decision branches are worked through, and only then can a real surveillance case of WMSD be identified. As for the diagnoses listed in the records used for passive surveillance, we recognize that different clinicians using different criteria might arrive at the same 'diagnosis'. While it would be best for all practitioners to use the same criteria, it is beyond the scope of this document to provide such criteria. Clinical diagnoses can be made only on the basis of a clinical exam, whereas it must be possible to classify WMSD symptoms according to symptoms data alone. Certain diagnoses included here (e.g. Raynaud's syndrome) are considered potential WMSD cases for surveillance purposes (Figure 5.1). Although these diagnoses have not been the focus of the WMSDs discussed in this book, since surveillance data must be very sensitive (i.e. have a high potential of identifying all true cases of WMSD), there is good reason to be over-inclusive of diagnoses at this stage of the surveillance process. Indications of severity (lost or restricted days, seeking treatment, duration of more than one week and aggravating factors) could also be used in the decision tree to increase specificity in case definition.

5.3.3 Active health surveillance: level 1 (using symptoms questionnaires)

5.3.3.1 Tools (active health surveillance: level 1)

The first level of active health (WMSD) surveillance is a relatively rapid appraisal of possible symptoms by means of a self- or group-administered symptoms survey or questionnaire (Figure 5.2 is only one example) that briefly addresses the type of symptoms, onset, duration, aggravating factors and severity. The issue of the individual's 'health perception and reporting behaviour' arises with the use of such symptoms questionnaires. For example, perceptions and reporting of pain are often dependent on health belief systems, gender, age, culture (Morris, 1991) and the respondent's level of trust about what will be done with the information provided.

Symptoms questionnaires are not intended to be clinically diagnostic, but instead allow standardization of information and surveillance data collection

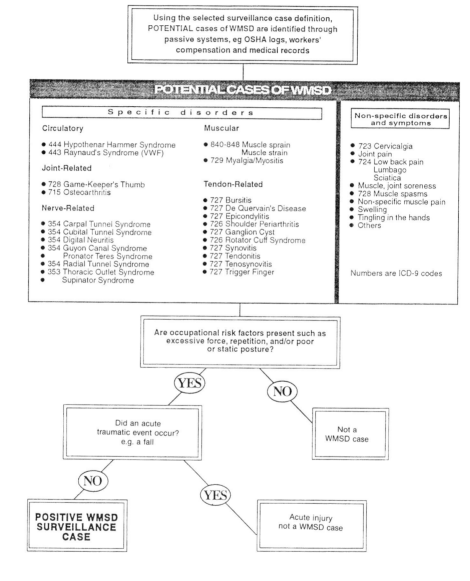

Figure 5.1 Decision tree for determining a WMSD surveillance case (using an example of passive surveillance with a record coded with ICD-9)

and help identify potential problems in the work environment. To date, none of the symptoms questionnaires available has been scientifically validated. In this context, 'validated' implies ensuring that we are truly measuring what we think we are, in our case WMSD, and that the symptoms we are measuring are related to the clinical outcome. Validation of some instruments is underway (Viikari-Juntura, 1988.

Self-administered symptoms questionnaires are the least expensive to administer but often provide incomplete data and the intent of the questions may be misunderstood. Literacy of the workforce should also be considered. Group-administered symptoms questionnaires require modest resources and have a health-care provider or ergonomics team member clarify the intent and provide individual assistance where needed. They increase the likelihood of obtaining reasonably reliable information, but may be influenced by lack of anonymity and co-worker pressure.

5.3.3.2 *Case definitions (active health surveillance: level 1)*

Symptoms surveys or questionnaires for level 1 active health surveillance, should contain, at a minimum, information by body part which will allow the symptoms case definition as set out in Box 5.2. While this definition may be adequate for level 1 active health surveillance purposes, it may also be necessary to incorporate more information at this stage to help determine priorities for level 2 resources (i.e. a detailed health interview or physical examination; see subsection 5.3.4). 'Severity of symptoms' can be used to determine these priorities. Thus the case definition in Box 5.2 could be used along with one or more of the following (if one is using the information obtained with the symptoms questionnaire suggested in Figure 5.2):

- current severity rating of 3 or greater (0-5 scale),
- sought medical attention,
- resulted in lost time (official or unofficial),
- light or restricted work (official or unofficial),
- changed jobs because of these problems.

5.3.4 Active health surveillance: level 2 (using health-related interviews or brief physical exams)

5.3.4.1 *Tools (active health surveillance: level 2)*

Health-related interviews and brief surveillance physical exams are two possible tools for level 2 active health surveillance. Health-related interviews are more labour-intensive than level 1 symptoms questionnaires, but can include probing to increase the sensitivity and specificity of the data. If done well and by a health team member trusted by employees, an interview can probably obtain as much

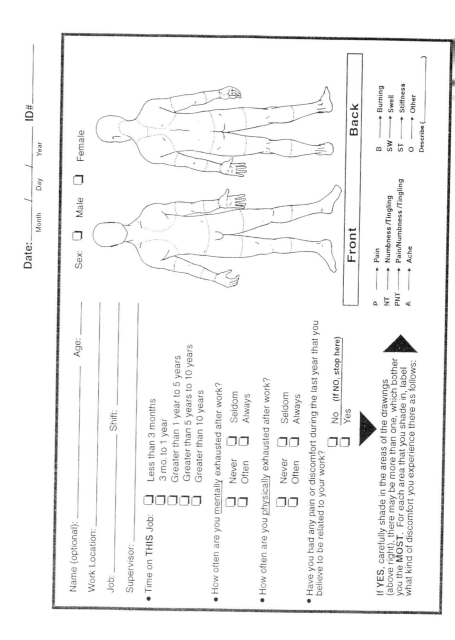

Figure 5.2 Symptoms questionnaire (Example of a tool for level 1 active health surveillance).

Please complete the column for each body part that bothers you   Name (optional) _____   ID # _____

	Neck	Shoulder	Elbow/forearm	Hand/wrist	Fingers
• 1 Which side bothers you?	☐ Left ☐ Right ☐ Both	☐ Left ☐ Right ☐ Both	☐ Left ☐ Right ☐ Both	☐ Left ☐ Right ☐ Both	☐ Left ☐ Right ☐ Both
• 2 What year did you first notice the problem?					
• 3 How long does the problem usually last?	A Less than 1 hour B 1 hr - 24 hrs C > 24 hrs - 1 week D > 1 week - 1 month E > 1 month - 6 months F More than 6 months ☐☐☐☐☐☐	A☐ B☐ C☐ D☐ E☐ F☐	A☐ B☐ C☐ D☐ E☐ F☐	A☐ B☐ C☐ D☐ E☐ F☐	A☐ B☐ C☐ D☐ E☐ F☐
• 4 How many separate times have you had the problem?	A Constant B Daily C Once a week D Once a month E Every 2-3 months F More than 6 months ☐☐☐☐☐☐	A☐ B☐ C☐ D☐ E☐ F☐	A☐ B☐ C☐ D☐ E☐ F☐	A☐ B☐ C☐ D☐ E☐ F☐	A☐ B☐ C☐ D☐ E☐ F☐
• 5 What do you think caused the problem?					
• 6 Problem in the last 7 days?	☐ Yes ☐ No	☐ Yes ☐ No	☐ Yes ☐ No	☐ Yes ☐ No	☐ Yes ☐ No
• 7 According to the scale of 0-5 shown, how would you rate this problem right now?	0 = no discomfort 5 = unbearable discomfort Rating = ☐	0 = no discomfort 5 = unbearable discomfort Rating = ☐	0 = no discomfort 5 = unbearable discomfort Rating = ☐	0 = no discomfort 5 = unbearable discomfort Rating = ☐	0 = no discomfort 5 = unbearable discomfort Rating = ☐
• 8 Have you had medical treatment for this problem?	☐ Yes ☐ No	☐ Yes ☐ No	☐ Yes ☐ No	☐ Yes ☐ No	☐ Yes ☐ No
• 9 Days of work lost in the last year due to this problem	_____ days	_____ days	_____ days	_____ days	_____ days
• 10 Days of light or restricted duty in the last year due to this problem	_____ days	_____ days	_____ days	_____ days	_____ days
• 11 Have you changed jobs because of this problem?	☐ Yes ☐ No	☐ Yes ☐ No	☐ Yes ☐ No	☐ Yes ☐ No	☐ Yes ☐ No
• 12 Please comment on what you think would improve your symptoms					

Figure 5.2 Symptoms questionnaire (Cont.).

Please complete the column for each body part that bothers you   Name (optional)   ID #

	Upper back	Low back	Thigh/knee	Low leg	Ankle/foot
• 1 Which side bothers you?	☐ Left ☐ Right ☐ Both	☐ Left ☐ Right ☐ Both	☐ Left ☐ Right ☐ Both	☐ Left ☐ Right ☐ Both	☐ Left ☐ Right ☐ Both
• 2 What year did you first notice the problem?					
• 3 How long does the problem usually last?	A Less than 1 hour B 1 hr - 24 hrs C > 24 hrs - 1 week D > 1 week - 1 month E > 1 month - 6 months F More than 6 months ☐ A ☐ B ☐ C ☐ D ☐ E ☐ F	☐ A ☐ B ☐ C ☐ D ☐ E ☐ F	☐ A ☐ B ☐ C ☐ D ☐ E ☐ F	☐ A ☐ B ☐ C ☐ D ☐ E ☐ F	☐ A ☐ B ☐ C ☐ D ☐ E ☐ F
• 4 How many separate times have you had the problem?	A Constant B Daily C Once a week D Once a month E Every 2-3 months F More than 6 months ☐ A ☐ B ☐ C ☐ D ☐ E ☐ F	☐ A ☐ B ☐ C ☐ D ☐ E ☐ F	☐ A ☐ B ☐ C ☐ D ☐ E ☐ F	☐ A ☐ B ☐ C ☐ D ☐ E ☐ F	☐ A ☐ B ☐ C ☐ D ☐ E ☐ F
• 5 What do you think caused the problem?					
• 6 Problem in the last 7 days?	☐ Yes ☐ No	☐ Yes ☐ No	☐ Yes ☐ No	☐ Yes ☐ No	☐ Yes ☐ No
• 7 According to the scale of 0-5 shown, how would you rate this problem right now?	0 = no discomfort 5 = unbearable discomfort Rating = ☐	0 = no discomfort 5 = unbearable discomfort Rating = ☐	0 = no discomfort 5 = unbearable discomfort Rating = ☐	0 = no discomfort 5 = unbearable discomfort Rating = ☐	0 = no discomfort 5 = unbearable discomfort Rating = ☐
• 8 Have you had medical treatment for this problem?	☐ Yes ☐ No	☐ Yes ☐ No	☐ Yes ☐ No	☐ Yes ☐ No	☐ Yes ☐ No
• 9 Days of work lost in the last year due to this problem	____ days	____ days	____ days	____ days	____ days
• 10 Days of light or restricted duty in the last year due to this problem	____ days	____ days	____ days	____ days	____ days
• 11 Have you changed jobs because of this problem?	☐ Yes ☐ No	☐ Yes ☐ No	☐ Yes ☐ No	☐ Yes ☐ No	☐ Yes ☐ No
• 12 Please comment on what you think would improve your symptoms					

Figure 5.2 Symptoms questionnaire (Cont.).

Box 5.2   Suggested symptoms case definition (by body part) for active health (WMSD) surveillance: Level 1

Perceived work related pain or discomfort (symptoms of pain, numbness, tingling, burning, swelling) occurring in the previous 12 months that lasted at least one week or occurred at least once a month and was not caused by an acute injury.

Box 5.3   Tools used in surveillance: summary

Approach	*Tool*
Health	
● Passive	● Existing records
● Active, level 1	● Symptoms surveys or questionnaires (self- or group-administered)
● Active, level 2	● Health-related interviews and/or brief physical exams
Risk factors	
● Active, level 1	● Quick checklists of risk factors
● Active, level 2	● In-depth job analysis

---

useful information as a brief surveillance physical exam, which is much more labour-intensive and probably no more reliable (Viikari-Juntura, 1988). Examples of tools which could be used in health-related interviews or physical examinations can be found in the Silverstein study (1985). Further, clinical diagnoses are probably beyond the scope of health-related interviews. In addition to health-related interviews, the use of functional disability scales (Viikari-Juntura, 1988) may be useful in estimating severity. Pain pressure threshold meters are another tool used by some investigators to 'quantify pain' (Takala *et al.*, 1992). It should be noted, however, that the variability between examiners and inter- as well as intra-subject variability may complicate their interpretation of the data obtained by interview or examination. As with all surveillance tools, employees should be assured of the confidentiality of information obtained. A summary of the tools used in surveillance, including health surveillance, can be found in Box 5.3.

5.3.4.2 Case definitions (active health surveillance: level 2)

A variety of case definitions for epidemiologic studies have been formulated and may be adequate for level 2 health surveillance (Waris, 1979; Silverstein and Fine, 1984). These may not, however, be appropriate for clinical diagnosis and case management (see chapter 9, on medical management). Examples of case definitions for some of the WMSD disorders (carpal tunnel syndrome, tendinitis, etc), that could be used for level 2 active health surveillance, are provided in the Silverstein study (1985).

5.4 Risk factor surveillance

The focus of this book is on workplace contributions to the development of WMSDs, and how to detect and prevent workplace problems. Accordingly, this chapter is concerned with workplace risk factor surveillance.

Passive surveillance for risk factors, although feasible, does not exist for WMSD risk factors, i.e. there are no records currently in existence that have been used to obtain information on risk factors *per se*. It is, however, possible that records may exist that could be useful for risk factor surveillance. Indeed, in order to assess the suitability of a job for the return to work of an injured worker, a very brief job analysis or physical demand analysis must be completed on that job in some provinces, states or countries. Such records could possibly provide some passive surveillance data on risk factors. Currently, however, all risk factor surveillance is active; there are two levels of active risk factor surveillance.

Risk factor surveillance tools, like health outcome surveillance tools, need to be simple and rapid rather than completely accurate. The more in-depth information on risk factors at work described in chapter 4 is intuitively appealing but may be difficult for inexperienced observers to rapidly apply to all jobs. Level 1 active risk factor surveillance will suggest quick measures of exposure, which will help determine whether a more in-depth analysis is needed. Level 2 active risk factor surveillance will suggest tools for a more in-depth job analysis. With increasing experience, it may be possible for workplace personnel to assess risk factors according to all the approaches described in chapter 4; these may lend themselves better to the search for strategies for solving workplace problems.

5.4.1 Active risk factor surveillance: level 1 (using checklists)

The first level of risk factor surveillance involves using a simple job checklist on all jobs or groups of jobs to quickly observe and characterize levels of risk factors present. Examples of potentially useful checklists are included in Tables 4.7 and 4.8 (RULA, Keyserling).

These checklists can be used by trained ergonomics team members, trained worker/supervisor teams or outside consultants. The benefit of a quick appraisal is that it does not rely on employees recognizing symptoms before action can be taken (action can be taken as soon as risk factors are found). Any time a new job, process or equipment or change in methods is instituted, a

Table 5.3   Example of a general assessment questionnaire (part of level 1 active risk factor surveillance)

───────────── General assessment questionnaire ─────────────

Please answer the questions listed below regarding your health and your work.
We want to know how you have felt this **past month** including today.
Do **not** put your name on this questionnaire, but be sure to indicate your job title and department at the bottom.
No one will know how you answered these questions.
This information will help us to improve your working conditions.

Thank you.

Please circle the number that indicates how often you have experienced each of the following during the last month.

		never	sometimes	often	always
1.	You had headaches	1	2	3	4
2.	Your hands or fingers got numb	1	2	3	4
3.	You had a cold or sore throat	1	2	3	4
4.	You had back pain	1	2	3	4
5.	Your wrists or hands hurt	1	2	3	4
6.	You woke up at night with hand paid	1	2	3	4
7.	You felt nervous or irritable	1	2	3	4
8.	Your arm hurt or felt numb	1	2	3	4
9.	You felt very tired at work	1	2	3	4
10.	You had pain in your neck or shoulder	1	2	3	4
11.	You felt pressured to work faster	1	2	3	4
12.	You felt you were in control of your job	1	2	3	4
13.	You had chest pains	1	2	3	4
14.	You liked your job	1	2	3	4
15.	Your legs hurt	1	2	3	4
16.	Your lower back hurt	1	2	3	4

Please fill in your job title, employee study ID# (if appropriate), and department below. Thank you for your help.

_____   _____   _____   _____/_____/_____

Job title   Employee study ID#   Department   Date

checklist should be used. Each workplace has unique characteristics that probably require specific checklists to be developed and scores or criteria decided upon. Use of a physical risk factor checklist may be combined with a general assessment questionnaire (Table 5.3) completed by employees whose jobs are being assessed. This general assessment questionnaire is an adaptation of a NIOSH health complaints checklist. It has been used by consultants and researchers and is thought to be a sensitive tool for identifying groups of employees with musculoskeletal discomfort and general job distress. It has content validity and its findings correlate highly with other self-reported indicators of discomfort (Smith and Zehel, 1992; Carayon *et al.*, 1987a, b). However, it has not been field-validated as an early indicator of WMSDs or predictor of WMSDs, nor has it been validated in medical/clinical trials.

The general assessment questionnaire, which includes employee perceptions of the effect of the work environment on them, should be compared with a physical risk factor checklist. If there is little concordance between these two instruments, it is likely that a significant organizational problem exists. If a potential organizational problem exits, the NIOSH Job Stress instrument (Hurrell and McLarey, 1988) or other survey instruments may be beneficial. Likewise, methods described in chapter 4 will provide more information on the actual stressors that could be the cause of the problem.

5.4.2 Active risk factor surveillance: level 2 (using job analysis)

Level 2 risk factor surveillance involves in-depth job analyses and should be implemented in those work areas identified as potentially high risk, based on level 1 risk factor surveillance. A more thorough workplace evaluation could be done, based on the information found in chapter 4. The extent of job analysis detail is aimed at finding the root cause of the risk factors so integrated solutions can be developed. Rarely does a 'problem job' occur in isolation. Usually, related jobs and the organizational context need to be assessed, and a thorough job analysis conducted. This facilitates integrated solutions. The surveillance component involves recording the status of the analysis and the risk factors identified, and processing the information with a view to developing solutions.

5.5 Determining which levels of health and risk factor surveillance to use

The full workplace surveillance flow chart is shown in Figure 5.6.

The importance of standardizing methods of surveillance between jobs, work areas, etc., or over time needs to be emphasized. It is extremely difficult to compare rates between jobs, departments and workplaces or across years unless the data have been collected in generally the same way with the same case definitions. It is also important to remember that in the surveillance process not all workers need to be examined. It can be quite appropriate to use samples of workers doing the same job. Obviously, if a problem is found in a particular job, all workers performing the job should be followed up.

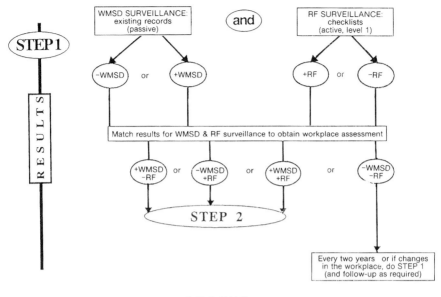

Figure 5.3 Step 1: workplace surveillance flow chart.

Ideally, every reported WMSD should have at least a quick look (checklist) of the job performed and problems fixed as soon as possible. At a minimum, where rates exceed those in Box 5.4, job observations must take place.

The surveillance flow chart presented in this section presents one ideal approach among many possibilities. However, in industry, there may be circumstances where compromises are necessary. In some industries, for example, where resources can be at a premium, it could be argued that identifying risk factors could be the priority in surveillance process.

5.5.1 Step 1

Passive health surveillance, using existing records and active risk factor surveillance (level 1) using quick checklists, should be performed by all firms initially to estimate the potential magnitude of the WMSD problem in the workplace (Figure 5.3).

5.5.1.1 Step 1: results of passive health (WMSD) surveillance

Does the firm have WMSDs (+ WMSD in Figure 5.3) or not (– WMSD in Figure 5.3)?

Equation 5.1   WMSD incidence rate (IR) i.e. new cases

$$IR = \frac{\# \text{ of new cases of WMSD during time}_x \times 200\,000 \text{ hours}}{\text{work hours during time}_x}$$

Where:

# of new cases of WMSD	=	WMSD data obtained using existing records meeting case definition
time$_x$	=	period of time used (i.e. during which new cases of WMSDs occurred e.g. 1 year)
200 000	=	estimated # of hours for 100 workers (based on 2000 hours worked per year per person)
work hours during time$_x$	=	cumulative # of hours worked by all employees during time$_x$

In order to interpret the WMSD data obtained through records, the incidence rate (IR) for WMSDs, i.e. the number of new cases, is calculated. To calculate the IR, the number of employees in each job, department or similar process should be determined. This is most often done by looking at the number of work hours for such employees. With both the health (WMSD) data obtained from records and the number of work hours, information on the incidence of WMSDs (the rate of new cases) can be obtained. This IR is equivalent to the number of new cases per 100 worker years. Workplace-wide incidence rates (IR) should be calculated for all WMSDs and by body part. The same rates should then be calculated for each department, process or type of job. Obviously, the data used for each calculation should be appropriate to the desired IR, i.e. for the IR of a specific department one would use the WMSD data for that department and work-hours for that department. Severity rates (SR) are based on the same calculations, except they traditionally use the number of lost workdays due to WMSDs in the numerator, rather than the number of cases.

5.5.1.2 Step 1: combining results (passive WMSD surveillance and level 1 active risk factor surveillance)

Based on passive surveillance data (medical logs, workers' compensation data and insurance data) and on checklist results for risk factors (level 1 active risk factor surveillance), no WMSD problem has been identified if the criteria shown in Box 5.4 are met. If no WMSD problem has been identified (–WMSD and –RF in Figure 5.3), then passive health surveillance using existing records and level 1 active risk factor surveillance using checklists should be done every two years or when changes in the work process, methods and equipment are introduced (Figure 5.3).

Box 5.4   Step 1: Criteria for deciding whether a WMSD problem can be identified (using WMSD passive surveillance data and RF level 1 active surveillance data)

No problem is identified

if   no WMSD problem is identified (−WMSD result in Figure 5.3),
        i.e. the workplace-wide IR is less than 1 per 200 000 hours
                and
        there is less than a twofold difference in IR between departments or work areas;

and   no risk factor problem is identified (−RF results in Figure 5.3),
        i.e. no obvious risk factors are identified during workplace walk-through as part of the level 1 active risk factor surveillance (see section 5.4).

Comparison with provincial or national workers' compensation rates over time may suggest the need to alter the criteria used in passive surveillance to help decide whether a WMSD problem is identified. For example, the present limits of an IR of 1 WMSD case per 200 000 hours shown in Box 5.4 would currently appear to be a reasonable limit when compared with workers' compensation data from Washington State, for instance. These compensation data indicate an industry-wide rate for WMSDs of the entire upper limbs of 0.82 per 200 000 work hours for 1988–1991 (cases resulting in lost time and/or medical costs).

5.5.2 Step 2

If the criteria listed in Box 5.4 for continuing with passive health surveillance and level 1 active RF surveillance are not met in step 1, then active surveillance of symptoms (level 1 active health surveillance) and job analysis (level 2 active risk factor surveillance) should be instituted promptly in step 2 (Figure 5.4).

5.5.2.1 Step 2: results of level 1 active health surveillance

Judging by the data obtained in step 2, does the firm have WMSDs (+ WMSD in Figure 5.4) or not (− WMSD in Figure 5.4)? Using symptoms questionnaires (level 1 active health surveillance), the prevalence rate (existing WMSD cases) can be calculated.

By the second year of using symptoms questionnaires (level 1 active health surveillance), new cases of WMSD (incidence rates, Equation 5.1) can be calculated in addition to existing cases (prevalence rates, Equation 5.2). The numerator for Equation 5.1 would then be the number of new cases arising between the first year and the second year and meeting the level 1 active health surveillance case definition.

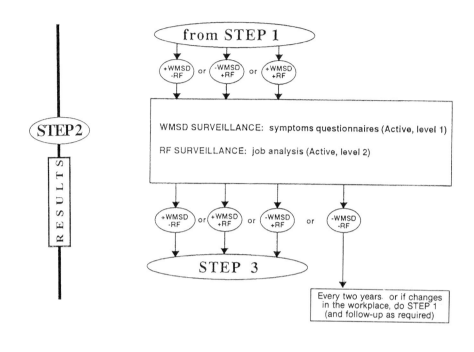

Figure 5.4 Step 2: workplace surveillance flow chart.

5.5.2.2 Step 2: combining results (level 1 active WMSD surveillance, and level 2 risk factor surveillance)

Based on the symptoms questionnaires (level 1 active health surveillance data) and on-job analysis (level 2 active risk factor surveillance), no WMSD problem has been identified (–WMSD and –RF in Figure 5.4) if the criteria shown in Box 5.5 are met. If no WMSD problem has been identified, then step 1, passive health surveillance using existing records, and level 1 active risk factor surveillance using checklists, should be done every two years or when changes in the work process, methods and equipment are introduced (Figure 5.4).

5.5.3 Step 3

If the criteria listed above (Box 5.5) are not met, then the following action should be taken (Figure 5.5):

Equation 5.2   WMSD prevalence rate (PR) i.e. existing cases

$$PR = \frac{\text{existing cases of WMSD during time}_x \times 200\,000 \text{ hours}}{\text{work hours during time}_x}$$

Where:

# of existing cases of WMSD	= WMSD data meeting case definition and obtained using level 1 health surveillance tools
during time$_x$	= time used (e.g. if the symptoms were recorded for the last month)
200 000 hours	= estimated # of hours for 100 workers (based on 2000 hours worked per year per person)
work hours during time$_x$	= cumulative # of hours worked by all employees during time$_x$

Box 5.5   Step 2: Criteria for deciding whether a WMSD problem can be identified (using active WMSD surveillance, level 1, and RF active surveillance, level 2)

No problem is identified

if   there is less than a twofold difference for a job or a work area when compared with the *prevalence* for the entire workplace (−WMSD result in Figure 5.4)

or

the workplace-wide *incidence* rate is less than 1 per 200 000 hours, and there is less than a twofold difference in IR between departments or work areas (−WMSD results in Figure 5.4);

and   no risk factor problem is identified when job analyses are done (−RF results in Figure 5.4).

- level 2 active health surveillance (using health interviews or brief physical exams)

and/or

- workplace/job interventions.

5.5.3.1 Step 3: WMSD active surveillance, level 2

Using symptoms questionnaires, i.e. level 1 active health surveillance, WMSDs are present (result + WMSD in Figure 5.4) when the following apply:

- twice the prevalence in a job or work area compared with the prevalence for the entire workplace (unlike incidence data, there are no normative

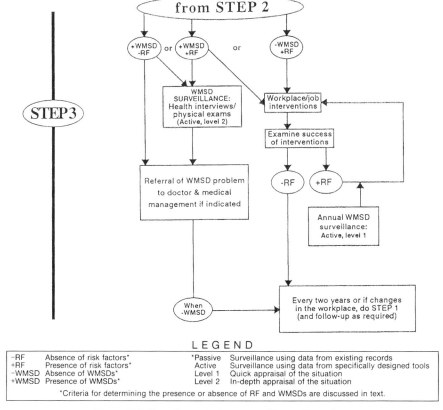

Figure 5.5 Step 3: workplace surveillance flow chart.

population data available for comparison where prevalence data are concerned – prevalence data can be compared only within the firm),

or

- an incidence rate of more than 1 per 200 000 hours and more than a twofold difference in IR between departments or work areas.

These above criteria are provided as a 'general rule of thumb' and are meant only to offer some guidelines.

If WMSD cases are found using symptoms questionnaires, such cases should be referred for medical attention, and health interviews and/or physical exams (level 2 active health surveillance) should be instituted (especially in the presence of risk factors). Level 2 active health surveillance is important in deciding the most appropriate action for the continued welfare of workers in such circumstances.

5.5.3.2 Step 3: Workplace/job interventions

If step 2 job analysis (level 2 active risk factor surveillance) reveals the presence of risk factors, then interventions to control such factors are essential in step 3. It is important that symptoms questionnaires (level 1 active health surveillance) and job analyses (level 2 active risk factor surveillance) be instituted prior to the workplace/job interventions, since these data will help establish baseline data from which to monitor the progress of ergonomics programmes (Figure 5.5). This is assumed in this flow chart, since these levels of surveillance are applied in step 2 and in fact helped determine whether the firm would proceed to step 3.

To evaluate the success of interventions, repeating job analyses (level 2 active risk factor surveillance) may be too costly and/or time consuming. Therefore, when workplace/job interventions have been introduced, active surveillance should include, at the very least, risk factor checklists (i.e. level 1 risk factor active surveillance) and annual symptoms surveys (level 1 active health surveillance). It would be best to collect baseline information throughout the workplace in order to determine priorities and monitor the effectiveness of ergonomics control programmes. However, it may be appropriate to begin in those areas of the workplace where:

1. the highest incidence rates were identified using passive surveillance,
2. the most people are affected,
3. the most severe cases were reported, or
4. large changes in processes or products are to take place regardless.

If the ergonomics programme is successful, one would initially expect an increase in reporting of symptoms to the medical department (due to both more accurate detection tools and increased awareness/understanding by all of WMSDs), followed by a decrease in the severity of cases and eventually a decrease in incidence rates. With very small numbers, WMSD rates can be very unstable and fluctuate widely. In such cases, rather than using incidence rates or prevalence rates, tools used in risk factor active surveillance might be useful and cost/time effective; indeed, in such situations, it might be more appropriate to concentrate on comparing risk factor scores and general assessment questionnaire scores over time. Scores for each risk factor should decrease over time if the interventions are successful. If interventions are successful, lower scores should be observed prior to a decrease in WMSDs. This approach offers a means of determining the potential effectiveness of the programme sooner. Other means of assessing the effectiveness of interventions and ergonomics programmes are mentioned in section 5.6.

5.6 Additional surveillance tools for evaluating ergonomics programmes

Possible tools for health and risk factor surveillance have already been discussed (sections 5.3 and 5.4). Other approaches do exist which, in this case, can be used

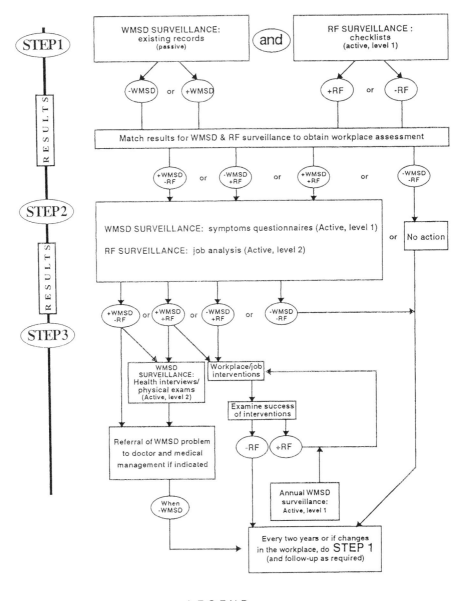

Figure 5.6 Complete workplace surveillance flow chart.

to evaluate the effectiveness of ergonomics programmes by replacing or complementing those tools already suggested for examining the success of such programmes (see 'examine success of interventions' in Figure 5.6). The discomfort survey (also known as 'localized fatigue survey') is one technique that can be used to examine the effectiveness of programmes (Figure 5.7) (Corlett and Bishop, 1978; Silverstein *et al.*, 1988). Using this instrument, employees rate current levels of fatigue or discomfort by body part. Approximately one month after the job has been modified, a repeat discomfort survey is conducted (same day of the week, same time of day as the initial survey, to ensure that the task/ job performed and general conditions are the same as those during the initial survey). A comparison of pre- and post-job change scores can provide an early indication of 'success' or of new problems that have been created by the job change. This type of risk factor surveillance does not wait for the 'WMSD body count' before assessing the ergonomic effectiveness of solutions.

More global measures of ergonomics programme effectiveness may include monitoring absentee rates, rates of requests for job transfers, types of employee grievances, types of employee suggestions for work improvements and job satisfaction surveys. One of the difficulties with using 'absenteeism' as surveillance data is that there are many reasons for absenteeism, both work related and non-work related. Patterns that differ substantially between departments or types of jobs may require more investigation. A similar caveat exists for job transfer request rates or turnover rates. Correlations between WMSDs and these other measures may be linear, synergistic or spurious. For example, a correlation between the level of absenteeism and the incidence or prevalence of WMSDs may mean that both are concomitant effects of the same 'unhealthy' work situation; in some instances people take more time off, in other instances they do not take time off and develop WMSDs. The same correlation may mean that poorly designed work produces WMSDs and those affected take time off work more often. Correlations are not directional, in that the way in which the sequence should be interpreted is not obvious. One must bear in mind, when viewing any negative outcome as a surrogate for a specific outcome (WMSD), that this negative outcome may be the result of other contributing factors having nothing to do with the case at hand, WMSD. For instance, acute injuries may be attributable to poor machinery design or maintenance, inadequate working procedures, etc. Absenteeism is a sensitive indicator of a number of problems, including a high level of stress, a poor work environment, a stringent management style, inadequate work organization and a lax collective bargaining agreement.

One difficulty in using attitudinal (job satisfaction) surveys in surveillance is that the baseline measure begins to change after a series of interventions so that the comparison is no longer with the original conditions, but rather with the most recent conditions. Nonetheless, instead of comparing the baseline with latest measures, looking for differences in patterns between groups of employees who have and have not experienced WMSDs may be a useful barometer of the perceived effectiveness of any ergonomics programme.

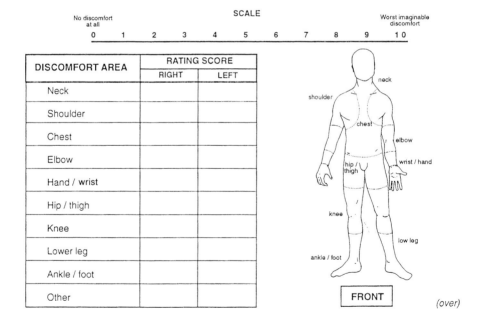

┌─────────────────── *DISCOMFORT SURVEY* ───────────────────┐

Last Name _____ First Name _____

Date _____ Time _____ am/pm

Department Name _____ Job Title _____

Shift _____

└───┘

Think about how you feel RIGHT NOW

1. *Shade in all the areas of discomfort on the figure.*

2 . *Rate the discomfort for each left and right side body part named in the the box below. Rate the discomfort using the scale below in which 0 = no discomfort at all and 10 = worst imaginable discomfort. Write the score in the box.*

SCALE

No discomfort at all   Worst imaginable discomfort

0   1   2   3   4   5   6   7   8   9   1 0

DISCOMFORT AREA	RATING SCORE	
	RIGHT	LEFT
Neck		
Shoulder		
Chest		
Elbow		
Hand / wrist		
Hip / thigh		
Knee		
Lower leg		
Ankle / foot		
Other		

FRONT

(over)

Figure 5.7 Discomfort survey (example of a tool to evaluate the effectiveness of workplace/job interventions).

SCALE

No discomfort at all										Worst imaginable discomfort
0	1	2	3	4	5	6	7	8	9	1 0

Think about how you feel RIGHT NOW

DISCOMFORT AREA	RATING SCORE	
	RIGHT	LEFT
Neck		
Shoulder		
Upper back		
Low back		
Elbow		
Hand / wrist		
Hip / thigh		
Knee		
Low leg		
Ankle / foot		
Other		

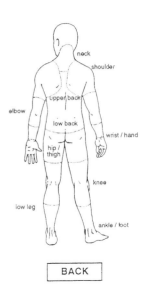

BACK

WHOLE BODY : Overall Discomfort Rating (use the same scale)	

Figure 5.7 Discomfort survey (Cont.).

5.7 Analysis and interpretation of surveillance data

Some examination and interpretation of surveillance data has already been done in section 5.5. In that section, we looked at using the data to determine which type and level of surveillance should be instituted. In this section, the surveillance data are analysed and interpreted to study possible associations between the WMSD surveillance data and the risk factor surveillance data. The question to be asked is whether we can observe an association, a link, between the number of WMSD cases found and a given job risk factor or a number of risk factors. There are many uses for such analyses of the data. Observing and comparing such associations may help, for example, in establishing which risk factor should be targeted first by interventions.

Analysis and interpretation of surveillance data to study associations can be a difficult task, especially for individuals with little training in biostatistics and epidemiology. The two principal goals of the analysis are: (1) to help identify patterns in the data which reflect large and stable differences between jobs or departments; and (2) to target and evaluate intervention strategies. This analysis can be done on the number of existing WMSD cases (cross-sectional analysis) or during time$_x$ on the number of new WMSD cases, in a retrospective and prospective fashion (retrospective and prospective analysis).

Selecting the appropriate method of analysis depends on the analyst's experience and access to statistical consultation, software and other resources. For the most part, the emphasis should be on simple methods and graphical methods. Attention should be directed toward reviewing the completeness of the data, and determining the response rate for each work area, department, etc. Response rates below 60 per cent may indicate that the data may not be representative of the actual situation for that work area/department, etc. Although all data can be tracked and analyzed by hand in a small workplace, a personal computer makes the process much more manageable. While elaborate software for surveillance systems exist, low-cost database entry and analysis programs are available, developed by the US Centers for Disease Control and the World Health Organization, specifically geared toward surveillance (e.g. EpiInfo 6.0).

The simplest way to assess the association between risk factors and WMSDs is to calculate odds ratios (Table 5.4). To do this, the prevalence data obtained in health surveillance are linked with the data obtained in risk factor surveillance. The data used can be those obtained with symptoms questionnaires (level 1 active health surveillance) and risk factor checklists (level 1 active risk factor surveillance). Each risk factor could be examined in turn to see whether it has an association with the development of WMSDs. In the example shown here (Table 5.4), one risk factor at a time is selected (e.g. overhead work for more than four hours). Using the data obtained in surveillance the following numbers of employees are counted:

Table 5.4 Examples of odds ratio calculations for a firm of 140 employees

		WMSDs are:		
		Present	Not present	Total
Risk factor (e.g. overhead work for more than 4 hours) is:	Present	15 (A)	25 (B)	40 (A+B)
	Not present	15 (C)	85 (D)	100 (C+D)
	Total	30 (A+C)	110 (B+D)	140 (N)

Numbers in each cell indicate the count of employees with or without WMSD and the risk factor.
Odds Ratio (OR)=(A×D)/(B×C)=(15×85)/(25×15)=3.4

- employees with WMSDs and exposed to more than four hours of overhead work (15 workers),
- employees with WMSDs and not exposed to more than four hours of overhead work (15 workers),
- employees without WMSDs and exposed to more than four hours of overhead work (25 workers),
- employees without WMSDs and not exposed to more than four hours of overhead work (85 workers).

The overall prevalence rate (PR), i.e. rate of existing cases, for the firm is 30/140 or 21.4 per cent. The prevalence rate for those exposed to the risk factor is 37.5 per cent (15/40) compared with 15.0 per cent (15/100) for those not exposed. The risk of having a WMSD depending on exposure to the risk factor, the odds ratio, can be calculated using the number of existing cases of WMSD (prevalence). In the above example, those exposed to the risk factor have 3.4 times the odds of having the WMSD than those not exposed to the risk factor. An odds ratio of greater than 1 indicates higher risk. Refer to Box 3.1 for help in interpreting the odds ratio. Ratios can be monitored over time to assess the effectiveness of the ergonomics programme in reducing the risk of WMSDs, and a variety of statistical tests can be used to assess the patterns seen in the data. Again, Box 3.1 may provide more information. Interestingly, some experienced epidemiologists suggest that statistical testing is overused and that data should be interpreted without very much reliance on formal statistical testing.

5.8 Example of the surveillance process

To illustrate how the proposed workplace surveillance programme could be used, a hypothetical example is presented, the case of Henry's T Shop.

Example 5.1:   Henry's T Shop

Henry's T Shop has 50 full-time employees:	Actual hours worked/year
3 work in the office	6000
2 are supervisors on the floor	4200
2 work in maintenance	4000
20 work in production area A	44 000
20 work in production area B	40 000
3 work in shipping/receiving	6000
Total hours	104 200

Table 5.5   1989-90 Incidence rates of WMSDs (per 100 worker years) for Henry's T shop

Work area	1990 Incidence rate	1989 Incidence rate
Office	0.0	0.0
Production A	13.6	5.0
Production B	5.0	0.0
Shipping/Receiving	33.3	33.3
Maintenance	0.0	0.0
Shop floor (supervisors)	0.0	0.0
Total	9.6	4.0

There were five WMSD cases (i.e. meeting passive health surveillance case definition) based on workers' compensation information during 1990. The plant-wide incidence (new cases) rate was (Equation 5.1):

IR = (5 cases × 200 000 hours) / 104 200 = 9.6 cases per 100 worker years.

Assessing this IR compared with the criteria provided in section 5.5 (1 case per 200 000 hours or 100 worker years), leads us to believe that active health surveillance is required on this basis alone. Further, if we look a little more closely, these five cases were distributed as follows: three cases of carpal tunnel syndrome were found in Production area A; one wrist tendinitis was found in Production area B; and one shoulder tendinitis was found in shipping/receiving. IR can therefore be calculated for each section/department of Henry's T Shop; results are shown in Table 5.5. Incidence rates were previously calculated for 1989 and these are also shown in Table 5.5.

As can be seen in Table 5.5, the total incidence rate rose between 1989 and 1990 (from 4.0 to 9.6). Also there was more than a twofold difference in incidence rates between work areas (criteria provided in section 5.5); although there was just one case in shipping in 1990, there were only three workers, so that there was 3.5 times (33.3/9.6) the risk of WMSD in shipping as in the plant as a whole. Because one case was also found in 1989 (IR = 33.3), it appeared that there might be an ongoing problem in shipping. The WMSD problem was increasing in both production areas, but was of particular concern in Production area A because the rate was more than twice that in Production area B.

By looking at other 'passive surveillance' data, including overtime and quality records, it was noted that Production area A had 4000 hours of overtime in 1990 (200 hours per employee), primarily because of the high scrap/reject rates and employee turnover (thus requiring more training time). On the other hand, there was no overtime in Production area B, which had much lower turnover and reject rates. Using a risk factor checklist, potentially important risk factors were identified. Both Production area A and shipping/receiving jobs exceeded a threshold score and required symptoms questionnaires (level 1 active WMSD surveillance) and job analyses (level 2 active risk factor surveillance) (step 2, Figure 5.4).

As part of the level 1 active WMSD surveillance program, a symptoms questionnaire was completed by employees (Figure 5.2). In Production area A, 60 per cent of employees had hand/wrist symptoms in the last seven days, 40 per cent had hand/wrist symptoms that lasted more than one month and 35 per cent of employees had no acute onset of symptoms; however, all these employees said that symptoms began after being on the current job. In Production area B, 25 per cent of employees had hand/wrist symptoms in the last seven days, and 15 per cent of employees had symptoms lasting more than one month. In shipping/receiving, all three employees indicated shoulder symptoms, and one (33 per cent) had low back pain. A more detailed evaluation of the jobs (level 2 RF surveillance) in these three work areas was completed and recommendations were made. These recommendations were discussed with employees after they had received ergonomics training. Employees thought some of the recommendations were impractical, and improved on other recommendations. For example, in shipping, a sheet metal slide was used to move boxes from one location to another, so that no boxes would be lifted from above shoulder height or below knee height. To evaluate the effectiveness of the interventions, discomfort surveys (Figure 5.7) were used before and after the changes. These indicated a reduction in shoulder discomfort from an average of six (maximum of 10) prior to the change to an average of three after the change.

Once the risk factor situation was controlled and workers with WMSDs had been properly attended to, the surveillance necessary in Henry's T Shop for these disorders and their risk factors was back at step 1 (Figure 5.6). That is, every two years, or whenever changes are introduced in the workplace, passive health surveillance using existing records and level 1 active risk factor surveillance using checklists should be conducted.

6

Managing solutions

The purpose of this chapter is to discuss approaches for eliminating the risk factors for WMSDs, or for mitigating the consequences of exposures to these risk factors. The common denominator of these approaches is that they are macroscopic in nature. Therefore they encompass strongly documented as well as less thoroughly studied risk factors. But because of variety and interrelatedness of risk factors at different levels, and because the most successful solution approaches are multifactorial, this chapter presents concepts and a model which can be used for managing solutions.

6.1 The systems concept: an overview for managing solutions effectively

The first part of the chapter will describe the interrelationships among various aspects of the work system, and how these can influence the success of intervention efforts and programmes. It is critical to recognize that solving ergonomic problems encompasses improvements at various levels of an organization, not only as concerns the job tasks, but also in the structuring of work organization, the supervision of work activities and the development of organizational policies and procedures. This is referred to as a 'systems' approach for improving ergonomics. It is often called 'macroergonomics'.

The chapter will begin with the philosophy of the systems approach and the role of organizational structure in successful intervention. The need to 'balance' various elements of the work system to achieve the best possible solutions will be discussed. This philosophical background will be followed by a discussion on issues involved in attempting to describe specific solutions for reducing the risk factors that can cause WMSDs, and issues on guideline development.

6.1.1 Definition of a system

'A system is a set of interrelated and interdependent parts arranged in a manner that produces a unified whole.' (Robbins, 1983, p.9). Organizational systems are made up of subsystems that act in concert to meet the overall objectives of the organization. Organizations themselves can be thought of as subsystems within larger societal systems. Of particular note is the fact that changing one part of an organizational system is likely to affect other system elements or subsystems (Smith and Sainfort, 1989). Often, if not planned for, these ripple effects of change are likely to have an adverse impact on both personnel and overall system functioning (DeGreene, 1973).

6.1.2 Characteristics of complex systems

Beginning with the classic research of the Tavistock Institute (e.g. Trist and Bamforth, 1951; Emery and Trist, 1960), organizations have been conceptualized as open systems engaged in transforming inputs into desired outputs. Organizations are viewed as open because they have permeable boundaries exposed to the environments in which they exist and upon which they depend for their survival.

Organizations most often bring two major components to bear on the transformation process:

- technology, in the form of a technical subsystem; and
- people, in the form of a personnel subsystem.

These are just two elements in a more complex work system that has many elements (see Box 6.1) (Smith and Sainfort, 1989).

The design of the technical subsystem defines the tasks to be performed; the design of the personnel subsystem determines the ways in which the tasks are performed. These two subsystems interact with one another at every human-machine and user-system interface. Thus they are interdependent and operate under joint causation, meaning that both subsystems are affected by events in the environment. The technical subsystem, once designed, is relatively stable and fixed. Prior to the advent of ergonomics, it was believed that because of the fixed nature of technology the personnel subsystem would have to adapt to environmental change. Ergonomics is a science that adapts the system to the individual to achieve the best possible fit, rather than adapting the individual to the environment. We now know that technology can and should be designed to be adaptive, or can be modified to meet personnel needs.

Joint causation between the technical and personnel subsystems gives rise to an important related systems concept of particular importance to ergonomics: joint optimization. Joint optimization means accommodating the needs of both technical and personnel subsystems at the same time. Optimizing just one subsystem and then fitting the other to it will result in suboptimization of the total system. In addition to the possible suboptimization of productivity, and especially production quality, employee stress and related physical symptoms are likely to be relatively high and job satisfaction low when just one element of the system is optimized. Joint optimization thus requires joint design of the technical and personnel subsystems, given the objectives and requirements of each, and understanding of the nature of the relevant external environment.

Box 6.1:   Important components of a work system

- organizational structure
- people, or the personnel subsystem
- technology, or the technological subsystem
- work tasks
- the relevant external environment

6.1.3 The synergistic nature of complex systems

A widely accepted view among systems theorists and researchers is that complex systems are synergistic in nature; that the whole is much more, or less, than the sum of the individual parts. Whether or not the whole is more or less than the sum of its parts depends on the extent to which there is harmonization among the parts. To the extent that there is true harmonization, conditions are likely to exist that minimize employee stress and enhance intrinsic job satisfaction. These are conditions that are optimal for minimizing WMSDs and related health symptoms. To the extent that there is not harmonization, the impact on employee health is likely to be more adverse than the simple sum of the parts would indicate. Achieving organizational harmonization requires successful completion of two major tasks. The first is harmonizing the key characteristics of the five interdependent major system components (Box 6.1). This involves the macroergonomic analysis and design of the system. The major requirement is following through and ensuring that individual jobs and related human-system interfaces are properly designed, and that these designs are harmonized with the macroergonomic design of the system. This in turn involves microergonomic analysis and design at the job level, which is then linked to the other components of the system in a harmonious manner.

Theoretically, effective harmonization of the system should lead to a synergistic improvement in productivity and reduction in stress-related symptoms, such as complaints of strain, accident rates, WMSDs and indices of job dissatisfaction. For example, instead of the 10 to 15 per cent improvements that typify successful microergonomic efforts (per cent quoted based on personal experience of several of the authors over three decades), greater improvements should be possible. Already there is some supporting evidence for enhanced improvement using a macroergonomic approach. For example, Nagamachi and Imada (1992) recently reported reductions in accident rates of 72 to 90 per cent as the result of macroergonomic interventions in both Japan and the United States. Although there is no guarantee that these findings in the area of accident reduction will generalize to reductions in WMSDs, they nevertheless are encouraging. Even more encouraging are the results of a macroergonomic approach to total quality management at the L.L. Bean Corporation in the United States (Rooney *et al.*, 1993). In this organizational change effort, injuries causing loss of time were reduced by 71 per cent and 76 per cent in the distribution and manufacturing divisions, respectively, from 1988 to 1992. Over 80 per cent of these reductions were in soft-tissue injuries.

6.2 Ergonomics and the work system

One way to define or to otherwise understand the nature of any field of science and practice is by noting the nature of its technology. The unique technology of ergonomics is human-system interface technology (Hendrick, 1991). Ergonomics as a science is concerned with developing knowledge about human performance capabilities, limitations and other characteristics as they relate to the design of the interfaces between people and other system components (see Box 6.1). As a practice, ergonomics concerns the application of human-system interface technology to the design or modification of systems to enhance performance, system safety, health, comfort, effectiveness and quality of life. At present, this unique technology has at least four identifiable major components:

- human-machine interface technology (hardware ergonomics);
- human-environment interface technology (environmental ergonomics);
- user-system interface technology (software ergonomics);
- organization-machine interface technology (macroergonomics).

6.2.1 Hardware ergonomics, or human-machine interface technology

Hardware ergonomics, or human-machine interface technology, began to develop during World War II, and represented the beginning of a formal science and practice of 'human factors' or 'ergonomics'. It primarily concerns the study of human physical and perceptual characteristics and the application of these data to the design of such areas as controls, displays, tools, seating, work surfaces and workspace arrangements.

6.2.2 Environmental ergonomics, or human-environment interface technology

Environmental ergonomics, or human-environment interface technology, had its origins in the 1930s as concern arose about the effects of noise, vibration, lighting, and other physical agents on human performance and health. During the last several decades, the importance of understanding the relation of humans to their natural and constructed environments has gained increasing focus. A very recent and quite different development has been the application of an ecological approach to human performance modelling and to classic ergonomic methods such as task analysis (Vicente, 1990). Increasing international awareness of the importance of ecological issues to human health and effectiveness is further highlighting the importance of human-environment technology research and application.

6.2.3 Software ergonomics, or user-system interface technology

Software ergonomics, or user-system interface technology, began to emerge in the third decade of ergonomics research (Hendrick, 1986a,b). It represented a shift in focus from the physical and perceptual to the cognitive nature of work. Much of this new focus was driven by the development of the silicon chip and the resulting work on computer systems. As more people began working on computers and computer-based systems, their manner of thinking and conceptualizing became more important to system design. Because user-system technology primarily is concerned with the cognitive aspects of human performance, it often is referred to as 'cognitive ergonomics'. There is growing awareness of the importance of effectively applied user-system interface technology in reducing operator frustration, stress and related medical symptoms and improving operator and system performance.

6.2.4 Macroergonomics, or organization-machine interface technology

Macroergonomics, or organization-machine interface technology, might be more appropriately labelled 'human-organization-environment-machine-tasks interface technology', for it does indeed involve consideration of all of these major system components. The central focus, however, is a human-centred approach to interfacing the overall organizational and work system design with the technology in the system to optimize human-system functioning.

The central focus of the first three ergonomics technologies, described above (subsections 6.2.1 to 6.2.3), has been the individual operator and operator teams or subsystems. Thus, the primary application of these technologies has been at the microergonomic level. In contrast, the organization-machine aspect (this subsection 6.2.4) of the human-system interface tends to be more macro in focus, because it deals with the overall structure of the work system as it interfaces with the system's technology; hence, it is referred to as 'macroergonomics' (Hendrick, 1986a,b).

Historically, organizational design and management (ODAM) factors have, at times, been considered in ergonomic practice, but macroergonomics, as a formally recognized technology, is relatively new. Its formal origin can be traced to a study in the United States by the Human Factors Society (HFS) of future trends relevant to ergonomics during the 20-year time frame between 1980 and the year 2000. As part of that study, Hendrick (1980) noted the following major trends of significance to ergonomics (and to the reduction of stress-related illnesses and accidents).

Technology. Recent breakthroughs in the development of new materials and the rapid development of new technology in the computer and telecommunications industries will fundamentally alter the nature of work in offices and factories during the 1980 to 2000 time frame, and will continue to do so into the next century. In general, we have entered a true information age and age of automation that will profoundly affect work organization and related human-machine interfaces.

Demographic shifts in the workplace. The average age of work populations in the industrialized countries will increase by approximately one-half year for each passing year during the 80s and will continue through most of the 90s. The major reasons for this 'greying' of the workforce are: (a) the ageing of the post-World War II baby boom demographic bulge that now has entered the work force; and (b) the lengthening of the average productive life span of workers because of better nutrition and health care. In short, the study noted that during these two decades the workforce would become progressively more mature, experienced, professional and, due to ageing, possibly more susceptible to WMSDs.

It is important for us to note that, as the level of professionalism increases, it becomes important for work systems to become less formalized, tactical decision-making to become decentralized and management support system designs to similarly accommodate the individual if the work system is to remain reasonably non-stressful and thus minimize the likelihood of WMSDs. As the workforce ages, it is also essential to provide ergonomic design improvements to reduce the risk of WMSDs.

The HFS study also noted the significant and progressive trend towards feminization of the workforce and the fact that this would have an impact on the anthropometric and biomechanical characteristics of the work population and related ergonomic design requirements. Although not a part of the study, similar conclusions could have been drawn about the progressive increase in the multi-ethnic make-up of the work force, particularly in the US, due, in large part, to the increase in immigrants from Southeast Asia.

Value changes. The report noted that, beginning in the late 60s, a fundamental shift in the value systems of the workforce had been occurring (Argyris, 1971). Briefly, workers valued and expected to have greater decision-making responsibility and more broadly defined jobs that offered a greater sense of both responsibility and accomplishment. Argyris (1971) further noted that, to the extent that organizational and work systems designs do not respond to these values, efficiency and quality of performance will deteriorate, and employee frustration and dissatisfaction will increase (and today we would add that a related increase in WMSDs could be expected). The HFS study noted that these value system changes were further documented in the United States in the late 70s by Yankelovich (1979), and in the 80s by Lawler (1986). Of particular note from these findings was the insistence that jobs must become less depersonalized, more meaningful and provide for greater opportunities for employee participation.

The limitations of traditional (micro) ergonomics. Results from the HFS study indicated that early attempts to incorporate ergonomics into the design of computer workstations and software have been disappointing. Although the design of workstations and software has improved since then, in general, the productivity gains are still less than expected, and continued physical and psychological symptoms of high job stress and poor intrinsic job satisfaction remain the norm rather than the exception (similar observations have been noted more recently by Sauter *et al.*, 1990). As noted several years later, it is entirely possible to do an outstanding job of ergonomically designing a system's components, modules and subsystems, yet fail to reach relevant systems effectiveness goals because of inattention to the macroergonomic design of the system (Hendrick, 1984, 1986a).

Collectively, the four above-mentioned trends indicated that, in the future, effective ergonomic design would require consideration of organizational design and management factors as an integral part of improved work systems design. Hendrick (1980) concluded that there was a strong requirement to integrate organizational design and management (ODAM) factors into ergonomics research and practice. Over the past decade this has been happening and a new technology, macroergonomics, has emerged.

Conceptually, macroergonomics is a top-down systems approach to organizational and work system design (subsection 6.2.4) and the design of related human-machine (subsection 6.2.1), user-system (subsection 6.2.3), and human-environment (subsection 6.2.2) interfaces. Herein, 'top-down' refers to the design concept that begins with consideration of the overall organizational structure and continues through to those of individual jobs and workstations.

6.3 Macroergonomics/work organization

In order to gain an understanding of macroergonomics, one must first have a grasp of the key dimensions of organizational structure.

6.3.1 Dimensions of organizational structure

An organization may be defined as 'the planned co-ordination of two or more people who, functioning on a relatively continuous basis and through division of labour and a hierarchy of authority, seek to achieve a common goal or set of goals' (Robbins, 1983, p.5). This concept of organization, with its division of labour and hierarchy of authority, implies structure. The structure of an organization can be conceptualized as having three major elements (Robbins, 1983):

- complexity,
- formalization, and
- centralization.

6.3.1.1 Complexity of an organization

Complexity refers to the degree of differentiation and integration within an organization. Three major kinds of differentiation are found in an organization's structure:

(a) vertical differentiation,
(b) horizontal differentiation, and
(c) spatial dispersion.

(a) *Vertical differentiation* is operationally defined as the number of hierarchical levels separating the chief executive's position from the jobs directly involved with the system's output. In general, as the size of an organization increases, the need for vertical differentiation also increases. The primary factor underlying this size–vertical differentiation relationship appears to be the span of control or the number of people that can be directly and effectively controlled by any one supervisor (Mileti *et al.*, 1977). As we shall see later, the optimal span of control is affected by other system characteristics of the organization.

(b) *Horizontal differentiation* refers to the degree of departmentalization and job specialization that is designed into the organization. Although it has the inherent disadvantage of increasing organizational complexity, the division of labour afforded by job specialization also has inherent efficiencies. Adam Smith (1870) demonstrated this over 100 years ago by noting that ten men, each doing a particular function (job specialization) could produce about 48 000 pins per day. Conversely, if each man worked separately, performing all of the production tasks, they would be lucky to make 200 pins per day (Smith, 1870). Division of labour creates groups of specialists. The way in which these specialists are grouped in organizational design is known as departmentalization. The optimal degree of specialization in the system depends upon various system factors, such as the characteristics of the workforce and the nature of the tasks. However, too much specialization of work functions and task activities can produce, for example, excessive repetitive movements of joints and tissues or invariability; this in turn, as has already been shown in chapter 4, could contribute to WMSDs.

(c) *Spatial dispersion* (another kind of differentiation) may be defined operationally as the degree to which an organization's facilities and personnel are dispersed geographically from the main headquarters. The three major measures of dispersion are: (1) the number of geographical locations within the organization; (2) the average distance to the separated units from the organization's headquarters; and (3) the number of employees in the separated locations in relation to the number at the headquarters location (Hall *et al.*, 1967).

Increasing any one of the three differentiation dimensions described above increases an organization's complexity.

Integration (the other aspect of complexity) refers to the extent to which structural mechanisms for facilitating communication, co-ordination and control across the differentiated elements of the system have been designed into its structure. Some of the more common integrating mechanisms include formal rules and procedures, liaison positions, committees, system integration offices, and information and decision-support systems. Vertical differentiation, in itself, also serves as a key integrating mechanism for horizontally and geographically differentiated units.

In general, there is a direct relationship between the complexity of an organization and the extent to which integrating mechanisms are required for optimal functioning: as the complexity increases, the need for integrating mechanisms also increases (Robbins, 1983). Incorporating the appropriate kinds and numbers of integrating mechanisms into the system's structure is a critical macroergonomic dimension of work systems design, and is important in managing solutions to WMSDs.

6.3.1.2 Formalization of an organization

From an ergonomics standpoint, formalization may be defined as the degree to which jobs within organizations are standardized (Hendrick, 1986a,b, 1987). In highly formalized organizations, jobs allow for little employee discretion over what is being done, when or in what sequence tasks will be accomplished and how tasks are performed. The management system includes explicit job descriptions, extensive rules and clearly defined procedures covering work processes (Robbins, 1983). Often, the design of the system's hardware, software and related human-machine and user-system interfaces restricts employee discretion.

Organizations having low formalization allow employees more freedom to exercise discretion; jobs and related human-machine and user-system interfaces are often designed to permit considerable autonomy and self-management. Employee behaviour thus is relatively unprogrammed and workers are able to make greater use of their mental capacities.

In general, the simpler and/or more repetitive the jobs, the greater the utility of formalization; the higher the level of professionalism of the jobs, the greater the utility of less formalization. It must be kept in mind that simple, repetitive jobs may lead to increased risks of WMSDs.

6.3.1.3 Centralization of an organization

Centralization refers to the degree that formal decision-making is concentrated in an individual, unit or level (usually high in the organization), thus permitting employees (usually low in the organization) only minimal input into decisions affecting their jobs (Robbins, 1983). Traditionally, centralization has been thought to be desirable: (a) when a comprehensive perspective is required, such as in strategic decision-making; (b) when operating in a highly stable and predictable environment; (c) for financial, legal and other decisions where they clearly can be done more efficiently when centralized; and (d) when significant savings clearly can be realized.

Decentralization is desirable: (a) when operating in a highly unstable or unpredictable environment; (b) when the design of a given manager's job will result in exceeding human information processing and decision-making capacity; (c) when more detailed 'grass roots' input to decisions is wanted; (d) for providing greater intrinsic job motivation to employees; (e) for gaining greater employee commitment to the organization and support of organizational decisions by involving employees in the process; and (f) for providing greater training opportunities for lower-level managers.

6.3.2 System considerations in macroergonomic design

As noted in section 6.2, the design of an organization's structure (which includes how it is to be managed and, obviously, management of solutions to WMSDs) involves consideration of the key elements of the other four major system components: (a) technology, or the technological subsystem; (b) people, or the personnel subsystem; (c) the work tasks; and (d) the relevant external environment. Each of these major sociotechnical system components has been studied in relation to its effect on the fifth component, organizational structure, and empirical models have emerged that can be used to optimize a system's organizational design (Hendrick, 1986b, 1987, 1991). These will be discussed in section 6.4.

6.3.3 Relation of macro- to microergonomic design

By using a macroergonomic approach to determine the optimal design of an organization's structure, one can describe many of the characteristics of the jobs to be designed into the system, and of the related human-machine and user interfaces. Some examples are as follows (Hendrick, 1991):

(a) Horizontal differentiation influences how narrowly or broadly jobs must be designed and, often, how they should be departmentalized.
(b) Decisions concerning the level of formalization and centralization will dictate: (1) the degree of routinization and employee discretion to be ergonomically designed into the jobs and attendant human-machine and user-system interfaces; (2) the level of professionalism to be designed into each job; and (3) many of the design requirements for the information, communications and decision-support systems, including what kinds of information are required by whom, and networking requirements.
(c) Vertical differentiation decisions, coupled with those concerning horizontal differentiation, spatial dispersion, centralization and formalization, will determine many of the design characteristics of the managerial positions, including the span of control, decision authority and nature of decisions to be made, information and decision-support requirements, and qualitative and quantitative educational and experience requirements.

In summary, effective macroergonomic design establishes much of the micro-ergonomic design of the system, and thus ensures optimal ergonomic compatibility of the system components with the system's overall structure. In system terms, this approach allows joint optimization of the technical and personnel subsystems from top to bottom throughout the organization. The result is greater assurance of optimal system functioning and effectiveness, including employee safety, health, comfort, intrinsic motivation and quality of work life.

6.4 Balancing the work system for ergonomic benefits

There are no 'perfect' jobs or 'perfect' workplaces that are free of all ergonomic hazards and provide complete psychological satisfaction for all employees. It is a poor strategy to believe that a 'perfect' workplace can be achieved within reasonable financial constraints, or even if there are no financial limits. Thus, we must consider the need for compromise among the competing needs for ergonomic improvements at the workplace. This requires that we establish a basis for identifying the most critical workplace conditions for design or redesign, and prioritize ergonomic needs. In any design or redesign process, trade-offs must be made among specific ergonomic improvements when attempting to achieve the best 'overall' ergonomic solution for a job. These trade-offs require us to think about how to best 'balance' the various ergonomic needs to achieve the solution that will have the greatest positive benefit for employee health and productivity (Smith and Saintfort, 1989).

6.4.1 Interaction among risk factors and the need for 'balance'

As we have discussed earlier in this book, there are many factors that can produce adverse ergonomic conditions that increase the risk of WMSDs. These can generally be classified as: (1) biomechanical, e.g. repetition, force, posture and vibration; (2) personal, e.g. anthropometry and health status; and (3) work organization, e.g. work stress, coping strategies and organizational practices. All of these classes of risk factors are interrelated, so that personal health status affects the extent of repetitions and forces that can cause tissue damage, and the nature of the supervisory relationship can affect personal behaviour such as exerting excessive force. Therefore, making modifications in one class of risk factors may not be beneficial if concomitant changes are not made in other related classes of risk factors.

For example, purchasing new keyboards that require less force to push the keys and are split to allow for better postural adjustment may not reduce ergonomic risk if the employees continue to strike the keys very hard as a behavioural response to high psychological job stress. In this case, to achieve positive results one must address both the biomechanical aspects of the keyboard design, and the work organization aspects of the job design to reduce the force, improve the posture and decrease the job stress. This example demonstrates the 'system' nature of ergonomic problems and the need to achieve a proper 'balance' of all related factors.

6.4.2 What is 'balance'?

There are two aspects of 'balance' to consider when addressing risk factors. These are: (1) system balance and (2) compensatory balance. System balance is based on the idea that a workplace or process or job is more than the sum of the individual components of the system. The interplay among the various components of the system produces results that are greater (or less) than the additive aspects of the individual parts, and determines the potential for the system to produce results. In our previous example of buying new keyboards, if the employer concentrates solely on the technological component of the system, then there is an 'imbalance' because the personnel and work organization risk factors are neglected. This can make for a very expensive 'mistake' since the costly new technology may not decrease the overall risk of WMSDs due to the excessive force used on the keyboard, in turn related to the psychological distress of the employees.

Experience in the telecommunications industry suggests that costly bio-mechanical risk reduction programmes for VDT operators such as new workstations and keyboards, without simultaneous consideration of job design and job stress factors, will not eliminate WMSDs and can have unsatisfactory results (LeGrande, 1993).

The second type of balance is 'compensatory' in nature. Experience in consulting with industry to solve ergonomic problems has shown that it is seldom possible to eliminate all risk factors. This may be due to financial considerations, or because it is impossible to remove some hazards that are inherent in certain job tasks. For example, a data-entry VDT operator depresses the keys many thousand times per hour. Often, the operator works continuously for one or two hours without a break. The operator spends eight or more hours per day, every day, at this task. This job puts the operator at high risk for developing WMSDs. The high repetition is an inherent aspect of the task. Yet, it is possible to reduce the exposure of this employee by rotating to other tasks with a low frequency of hand/finger motions each hour. In addition, these tasks may also provide more variety in job content, which enhances job satisfaction, lowers stress and decreases the risk of psychological aspects of WMSDs. This is a form of 'compensatory' balance that provides a better work organization environment to reduce biomechanical exposure and psychological distress.

In this same VDT operator example, it would be important to provide proper workstation design to enhance shoulder, arm, wrist, hand and finger postures. This would also provide a better system balance, since research has shown that postural factors interact with the frequency of motions in producing WMSDs. The ergonomic considerations for reducing biomechanical strain also serve to enhance the work organization environment by making employees more comfortable, demonstrating their importance to the company, and showing employers' commitment to the welfare of employees.

6.4.3 Critical components of a system to consider for achieving balance

Figure 6.1 illustrates a model for conceptualizing the various elements of a work system, that is, the loads that working conditions can exert on workers. At the centre of this model is the individual with his/her physical characteristics, perceptions, personality and behaviour. The individual has technologies available to perform specific job tasks. The capabilities of the technologies affect performance and also the worker's skills and knowledge needed for their effective use. In addition, the task requirements affect the skills and knowledge needed. Both the tasks and technologies affect the content of the job and the physical demands. The tasks (along with the use of their technologies) are carried out in a work setting that comprises the physical and the social environment. There is also an organizational structure that defines the nature and level of individual involvement, interaction, control and supervision. This model can be used to establish relationships between job demands, job design factors and ergonomic loads, and shows that these various elements interact when work is being done. The overall physical and psychological effects are defined by the ways in which these elements are integrated. This is a systems concept in that any one element will have an influence(s) on other element(s). Demands are placed on the individual by the other four elements, which create loads that can be healthy or harmful. Harmful loads lead to physical and psychological stress responses that can produce adverse health effects such as WMSDs. Each element in this model and some examples of the potential adverse aspects are described below.

6.4.3.1 Relevant external environment

'Relevant external environment' refers to that part of the organization's external environment that is made up of the firm's critical constituencies (i.e. those aspects that can positively or negatively influence the organization's effectiveness). Negandhi (1977), who reviewed field studies of 92 industrial organizations in five different countries, has identified five external environments that significantly impact on organizational functioning. These are:

- socioeconomic (including the nature of competition and the availability of raw materials);

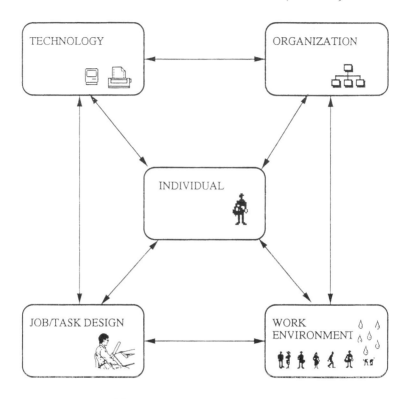

Figure 6.1 Model of the work system (Smith and Sainfort, 1989).

- educational (including both the availability of educational programmes and facilities and the aspirations of workers);
- political (including governmental attitudes towards business, labour, and control over prices);
- legal; and
- cultural (including the social class or caste system, values and attitudes).

Of particular importance in managing solutions to WMSDs is the fact that relevant external environments vary in two dimensions that strongly influence the effectiveness of an organization's design: the degrees of environmental change and complexity. The degree of change refers to the extent to which a specific environment is dynamic or remains stable over time; the degree of complexity, in this context, refers to whether there are few or many relevant environments. As illustrated in Table 6.1, these two environmental dimensions combine to determine the environmental uncertainty of an organization.

Of all the system factors that impact on the effectiveness of an organizational design, environmental uncertainty repeatedly has been shown to be the most important (Burns and Stalker, 1961; Emery and Trist, 1965; Lawrence and Lorsch, 1969; Duncan, 1972; Negandhi, 1977). With a high degree of

Table 6.1 Environmental uncertainty of organizations

| | | Degree of change | |
		Stable	Dynamic
	Simple	Low uncertainty	Moderate/high uncertainty
Degree of complexity			
	Complex	Low/moderate uncertainty	High uncertainty

uncertainty, a premium is placed on an organization's ability to be flexible and respond rapidly to change. With low uncertainty, maintaining stability and control for maximum efficiency and effectiveness become the more important criteria for survival.

Accordingly, the greater the environmental uncertainty, the more important it is for the organization's structure to have relatively low vertical differentiation, decentralized tactical decision-making, low formalization and a high level of professionalism among its work groups. In contrast, if adequate intrinsic motivational characteristics are incorporated into the design of the jobs, highly certain environments are ideal for relatively high vertical differentiation, formalization and centralized decision-making, such as is found in classical bureaucratic structures. Of particular note is the fact that, today, most high-technology corporations are operating in very dynamic and complex environments. Failure to respond to this uncertainty in terms of the work system design is likely to result in employee stress, poor job satisfaction and suboptimal productivity – conditions that increase the likelihood of WMSDs.

6.4.3.2 Work tasks

Several task characteristics can influence the risk of WMSDs. For instance, repetitiveness (Cox, 1985) increases the frequency of motions and induces boredom, which is a psychological stressor. The complexity of the content of the work is related to repetitiveness in that invariant tasks lead to repetition. Content also affects the perceived meaningfulness of work, which can cause psychological stress (Margolis *et al.*, 1974; Caplan *et al.*, 1975; Hackman *et al.*, 1975).

Workload issues such as overload and underload (Frankenhaeuser and Gardell, 1976; Coburn, 1979) can create adverse conditions for WMSDs and psychological stress. Overwork causes overloading of the muscles, tendons, ligaments and other connective tissues, creates 'fatigue' and produces psychological stress. Underload creates boredom, fixed static postures, anxiety over job loss and psychological stress.

Machine-paced work tasks are more likely to be repetitive, invariant and short-cycle. Machine pacing has been shown to be more stressful than non-paced tasks (Salvendy and Smith, 1981). Lack of participation (French, 1963; Caplan *et al.*, 1975) and lack of control (Coburn, 1979; Karasek *et al.*, 1981;

Table 6.2 Knowledge-based technology classes (adapted from Perrow, 1967)

		Task variability	
		Routine with few exceptions	High variety with many exceptions
	Well-defined and analysable	Routine	Engineering
Problem analysability			
	Ill-defined and unanalysable	Craft	Non-routine

Fisher, 1985) in task activities can produce emotional problems and increase the risk of circulatory diseases. More information on the various task characteristics that can influence the risk of WMSDs can be found in chapter 4.

6.4.3.3 Technology, or the technological subsystem

Technology, as a determinant of organizational design, has been operationally defined in several distinct ways. Perhaps the most thoroughly validated and generalizable model of the technology-organization design relationship is that of Perrow (1967), which utilizes a knowledge-based definition of technology. In his classification scheme, Perrow begins by defining technology as the action one performs upon an object in order to change that object (in sociotechnical systems terms, this is referred to as the transformation process). Perrow notes that this action always requires some form of technological knowledge; hence, technology can be categorized by the required knowledge base. Using this approach, he has identified two underlying knowledge dimensions of technology. The first of these is task variability or the number of exceptions encountered in one's work. For a given technology, these can range from routine tasks with few exceptions to highly variable tasks with many exceptions.

The second dimension has to do with the type of search procedures one has available for responding to task exceptions, or task analysability. For a given technology, the search procedures can range from tasks being well-defined and solvable, by using logical and analytical reasoning, through to being ill-defined with no readily available formal search procedures for dealing with task exceptions. In this latter case, problem solving must rely on experience, judgement and intuition. The combination of these two dimensions, when dichotomized, yields a 2×2 matrix as shown in Table 6.2. Each of the four cells represents a different knowledge-based technology.

1. Routine technologies have few exceptions and well-defined problems. Mass production units most frequently fall into this category. Routine technologies are best accomplished through standardized procedures, and are associated with high formalization and centralization. (However, care must be taken to

ensure that the design of specific jobs includes adequate intrinsic motivational characteristics.)
2. Non-routine technologies have many exceptions and difficult-to-analyse problems. Aerospace operations often fall into this category. Most critical to these technologies is flexibility. Thus they lend themselves to decentralization and low formalization.
3. Engineering technologies have many exceptions, but they can be handled using well-defined rational-logical processes. They therefore lend themselves to centralization, but require the flexibility that is achievable through low formalization.
4. Craft technologies typically involve relatively routine tasks, but problems rely heavily on experience, judgement and intuition for decision. Problem-solving thus needs to be done by those with the particular expertise. Consequently, decentralization and low formalization are required for effective functioning.

Perrow's model suggests that when the organizational and work systems designs are not compatible with the technology, productivity may be suboptimal and symptoms of employee stress, WMSDs, and job dissatisfaction are likely to appear.

The ability to accomplish tasks and the extent of physiological and psychological load are often defined by the effectiveness of the technology used by the worker. When the technology produces too much or too little load then increased biomechanical and psychological stress can occur, as well as adverse physical health outcomes, including WMSDs (Smith *et al.*, 1981; Johansson and Aronsson, 1984; Östberg and Nilsson, 1985). One of the main influences of technology concerns worker fears of lacking adequate skills to use the technology. Another is the fear over job loss due to increased efficiency of technology (Östberg and Nilsson, 1985; Smith *et al.*, 1987). On the other hand, new technology, such as lifting aids or powered hand-tools can reduce biomechanical loading, while others, such as new computers, can enhance job content by providing positive feedback that is unavailable with other technologies (Kalimo and Leppanen, 1985).

6.4.3.4 Organizational structure

The work organization establishes the level of work output required (work standards), the work process (how the work is carried out), the work cycle (work/rest regimens), the social structure and the nature of supervision. Thus, the organizational context in which work tasks are carried out has major influences on the worker's physical and psychological stress and health. As mentioned earlier, technology often requires new skills from workers for proficient performance. The way in which workers are introduced to new technology and the organizational support they receive, such as training and time to become acclimatized, have been related to stress and emotional disturbances (Smith *et al.*, 1987). Employees often have to be 'retrained' to acquire better work practices to reduce biomechanical strain. The effort that the organization puts into this 'retraining' and in supervising to ensure compliance with proper procedures will affect the acceptance and successful use by employees.

The ability to grow in a job and to be promoted (career development) has a stress connection (Arthur and Gunderson, 1965; Smith *et al.*, 1981). The chance to move from one job to another helps reduce the effects of long-term exposure to specific loads on particular muscles, tendons, ligaments and joints. The possibility of job loss also can create psychological stress that often has concomitant muscular strain (Caplan *et al.*, 1975; Cobb and Kasl, 1977; Kasl, 1978; Smith *et al.*, 1981).

Other organizational considerations such as work schedule (shiftwork) and overtime can have negative effects on the extent of exposure, biological susceptibility to stressors, and mental and physical health (Breslow and Buell, 1960; Margolis *et al.*, 1974; Monk and Tepas, 1985). Other aspects such as role conflict and ambiguity also have negative emotional consequences (Caplan *et al.*, 1975).

6.4.3.5 *People, or the personnel subsystem*

A number of personal considerations also contribute to the physical and psychological effects that the previously mentioned elements of the model will produce. These include the strength and health of the individual, previous musculoskeletal or nerve injury, personality, perceptual-motor skills and abilities, physical conditioning, prior experiences and learning, motives, goals and needs and intelligence (see for example Levi, 1972).

6.4.4 Achieving balance in a work system

The essence of this approach is to reduce physical and psychological stress and their negative health consequences by 'balancing' the various elements of the work system. Proper job design can be achieved by providing those character- istics of each work element that meet recognized criteria for physical loads, work cycles and job content (i.e. that meet individual physiological and psychological needs). The ideal design would eliminate all sources of stress. However, since such a perfect job cannot often be achieved, the use of positive work elements to compensate for poor work elements can balance the stress by moderating those factors to reduce the load and potential health consequences.

The five components of the work system function in concert to provide the resources for achieving individual and organizational goals. We have described some of the potential negative attributes of the elements in terms of job stress, but there are also positive aspects of each that can counteract the negative influences of others. For instance, the negative influences of inadequate skill to use new technology can be offset by training employees. The adverse influences of low job content that creates repetition and boredom can be balanced by an organizational supervisory structure that promotes employee involvement and control over tasks and job enlargement that introduces task variety. This approach proposes to balance the loads that are potentially stressful by considering all of the work elements. For instance, organizational structure could be adapted to enrich jobs in order to provide greater task content and subsequent training to enhance skills. Alternatively, new organizational structures could provide shared responsibilities or increased financial resources to accomplish tasks. Interventions should look at the total work system instead of focusing on one element if they are to be more successful, not only from an organizational and performance standpoint, but more importantly from an individual perspective. This model defines the process by which working conditions at different levels (i.e. individuals or people, task, environment, technology, and organization) can produce loads that can lead to worker stress. It also proposes a system that helps balance these loads to reduce worker stress. When balance cannot be achieved through changing the negative elements of the job, then it can be improved by enhancing the positive elements of the job. Thus, the positive aspects of work can be used to counterbalance the negative.

A major advantage of this model is that it does not highlight any one job factor such as frequency of motions, or a small set of factors such as workload and posture. Rather, it examines the design of jobs from a holistic perspective to emphasize the potential positive elements in a job that can be used to overcome the adverse elements when such adverse elements cannot be eliminated. Thus, all aspects of the job can be considered in developing a proper design or redesign.

6.5 Solutions and guidelines: issues for consideration

Thus far, chapter 6 has discussed the management of solutions in more general terms. This section will examine a number of issues involved in attempting to identify specific solutions and formulate guidelines.

6.5.1 Introduction

Chapter 4 described a wide variety of risk factors, ranging from anthropometric considerations of being able to fit, reach and see within the work area to environmental stressors (cold, heat and abrasion), musculoskeletal load, static load, perception and cognition, and psychosocial effects. We can say that: (1) risk factors are heavily interdependent and interact strongly; and (2) their predictive ability (i.e. if the level of the factor is changed, does the level of risk and WMSDs change?) is largely undetermined.

Despite a lack of epidemiological support, the best available evidence supports the reduction of risk in general as a viable and practicable approach to reducing the risk of WMSDs. There is a large body of literature describing controls for the reduction of WMSDs and their application in a wide variety of work settings. It is not our purpose here to duplicate the extensive literature on these preventive approaches, but rather to approach the issue of solutions from the viewpoint of individuals or groups either planning or attempting to develop guidelines and legislation aimed at reducing WMSDs.

It is our belief that co-ordinated intervention using different types of controls is required to produce a healthy workplace. Ill-health in the sense of work related musculoskeletal disorders may be one outcome of ignoring the broad nature of the risk factors. The aim of any guideline should be to minimize the risk of WMSDs, usually by reducing the magnitude of the risk factors.

6.5.2 Approaches to reducing WMSDs: information available

6.5.2.1 Information on controls

In the model described in chapter 2, we identified a number of workplace features, such as workstation design or supervision or computer monitoring, which are potentially modifiable. It is through the design and modification of these features that reduction of WMSD risk is thought to occur. Controls for the reduction of WMSD risk are usually categorized as engineering controls, personal protective equipment and administrative controls. Engineering controls are those changes to the workplace that modify its structure and dimensions, such as the tools, fixtures and technology used, as well as the temporal nature of the task and job design. Classic examples include workstation dimensions and design, changes to tool design, to reduce weight for instance, or the design of a product to reduce insertion force for a part. We wish to stress the structural element of engineering controls, including not only physical elements such as workplace dimensions and forces, but also work organization, cognitive demands and autonomy.

Administrative controls refer to methods of reducing either the magnitude or the duration of exposure to risk factors by management or personnel methods: methods directed at the worker. Classic examples include job rotation, two persons handling heavy objects and rest breaks.

Although of limited applicability and of unknown efficacy, personal protective equipment is sometimes proposed where a risk factor cannot practically be reduced to arrive at a minimal level of WMSD risk. Protective equipment is used to shield the individual from the risk factor, thus reducing WMSD risk. A common example is the use of gloves to reduce abrasion or cuts or local mechanical stressors. However, considerable debate surrounds the use of protective equipment such as wrist splints, anti-vibration gloves and arm supports for the reduction of WMSDs.

Engineering controls that change the structure of the job are often recommended as primary interventions, with administrative controls and personal protective equipment being regarded as interim or less-effective approaches.

6.5.2.2 Information from specifications

One may broadly identify two extremes in the way that WMSDs can be reduced: through the use of performance or outcome specifications or design specifications. An example of the difference between the types of specifications can be seen in two approaches to regulating supermarket cashier's work-stations.

A Swedish publication on work at checkout counters, ASF (1986), takes an 'outcome' approach, requiring that the entire cashier 'environment be accep-table' in general terms with very little in the way of proscribed data, for example on dimensions. On the other hand, in Germany, a design specification approach was adopted for each separate component of the cashier workstation, and thus specific data were available for the chair, table, etc. (Federal Agency for Industrial Safety, 1985). This document does note, however, that the separate elements must be combined in a skilful manner to come up with a satisfactory solution, which is, of course, the major problem with a complex interaction between individuals and the environment.

Recent documents in the United States, such as the Red Meat Guidelines distributed by the OSHA (1991), have adopted a performance approach to workplace regulation. More recently, the ANSI Committee on Cumulative Trauma Disorders (ANSI Z-365) explicitly rejected specifying levels of causes of cumulative trauma disorders and opted instead for a performance-based approach.

6.5.2.3 Data from intervention studies

Chapter 4 demonstrated that the literature supports the identification of a range of risk factors. Interventions are usually centred on the reduction of workplace features that produce these risk factors. Although many changes have been suggested for the reduction of risk factors leading to WMSDs, and many of these make sense from a physiological, psychological or biomechani-cal viewpoint, there has unfortunately been little evaluation of their effective-ness and even less of their cost-effectiveness. Reviews by Silverstein (1987) and Kilbom (1988) have shown the lack of strong evidence of the effectiveness of these interventions. Even more recently, Cole *et al.* (1992) reviewed more up-to-date studies, subjecting them to epidemiological criteria for validity, and proposed that none of the studies in the literature achieve a satisfactory validity rating.

This lack of epidemiological support may be due to the difficulty in demonstrating effectiveness in occupational settings because of problems includ-ing sample size, rapid turnover and multiple simultaneous co-interventions (Silverstein, 1987). It could, however, also be due to issues associated with the relationship between work and the development of WMSDs. These issues were examined in chapters 2 and 4, and include the problem that the risk factors are heavily interdependent and interact strongly and that the exposure-response relationship may be non-linear or 'U' shaped.

6.5.3 Issues to be considered in formulating guidelines

Worldwide there is a desire to formulate guidelines or legislation to control the development of WMSDs. In some jurisdictions, the specification of workplace parameters is being considered. A comparison is often drawn between ergonomics and occupational hygiene. In occupational hygiene, the setting of threshold limit values (TLVs) is a common practice. It should be recalled, nonetheless, that in that field the values are for each agent or substance individually, with the problem of multidimensional exposures still in its infancy. As discussed in chapter 4, physical exposure is inherently multidimensional.

In addition, the requirements of defining threshold or criterion values are relatively difficult. Exposures have to be measured in closely defined ways and have demonstrated epidemiological validity. For WMSD risk factors, the difficulties involved in defining criterion values include the unknown form of the cumulutative exposure-response relationship, the presence of multiple injury sites and injury mechanisms, the interaction between risk factors and the importance of time variation of a risk factor (e.g. peak or average values). Furthermore, there is a lack of detail on exposure and of data on different workplaces in a number of epidemiological studies.

6.5.4 A strategy for operationalizing WMSD guidelines

In the absence of general applicable guidelines and criteria on minimizing and/ or optimizing risk factors, two complementary approaches have merit for the prevention of WMSDs:

1. General guidelines
    Such guidelines describe in general terms the principles and policies to be adopted in preventing WMSDs. They may be actual policy documents or in-house standards and other material intended to outline the principles of prevention and raise awareness. General guidelines are usually constructed on the basis of scientific knowledge and general experience with respect to WMSDs.
2. Specific guidelines
    Specific guidelines aim at the design and redesign of work and tasks that are known in detail. Specific guidelines draw on both scientific knowledge and the collective experience of an industry or company. For this reason, specific guidelines may be much more detailed and may contain quantitative and

recommended data and even limit values. Information from workplace surveillance can be useful in the design of these guidelines.

6.5.4.1 Guideline development

The elements that could go into a guideline are outlined below. It was demonstrated earlier that risk factors interact in various ways. A good guideline makes allowance for this interaction, even in cases where the focus is on single risk factors. In formulating guidelines, it should be remembered that several risk factors – a typical example is musculoskeletal load – cannot and should not be reduced to a minimum. For instance, a continuous immobile posture can be as deleterious as to the worker as repeated difficult postures. Guidelines should therefore stipulate variable and moderate loads and avoid extremes in any direction; the key word here is optimization.

FIT, REACH AND SEE
Dimensional and other factors related to 'fit, reach and see' have been described in detail in the classical ergonomic literature. Being concrete and illustrative, these factors often form the foundation of guidelines on WMSDs.

ENVIRONMENTAL RISK FACTORS
Environmental risk factors, especially cold and vibration, are fairly easy to integrate into guidelines. Environmental physiological data are available on the effects of cold, and the ISO (1986) standard offers guidelines for evaluating the effects of vibration. Other local stressors, such as pressure, chafing, minor injuries, etc. are usually known to workers and supervisors in a given workplace and can thus be integrated into guidelines.

POSTURES
Postures – either of the trunk or the limbs – are linked to extremes of range of movement or to muscle effort while resisting a load. They are easy to observe and are therefore suitable for use in guidelines. Also, since it is easy to record the duration of postures using standard ergonomic or work study analyses, a quantitative idea of the potential risk may be obtained.

MUSCULOSKELETAL LOAD, STATIC LOAD AND INVARIABILITY
Various aspects of risk factors related to musculoskeletal load should go into a guideline. Reduction of peak forces – resulting from either the working methods or the use of hand tools – are elements that can be easily included in guidelines. Static load – either as a general concept or in terms of static muscle effort – may be less easy to operationalize as a guideline item. As explained in chapter 4, it may be linked to a lack of variability in postures, movements, etc. Repetition and duration of musculoskeletal load may be expressed in quantitative form and thus incorporated into guidelines.

COGNITIVE DEMANDS

Cognitive demands related to WMSD risk factors are closely linked to job content. The continuous attention required by the task, precision and the complexity of the task are among the cognitive elements that could be covered in guidelines.

ORGANIZATIONAL AND PSYCHOSOCIAL ASPI CTS

Organizational and psychosocial aspects have been seen by practitioners as being difficult to operationalize and thus complicated to adopt in a guideline format. However, there are examples of well-designed guidelines/checklists where these items have been successfully integrated in an industrial context (Régie nationale des usines Renault, 1976). Some examples of organizational and psychosocial issues that could be included in guidelines are as follows:

- Repetition. In addition to being an essential component of the physical risk factors, repetition is also a strong psychosocial element that can be expressed in quantitative terms.
- Job content covers various aspects that may be expressed in semi-quantitative terms. Such aspects may include: (a) to what extent subsidiary tasks are part of the job (job enlargement); (b) to what extent workers can control their work (paced vs non-paced work); and (c) the possibility of personal contacts.

PERSONAL PROTECTIVE EQUIPMENT

The requirement to use personal protective equipment (although jobs ideally should not require personal protection) can be easily formulated and adapted for guideline purposes.

MACHINES AND TOOLS

The ergonomic/biomechanical literature gives ample information on and examples of design principles and the use of hand tools to reduce WMSD risk factors. Because of their nature these elements are often easy to incorporate into guidelines.

7

Managing change

7.1 Introduction: the change process

Any attempt at reducing WMSDs in an organization involves some kind of change in tasks, work organization, environment, tools and technologies, and individual employee behaviours and attitudes. Change does not occur in a vacuum. It is an organizational phenomenon that has an impact on various subdivisions and members of the organization. In this chapter, we will discuss methods of planning for change (how to identify problems and design solutions), implementing change and monitoring the success of change. It is important to understand that change itself, if not well managed, can be a source of employee stress. In fact, it can be considered a factor that can diminish or intensify health outcomes (WMSDs). Change can produce uncertainty, feelings of lack of control and increased workload, all of which are well-known psychosocial sources of stress. Therefore, it is important to develop mechanisms and methods of planning for, implementing and monitoring change.

Controlling WMSDs can be a very complex process because of the many diverse factors involved in their causation. It is a serious mistake to believe that ergonomic improvements are easily made and that the results of ergonomic improvements will be substantial and immediate. We do not know everything we need to know about some WMSD risk factors. Adhering to the suggestions made in this document does not ensure that the WMSDs discussed here will be completely controlled. In fact, it is possible that the process of conducting an ergonomic analysis, training the workforce about ergonomics and other general factors that sensitize the workforce to WMSDs may initially produce an increase in the reporting of these disorders. In addition, personal risk factors such as individual work methods (see subsection 4.2.10), chronic illness, off-the-job activities, the ageing process and individual health status (see section 9.8) can contribute to WMSDs. These personal risk factors are very difficult to modify.

Even when good ergonomic practices are adopted, it is very possible that the extent of WMSDs may not subside and may actually increase for a short time. Proper ergonomic redesign involves changes in work organization, job content and socialization at work as well as workstation redesign, task improvements and changes in job demands. Planning appropriately for change is an important part of the process in order to ensure that WMSD risk factors are reduced or eliminated from the work system. In addition, adequate planning can facilitate the implementation of change and reduce the stress generated in the change process.

In order to ensure that change is well managed, it is important to recognize that the change process takes time: any change should be first planned, then implemented, and finally monitered. In addition to time, other resources (e.g. personnel, financial) often need to be invested in the change process.

7.1.1 Change process models

There are two well-known and widely used empirically developed models of organizational change processes that are internally driven by the organization. We feel that each of these provides practical guidance for effectively implementing such changes as ergonomic improvements and programmes.

Lewin's three-phase model. This is a well-known model developed just after World War II that focuses on the social and personal learning process by which employees unlearn old patterns of behaviour and adopt new ones (Lewin, 1947). In this change process Lewin identified three phases which he termed *unfreezing*, *changing* and *refreezing*. The unfreezing phase involves making persons aware, both cognitively and emotionally, of the need to change old ways of doing things. Management and consultants can accomplish this through such things as educational programmes (e.g. briefings, discussion groups) which enable employees to fully understand what is wrong, what needs to change and the consequences of not changing. In this regard, employee ergonomic education would be critical for informing the workforce of the reasons for making changes in work processes, jobs and the physical plant. A common characteristic of failed change efforts is for management to skip this step and, instead, intensely push the change (and thus attempt to force employees to accept it, despite their frustration and lack of understanding or appreciation of the need for change). The educational programmes give employees needed information and also provide opportunities for them to ask questions about why the change is necessary, what will be changed, how the change will be accomplished, when it will occur and what will be expected of them. Such educational efforts increase employees' acceptance of change, enhance positive employee responses, provide employees with the necessary knowledge to carry out their role appropriately and help reduce anxiety concerning change. It is well known that employees become apprehensive before change occurs, even when the change will provide better working conditions. Reducing employees' apprehension creates the proper 'climate' for the benefits of change to be successful.

The second phase is changing, or implementation of the actual change strategy. As described later in this chapter, the change strategy may be structural, technological and/or people oriented. This implementation phase requires both careful planning and selection of the change methods. At this stage it is critical to have an organized plan of action that spells out the specific changes to be made, the process through which the changes will be implemented, the timetable for specific changes and the roles of managers, supervisors, employees and consultants. The process of change will be discussed in detail later in the chapter.

The final phase is refreezing. Critical to the refreezing phase is the reinforcement of the new employee skills, knowledge, attitudes or work procedures by supervisors and management. The very best changes cannot succeed to their fullest extent unless employees utilize technological and workstation changes properly and/or follow the necessary procedures for safe behaviour. This requires employee commitment to successfully making the changes, motivation to take appropriate action and the knowledge and skills to perform well. Critical to ensuring proper employee reinforcement is the training of supervisors and managers on both what to reinforce and how to do it effectively, and ensuring that they, in turn, are reinforced by their superiors.

Dalton's sequential model. This model of organizationally induced change emphasizes two critical conditions that must be met if a change process is to be successful. First, *tension* resulting from the current situation must exist within key persons or groups within the organization. Simply stated, people must feel the need for change and want to do something about the need. Secondly, the forces for change represented by this tension must be mobilized and given direction by a *prestigious influence agent*. Employees need to believe not only that the change is valid, but that the key 'change agent' within the organization has the necessary commitment, knowledge, and power to successfully bring it about (Dalton, 1969). Often, this 'change agent' is the chief executive officer (CEO) of the company.

Dalton has identified four major learning subprocesses in successful change efforts. These are: (1) movement from generalized to specific objectives; (2) altered social ties, characterized by a loosening of old relationships and the establishment of new ones that support and reinforce the change; (3) growth in self-esteem, brought about by employees experiencing a sense of personal growth during the change process; and (4) internalization of employee commitment to the change. Internalization is most often brought about by employees finding that the change helps them to solve problems and/or that it is congruent with their own orientation (Dalton, 1969; Szilagyi and Wallace, 1990). These learning subprocesses do not simply occur as a part of change; rather, as described later in this chapter, they require careful planning.

7.1.2 Resistance to change

The most serious threat to successful change often comes from 'resistance to change' by various persons or groups in the organization. Individuals resist change for a variety of reasons, some related to the way they perceive the change, others related to the change process itself (e.g. imposed versus participatory process) and yet others related to individual characteristics. Resistance to change is an attitude that is determined by many different factors: individual characteristics (e.g. propensity to avoid change, need for status quo, past experiences, perceptions), organizational factors (e.g. inadequate/insufficient information, lack of participation) and content of the change (e.g. losing something valuable – the difference between the existing situation and the future situation). The concept of resistance to change can be examined as the sum of losses and gains. Employees will resist change depending on their perception of the losses and gains related to the change, and the value attached to these losses and gains. Bolle de Bal (1982), a Belgian sociologist, has crystallized the resistance to change in three paradoxes with respect to participation.

To understand why people resist, one must be aware of the major causes which are likely to be present when change is introduced. Szilagyi and Wallace (1990) note the following.

1. *Fear of economic loss.* Any significant change is likely to create the feeling that some positions may be eliminated or require skills which the incumbent does not possess, and thus make the incumbent vulnerable to firing, replacement or transfer to a lower-paying position. Accordingly, it is extremely important for management to communicate clearly how any change will affect job security, retraining opportunities, etc. at the very beginning of the change process. Otherwise, the vacuum created by a lack of information will be filled with rumours, heightened fear and defensive behaviour.
2. *Potential social disruption.* A major source of both job comfort and satisfaction are the social relationships that one has developed at work. Almost any change, be it in structure, technology or personnel, can affect these social relationships and thus be seen as undesirable.
3. *Inconvenience.* The introduction of change invariably involves some change in the way of doing one's work. Simply because of this disruption, the change is likely to meet with at least some resistance. To the extent that the change makes one's job easier, safer, healthier or more intrinsically rewarding, it is likely to blunt the resistance due to inconvenience.
4. *Fear of uncertainty.* Any change disrupts that which is known and creates some degree of uncertainty. Uncertainty tends to cause people to engage in defensive behaviours aimed at reducing the uncertainty. The most frequently observed means of accomplishing this is simply to resist changing. In order to create more certainty and thus reduce resistance, managers need to provide as much information and structure as possible to accompany the change process. Actively involving employees in the change process can give them a greater sense of control over uncertainty and thus reduce resistance. More will be said about participatory strategies later.

5. *Resistance from groups.* If management initiates changes that threaten to disrupt a group's norms or sense of importance, they are likely to meet with group resistance. Similarly, group pressures are likely to be placed on individual group members to resist, even if they support the change.

7.1.3 Overcoming resistance to change

Resistance to change can be prevented or diminished by using several methods for determining the need for change and for implementing it. Some methods emphasize the need for employees to know more about the nature of the change and how it can be accomplished. Misunderstandings often occur because employees are not sufficiently or adequately informed about the change or because employees have had bad experiences involving past changes. Other methods rely on the benefits of employee involvement in the change process. Employees who are involved in the change process are more informed about the change and develop positive feelings toward it. This means they are less likely to resist change.

Although there are no formulas for overcoming resistance to change, there are some useful guidelines (Szilagyi and Wallace, 1990). First, any change pro-gramme needs to integrate the needs and goals of the organization with those of the employees. This will enable employees to see the personal benefit of the change and will thus reduce employee resistance. Second, as Dalton's model indicates, a prestigious individual should introduce the change (Dalton, 1969). This lends credibility to the need for change and confidence that it can be carried out successfully. Third, employees should be involved in the change process. This can reduce employee fears and gain their active support. Fourth, frequent feedback on progress in the change process and on what can be expected should be provided. This kind of feedback can go a long way toward reducing employee fears.

The following are among the most frequently used approaches for overcoming resistance to change:

1. *Knowledge and communication.* This approach is critical to enabling employees to understand the need for change, providing knowledge about the required change, understanding the implementation process and reducing fears about the consequences of the change.
2. *Participation and involvement.* Use of some form of employee participation in the change process provides them with a greater sense of control over the process and can help gain their proactive support. Because this approach takes advantage of employees' first-hand knowledge, it may also result in better implementation decisions.
3. *Provide incentives for compliance.* This might be financial, such as salary or fringe benefit increases. Alternatively, it might involve non-economic incen-

tives, such as greater flexibility in work hours or greater job autonomy. In any event, positive reinforcement from supervisors and managers should always be part of the change strategy.

4. *Provide empathic support.* Perhaps the simplest, most direct way for managers to overcome resistance to change is simply to be empathic and supportive. Such activities as simply being a good listener, making an extra effort to maintain close personal contact, giving time off after a particularly heavy work period, and assisting employees in their personal growth via tutoring and providing training opportunities can go a long way toward reducing resistance.

7.2 Planning for change

Of all the characteristics that have distinguished successful from unsuccessful change efforts in organizations, the most striking has been the extent to which the change effort has been carefully planned. Some of the reasons for the critical nature of planning that have been noted empirically are as follows.

Avoiding unintended ripple effects. As noted in our discussion of organizational systems in section 6.1, a change to any one element of the system affects others, often in unintended and dysfunctional ways. Careful planning is essential to anticipating possible interaction effects and developing strategies for enhancing those that are desirable and, especially, for blunting those that are dysfunctional.

Overcoming resistance to change. As noted in section 7.1, for a variety of reasons, change efforts invariably meet with employee resistance. Careful planning is essential to reducing people's needs to resist change and to gaining their proactive support. In fact, the very existence of a carefully thought-out plan that is clearly understood by all affected employees can reduce uncertainty and thus lower resistance.

Harmonizing change among organizational elements. When an organization has either no plan, or a poorly prepared plan for its change efforts, there is invariably confusion among the various organizational elements as to specific goals, roles, responsibilities and/or strategies for effecting change. The result often is conflicting and counterproductive behaviours – even when everyone sincerely is trying to be supportive of the overall change effort. Careful planning can clarify goals, roles, responsibilities and strategies, and thus ensure an integrated, harmonized change effort.

Determining what should be changed. Failure to plan adequately for change may result in not focusing on the right 'what' of change. In particular, there is a tendency to focus on that which is most obvious, quantifiable or easy to change, and to overlook or avoid what may be the most relevant, or at least equally important, areas requiring change.

7.2.1 Determining the most appropriate strategy for change

Failure to carefully plan how to go about implementing change often results in an ineffective strategy. Carefully working out the most effective strategy for change is as in portant to its success as knowing what to change. As part of this strategy, determining the appropriate tempo of the change effort is also critical to its success.

Adequate planning assumes that there is a need for change, an impetus for change. This impetus may come from various members of the organization or from external sources (e.g. government agencies). In planning for change, it is first important to understand the sources of the impetus for change. The sources can come from different levels. Change experts emphasize the need for involvement of the various parties concerned about WMSDs. The timing and nature of involvement will be discussed later. The success of change depends not only on the involvement of all parties concerned, but also on senior management's commitment. Several methods will be discussed that can be applied to planing for change aimed at reducing WMSDs: participatory ergonomics and inspection programmes. Controlling WMSDs is a long, uncertain process. The process of improving quality is very similar to the process of preventing WMSDs because of the many factors influencing it and the need for involvement by different members of the organization.

7.2.2 Levels of planning

There are traditional WMSD risk factors, such as the forces and loads encountered, the frequency and duration of repetitive motions, the postures of the body parts when working, vibration exposure and, to some extent, personal susceptibility. But there are also systemic risk factors, such as sociotechnical issues, organizational policies and procedures and job design features, which can cause psychological stress and dysfunctional coping. These can have serious health consequences, including increased susceptibility to WMSDs. They can also cause maladaptive employee behaviour, which can intensify the risk of WMSDs and undermine the benefits of positive ergonomic improvements. To develop effective planning, we must first recognize the interrelationships among the various levels of the work system that can have an influence on WMSDs.

These levels of influence can be conceptualized as a hierarchy with sociotechnical factors at the top and job design considerations at the bottom. Sociotechnical factors define the relationships between people's jobs and societal norms, attitudes and rules. Societal laws, rules and regulations establish the legal basis for defining and controlling workplace health and safety hazards. In the case of WMSDs, this means the criteria for acceptance of workers' compensation claims, development and enforcement of ergonomic regulations for controlling employee exposures to WMSDs risk factors and the legal process for pursuing claims against employers and product manufacturers.

In North America there is a substantial difference among the Canadian provinces in the awarding of workers' compensation for WMSDs and in the enforcement of ergonomic regulations. This is also true of the workers' compensation and safety regulatory programmes in the various states in the United States. However, at the federal level in the United States there has been vigorous enforcement of ergonomic hazard abatement in select industries such as meat packing. Beyond the legal basis for acceptance of WMSDs are the background of societal attitudes toward workplace injuries in general and WMSDs in particular. There are times when the general public (society) feels that some WMSDs are a recognized part of a particular job and that employees accept the risk of injury when they take that job. Other times, public sentiment and activism bring about a change in what are considered acceptable and reasonable employment conditions and put pressure on government and employers to make improvements. This change in what is socially acceptable can itself produce increases or decreases in the level of reporting of injuries and in WMSD-control efforts by unions, companies and government agencies.

A good example of this particular type of sociotechnical level of influence is the substantial outbreak in reporting of WMSDs in Australia ('RSI' as they were known in Australia) in the mid- and late 1980s (see e.g. Hocking, 1987). Prior to this time there had always been a low level of reporting of upper-extremity disorders by keyboard operators in Australia. In most cases, these employees were not compensated for their RSI because it was not accepted that keyboard work could produce such disorders. Society (the workers' compensation system) had determined that such disorders were work related only for other kinds of jobs where such disorders had a historical basis, such as assembly-line work.

When research findings from Europe suggested a possible connection between keyboard work and precursors to WMSDs, a group of political activists, unions and health researchers launched an effort to change the workers' compensation rules to allow keyboard operators to be compensated for RSI. This was accomplished when one of the activists was appointed to a major government post and could influence government policy in this area. Once the compensation system started approving RSI claims, the 'epidemic' began and grew to become a national crisis. Later, when the government reverted back to the older, more stringent rules for compensation, the 'epidemic' subsided and reported cases of RSI returned to the former low level. Does this mean that the WMSD problems were miraculously resolved? Such a conclusion is unlikely and illogical. However, it is clear that the government's action did have a substantial influence on what type of injuries were accepted and compensated for.

This example illustrates how the sociotechnical level influences WMSD reporting and control. For society, WMSDs have many ramifications such as economic and international competitiveness, access to and use of health care resources and employee safety and security issues, to name just a few. Society wants to protect employees from workplace injury, but also to control rising health care costs. When RSI for keyboard jobs became a 'legitimate' and therefore compensable injury, the level of reporting in Australia increased substantially. But when the government (society) no longer accepted such injuries for keyboard operators, compensation reporting dropped to a low level (however this may have only shifted the problem to medicare or personal medical benefits).

The lesson here is not that WMSD injuries to keyboard operators in Australia were not legitimate or real, but to illustrate how societal acceptance and government regulations influence employee claims of these injuries. This societal influence in no way defines the actual prevalence of such injuries in the workforce, but rather defines what 'society' will accept and take action on. In defining strategies to control WMSDs in the workplace, societal beliefs, attitudes, rules and government regulations have potential consequences that define the legitimacy of an injury or a problem solution. We must realize that changes must be made in society's perceptions, beliefs, attitudes and rules before substantive improvements can occur at lower levels in the hierarchy.

The next level of the hierarchy is the organizational policy and procedures that reflect the attitudes, rules, goals and procedures of a company in dealing with occupational safety and health in general, and specifically with WMSDs. There are various organizational methods that can be used to successfully implement solutions for controlling WMSDs. These will be reviewed in the following section.

The final level of the hierarchy is the job itself. This is where there is direct interaction between the employee and the workplace and includes elements of job task characteristics, job demands and pressures, supervision, scheduling, social interaction and career development. This is where the traditional task risk factors of force, repetition and posture come into play. It is also where job stress factors have a role. Chapter 2 has introduced the theoretical mechanisms by means of which stress could influence WMSDs. There is growing evidence that the psychosocial environment in the workplace has a substantial effect on the occurrence and control of WMSDs. Recent studies of telecommunications workers doing typing tasks indicates that high levels of job stress are related to increased levels of upper-extremity health complaints (Hales *et al.*, 1992; Smith *et al.*, 1992) and medically diagnosed upper-extremity WMSDs (Hales *et al.*, 1992).

An example will illustrate the importance of considering all levels of the ergonomic hierarchy in implementing controls. It shows the serious consequences of focusing only on physical ergonomic factors to the exclusion of organizational and job design factors. A major telecommunications company in the United States began to experience a substantial number of upper-extremity WMSDs cases in telephone operators in the late 1980s. The union complained, and when the company did not respond positively to its request, OSHA was asked to investigate. OSHA issued a citation for ergonomic hazards which was later withdrawn. During the OSHA investigation, the company hired ergonomics consultants to provide advice on ergonomic improvements to reduce the level of upper-extremity WMSDs. The consultants provided advice on work-station and environmental design, and similar advice was given by OSHA. This advice conformed to the ANSI/HFS 100 - 1988 standard for VDT workplaces.

At the time, there were about 150 telephone operators out of a workforce of 500 reporting some type of upper-extremity problem. At least several dozens had undergone surgical treatment. Up to that time, the company had spent about $2 000 000 on workers' compensation medical treatment and lost wages. The company decided to implement the consultants' and OSHA's recommendations. Approximately the same amount was spent on improving the environmental lighting and glare, reducing noise levels, purchasing dual height-adjustable workstations, height-adjustable chairs and new computer equipment. Given the workers' compensation costs, this investment was considered justified.

Two years after this investment, the level of upper-extremity WMSD cases remained substantially the same, as did the workers' compensation costs. In fact the problem spread from the original location to other locations of the company in other states. At this point, the company and the union agreed that more consultative advice was necessary. They agreed to jointly invite NIOSH to conduct a study into the causes of the telephone operators' WMSDs (Hales *et al.*, 1992). The NIOSH investigation indicated that 80 per cent of the computer workstations at the several sites examined complied with recognized ergonomic design principles. At the site where the investment had been made, there was almost total compliance. Yet for all sites studied the prevalence of upper-extremity WMSDs was 21 per cent for directory assistance operators, with a 20 per cent prevalence at the ergonomically improved site. NIOSH further found that the most predictive factors of employee-reported upper-extremity health complaints were psychosocial job design factors.

This example shows that effective efforts to solve WMSDs must address the total work system (see for example Smith and Sainfort, 1989). This 'systems' approach can serve as the basis for ergonomic improvements. It looks at the concept of properly fitting the employee into the workplace, as well as issues of adequate employee knowledge and skills (training), establishing proper job demands (workload and work pressure), providing opportunities for participation and ensuring an adequate psychosocial work environment.

Planning for change should include an examination of all three levels, sociotechnical, organizational and job. An organization should first understand its sociotechnical environment (e.g. laws and regulations and workers' compensation system). At the next level, organizations can start elaborating strategies for identifying the problems, designing solutions and examining methods for implementing change. Several methods for implementing change will be discussed in the next section. At the final level (job) of the planning phase, organizations can define the specifics or technical content of the change.

7.2.3 Timing and nature of involvement

Planning for change in an organization requires the co-ordination of the various subsystems, e.g. technological, administrative, social and personal. This creates the need to foster interaction among the subsystems and to gather information from each for proper co-ordination. This duty falls to the administrative functions in the organization because there are often differences of opinion among subsystem managers about the relative importance of strategic aspects of change. Organizational change must be directed by the chief executive officer of the company to ensure agreement (and compliance) among subsystem managers. Without the involvement of senior management it is virtually impossible to gain the necessary co-ordination among subsystems, and lack of co-ordination can lead to serious implementation problems. Senior management involvement also demonstrates the importance of the proposed changes to managers and employees.

There is agreement among change experts that the most successful strategies for workplace change include aspects of involvement by all subsystems that will be affected by the change. Involvement assumes that there is an active role not only in providing strategic information, but also offering opinions about the implications of the change for the subsystem. Different aspects of involvement are important, such as the timing, nature and method. In the planning phase, it is important for senior management to consider the timing of involvement.

Organizations need information for effective decision making, and to achieve this they keep many sources of information about the subsystems in centralized files. Therefore, many types of information about the current status of subsystems can be obtained without having to solicit input from the subsystems themselves. Because of this centralized information resource, the administrative subsystem often starts planning for change without participation from the other subsystems. Preliminary planning takes place and the results of this preliminary analysis and recommendations are sent to the subsystems for response. This puts the subsystems in a 'reactive' response mode. Such a response mode is reflected in lowered motivation for the proposed change and resistance to the proposed change and can lead to 'political' actions against the proposed change.

In contrast, when subsystems are brought into the planning for change at the beginning of the process there is a more positive subsystem response. Participation from the outset builds ownership of the recommendations, enhances motivation for the success of the change and 'political' actions are brought to light immediately for quick resolution.

There is always the issue of the sharing of power when subsystems are involved in the change process. Power can come from several sources: (1) critical information; (2) opinions; (3) beliefs, policies, rules and procedures; and (4) authority. The sharing of power means that senior management relinquishes its control over some or all of these sources to others. Effective involvement requires the sharing of critical information among subsystems to identify the best possible strategies and potential problems. Active participation always generates greater motivation and better acceptance of solutions than do passively providing information and taking orders. Active participation is achieved by soliciting opinions and sharing authority to make decisions about solutions. One drawback of active participation is the need to develop a consensus among participants who have differing opinions and motives. This usually takes more time than traditional management decision making and can lead to conflict among subsystems.

7.2.4 Determining the success of change efforts

One of the most frequently overlooked areas in poorly planned change efforts is that of evaluation. Unless careful attention is given to determining how to evaluate the effectiveness of the various aspects of the change efforts, it will be extremely difficult to really know: (a) what progress is being made; (b) what aspects of the effort need modification as the effort progresses; and (c) upon completion, the overall success and what 'lessons' have been learned from the efforts. These data are critical not only to correcting the change efforts as they progress, but also to planning for any follow-up change efforts. Because planning for and conducting evaluations is so critical to ensuring successful change efforts, it is treated as a separate section at the end of this chapter.

7.2.5 Determining the goals of change efforts

The first consideration in any change effort should be to determine just what it is that one wants to accomplish as a result of the change effort. Successful change efforts typically begin by specifying generalized goals for the organization and then moving progressively to more specific and concrete objectives. This process moves from the overall organization down through each vertically, horizontally and geographically differentiated element to be involved in the change effort. It is also an iterative process: as the change efforts progress, the goals should become both more immediate and concrete. Invariably, this results in modifying or resetting some of the initial goals, or in making deletio₁ s and additions as experience is gained with the actual change process.

Avoiding tunnel vision. A major pitfall to effective change efforts is seeing symptoms of a problem, and then focusing too narrowly on its probable cause. For example, as noted earlier, the research literature on WMSDs has shown time and again that there is usually more than a single cause, or even a simple combination of a few causes, for the observed symptoms. Yet the literature on organizational decision making has repeatedly documented the tendency of decision-makers to concentrate on one or two probable causes and thus develop change effort goals that are too narrow in scope. Simply stated, effective goal setting for change efforts requires a true systems analysis of the problem and a systems approach to the goal-setting process.

Operationalizing goals. Critical to the initial goal-setting portion of the planning is not to begin by focusing on what can be readily seen and measured. More often than not, that which is obvious and easily measured is not what is most relevant, and can lead to the tunnel vision described above. Only after the goals are determined should attention be given to how to operationalize them. Ultimately, developing ways to qualitatively or quantitatively measure goals, and to provide feedback to the affected employees on progress toward goal attainment, is an essential part of the planning process.

7.2.6 Determining what to change

As suggested by our discussion of the nature of organizational systems in section 6.1, and of ergonomics interfaces in section 6.2, the 'what' of change can be categorized into five distinct areas: technology, the work environment, people, jobs and organizational structure. These five, either singly or in combination, are potentially relevant to reducing WMSDs.

Technology. Technological change to reduce WMSDs can take several forms. Among these are: (a) changing the design of hardware to improve human-machine interfaces; (b) changing the design of system software to improve user-system interfaces; and (c) introducing new technology to relieve employees from tasks that directly or indirectly induce WMSDs.

Environment. Modifying those aspects of the work environment that are known to aggravate WMSDs can help, e.g. (a) reducing exposure to cold or heat through changes in the facility's temperature and humidity; and (b) providing protective clothing to guard employees from environmental exposure.

People. Changing employees can focus on attitudes, values, motivation, knowledge or behavioural skills. In a given situation, any or all of these may be relevant to reducing WMSDs. Changing people invariably involves some form of training, and transfer of that training to the actual job situation.

Jobs. Changing the design of jobs can involve any or all of the following: (a) reallocation of functions and tasks between persons and machines or software to make more effective use of human capabilities and reduce stress on human limitations; (b) moving employees between different positions (job rotation); (c) increasing the variety of tasks to reduce repetitiveness (job enlargement); (d) recombining tasks into more effective work modules to reduce frustration; (e) increasing job control and the content of tasks to allow for greater employee responsibility and discretion (job enrichment); and (f) redesigning tasks to provide better sequencing, a more optimal difficulty level or more appropriate workload.

These job design changes can not only reduce physical and psychological stress responses and WMSDs, but often lead to improved job satisfaction, sense of self-worth and motivation. In some instances, these same changes may lead to increased productivity and quality (Bammer, 1993). In redesigning jobs, care should be taken to ensure that the resultant jobs are compatible with the overall organizational design.

Organizational structure. Changing the structural design of an organization involves modifying one or more aspects of its design complexity, formalization and/or centralization of decision making. Effective organizational redesign (to optimize conditions for minimizing WMSDs and related symptoms) requires a macroergonomic analysis of the major system components. Based on this analysis, the structure can be modified to ensure harmonization with the critical characteristics of the organization's technology, personnel subsystem and external environment. (See section 6.4 for a discussion of macroergonomic analysis).

7.2.7 Determining how to change

Determining how to go about implementing change requires careful analysis and consideration of how power will be utilized, how personal or impersonal the relationship approach will be, and the tempo or speed and depth of the process to be followed (Szilagyi and Wallace, 1990).

7.2.7.1 Power utilization

Greiner (1967) identifies three approaches to change, based on the extent to which power is shared via employee participation in the change process: unilateral, shared or delegated.

Unilateral power occurs when the superior does not share power in the change process. This can be accomplished by: (a) decree, in which the superior simply announces what change will occur and what is expected from each subordinate in implementing the change; (b) replacement, in which one or more individuals are replaced as the means of effecting change; and (c) structure, in which the superior unilaterally changes the organizational structure as the means of achieving goals. The unilateral approach is most likely to be appropriate when the workforce is relatively unskilled or inexperienced.

Shared power. Shared power assumes that authority must be used with discretion. When the organization has capable employees, power can be shared to set change goals, reach important change decisions and determine implementation strategies. Although there are different variations to the shared approach, they can be grouped into two general categories: (1) group decisions and (2) group problem solving.

1. The group decision. In this approach, group members select a solution from among several alternatives offered by superiors. This approach usually does not involve actually identifying the problem or solving it. Instead, its primary objectives are to give the group some sense of control and to obtain group agreement.
2. Group problem solving. This approach involves the group actually solving the problem through group discussion and analysis. Usually, the group is

given considerable latitude in identifying and analysing the problem and in developing alternative solutions. However, management may serve as an information resource, set certain guidelines, make input to the process and may even make the final decision from among the alternatives proposed by the group. One form of group problem solving that has become widely and successfully used in change efforts in recent years is participatory ergonomics (e.g. Noro and Imada, 1991; Nagamachi and Imada, 1992). This approach is discussed in section 7.4.

Delegated power. With delegated power, subordinates are given complete control over the decision-making process. From the outset, employees are given responsibility for identifying problems, analysing them, developing proposed solutions and making the final decision. This approach assumes a highly experienced and responsible workforce. A primary reason for using this approach is to enhance employee self-management and autonomy, and thus gain greater organizational commitment, motivation and productivity. Its major drawbacks are that management has less control and the process ignores the experience and perspective that management can bring to the decision-making process.

Based on a survey of several cases of organizational change, Greiner (1967) found the shared approach to be the most successful. However, the level of employee training and experience, as well as situational factors, may sometimes make either the unilateral or delegated power approach the most appropriate. In particular, shared strategies that come close to being a delegated power approach have been seen more frequently in recent years (e.g. self-managed work teams).

7.2.8 Other considerations in planning for change

Szilagyi and Wallace (1990) have noted several other considerations to be taken into account when planning for change. These include: (a) altering social ties; (b) enhancing employee self-esteem; (c) gaining employee internalization; and (d) ensuring transfer of learning.

Altering social ties. Successful change programmes are characterized by a loosening of old social ties and the establishment of new relationships that support and reinforce the change. Breaking down and loosening existing relationships is an essential part of the 'unfreezing' process referred to earlier, but this alone does not ensure that the resulting change will be in the desired direction, or will be lasting. What seems essential is to establish new relationships that reward desired behaviours and support appropriately changed attitudes. Accordingly, as part of the planning process, careful attention needs to be given to developing and implementing appropriate rewards and other reinforcements. Several approaches for accomplishing this reinforcement are discussed in section 7.4 on implementation methods.

Enhancing employee self-esteem. Beginning with the classic Hawthorne studies at Western Electric (Roethlisberger and Dickson, 1939), abandonment of previous behaviour patterns has been found to occur more readily when conditions are created that increase employees' sense of self-worth and importance. Attention to creating these conditions should be an integral part of the planning process.

Gaining employee internalization. If a change effort is to be successful, employees must first come to internalize the motive and rationale for the change. Simply stated, employees will adapt new behaviour because they believe it will help solve problems or is congruent with their own orientation. Planning efforts thus should include means for: (a) increasing employee awareness of the need for change; and (b) ensuring that both the organizational change goals and the change process itself will be congruent with employees' personal needs. Ultimately, the actual testing of the new change through personal experience will determine whether it will become fully internalized. The bottom line is that the change – be it to organizational structure, jobs, people, technology or some combination of these – must be found valid when tested against real organizational life.

Transfer of learning. Most major change programmes involve at least some form of education or training. Frequently, this learning takes place away from the actual job. For example, skills training may first occur on simulators or part-task trainers, or as some form of simulation exercise (often computer-based). Similarly, some of the learning may take the form of classroom study. In addition, a number of widely used intervention strategies, such as sensitivity training, goal setting and team building take place away from the job and, indeed, often away from the organization's facilities. Care must thus be given in the planning process to ensuring that what is learned is likely to be transferred to the actual job situation. One primary way to help ensure that transfer of learning will occur is to design the education and training programmes (see chapter 8) so that the elements of behaviour learned away from the job are as similar as possible to those required for acceptable on-the-job performance. A second major consideration is to ensure that the new employee behaviour, once learned, is reinforced back on the job. Often, this requires training superiors in both what behaviours to reinforce and how to go about reinforcing the desired new behaviours.

7.3 Implementing change

Through a thorough process of planning for change – one that begins with a careful systems analysis and then follows through with goal setting, determining what to change, how to change and other considerations as outlined above – many of the decisions concerning implementation already will have been determined. However, there still are other aspects of the actual implementation process to be considered. These include: (a) the change agent; (b) the change model; and (c) constraints on implementation. Finally, the actual methods to be used to effect change must be selected.

7.3.1 The change agent

Organizational change efforts frequently require the assistance of someone with an outside perspective and/or special expertise to facilitate the change process. This change agent thus is able to bring in new ideas, perspectives, approaches, implementation skills and specialized knowledge to help solve the problem. At least five types of organizational change agents can be identified, as follows.

Outside pressure type. These change agents work outside the organization and use various pressure tactics to create change. These include such tactics as consumer-advocacy activities, political lobbying, litigation and other forms of legal pressure, and public demonstrations.

People change type. These change agents bring a strong psychological orientation to the change process and focus on the individual employee. Typical change methods include training, behaviour modification, behavioural modelling, specific goal setting and counselling.

Organizational development type. Organizational development change agents focus primarily on organizational processes. They typically concentrate on methods that enhance team building, problem solving, conflict management, leadership behaviour and intra- and intergroup communication.

Analysis from the top type. The focus of this type of change agent is on identifying needed structural changes. Analytical approaches such as operations research, systems analysis, risk analysis and computer modelling are used to advise senior management of needed change.

Human-system interface type. Recent years have seen the emergence of this new type of organizational change agent. Typically, this type has a combined professional background in the behavioural sciences and industrial engineering with specialized training in ergonomics or human-system interface design. Others may combine a background in medicine, physical therapy or other related fields with ergonomics. Often they are known as ergonomists, human factors specialists or human factors engineers specializing in organizational change. Their primary focus is on identifying deficiencies in the various human-system interfaces that exist in the organization and employing ergonomic methods, design specifications and design guidelines to correct them. This type of change agent can be particularly useful for identifying and correcting interfaces that may be directly or indirectly inducing WMSDs. Section 6.2 discusses the various types of human-system interfaces and related ergonomics technologies.

Although most organizational change agents tend to have one of the above profiles, in reality there is overlap in the techniques that they will actually use in a given situation.

7.3.2 Change agent implementation models

In assisting organizations with the change process, a variety of models can be employed by change agents. Some of the most popular models are: (a) the medical model; (b) the doctor-patient model; (c) the process model; and (d) the engineering model (Szilagyi and Wallace, 1990).

Medical model. In this model, management, much like the attending physician, seeks assistance from an outside specialist (the change agent) in diagnosing problems, clarifying issues and recommending possible courses of action; but management, much like the attending physician, makes the final decision as to the course of therapy.

Doctor-patient model. Much like the medical model, the change agent makes a diagnosis and recommends a course of action to management. In this case, the change agent is the physician and management is the patient. The final decision rests with the 'patient'; but, because of the established relationship, the recommendations are usually adopted.

Process model. In the process model, management and the change agent work together to diagnose, plan, implement and evaluate the change effort. However, final responsibility for the change remains with management.

Engineering model. In its purest form, the diagnosis and selection of an approach have already been made. The change agent is then called in as the 'engineer' to design the change effort to fix the problem.

7.3.3 Constraints

Although there are constraints that must be considered with each change and development technique, there are four extremely important ones that affect any implementation strategy. The first of these is resistance, which can take the form of sabotaged performance standards, passive-aggressive behaviour, absenteeism, unfounded grievances and reduced productivity (Goldstein, 1988). The psychological basis for resistance to change and strategies for overcoming this resistance were discussed in section 7.1. The other three constraints are: (a) leadership climate; (b) individual characteristics; and (c) formal organizational design (Szilagyi and Wallace, 1990).

Leadership climate. Leaders can exercise considerable influence on subordinates either to accept or reject changes implemented by senior management. Accordingly, leaders' values, attitudes and perceptions can all be constraining factors. Thus it is essential, when implementing any change, to first ensure that the leaders fully understand and are committed to the change. Perhaps the best way of accomplishing this is by involving them in the change effort during the planning stage. Alternatively, a carefully developed seminar or series of seminars for the leaders prior to implementation could be considered. Although such seminars can be developed by the change agent, direct senior management participation is essential to give the change effort credibility (see Dalton's model in section 7.1).

Individual characteristics. Among those characteristics that have been found important are learning abilities, current skills, attitudes, expectations and personality. For example, if one introduces new technology as part of the change effort, then it is essential to: (a) ensure that workers get the necessary skills training; (b) take steps to allay their fears about such things as job security or having adequate time to achieve performance standards; and (c) ensure that reinforcement mechanisms are in place to provide feedback and bolster self-worth. Failure to take such steps to adequately deal with these individual constraints has torpedoed many change efforts.

Formal organizational design. The formal organizational structure must be compatible with the proposed change, or it will prove a major obstacle to successful change implementation. For example, in a highly hierarchical, formalized and centralized structure, introducing participatory decision making without first changing at least some elements of the structure would be unrealistic. Conducting a macroergonomic analysis of the organization as part of the planning process can be an excellent first step in coping with this constraint. As noted in chapter 6, when there is effective harmonization of the organizational elements, synergistic improvements are possible.

7.4 Implementation methods

7.4.1 Introduction: implementation and management – employee involvement

There is a substantial body of safety management literature that supports the belief that management commitment is a necessary condition for successful injury-control programmes (see e.g. Simard *et al.*, 1993). This commitment can be generated by external sources such as government regulatory action and/or by senior management attitudes toward WMSD remediation. In the United States, the recent emphasis by OSHA enforcement on improving ergonomic conditions in selected industries such as meat packing and automobile manufacturing has led to a number of million-dollar fines. This has created attitude changes in the senior management of these industries, with a greater emphasis on ergonomic improvements. OSHA's production of an ergonomic guideline for the red-meat industry has led to increased recognition of ergo-nomic problems in many industries and provided a basis for industries to take some positive action. In addition, the document has had a substantial impact on specific policies and procedures regarding WMSDs and ergonomic improve-ments in companies in affected industries.

Proper attitudes toward occupational safety and health start with senior management and must flow downward through the entire organization to middle and lower-level management and to the workforce. A positive attitude throughout the organization is necessary for successful implementation of ergonomic improvements and WMSD control. This requires recognition by everyone that the organization is committed to ergonomic improvements. Such commitment starts with a policy statement from senior management addressed to all employees regarding the importance of and commitment to ergonomic improvement. This action needs to be followed up by meetings between supervisors and employees regarding specific rules and procedures, and a discussion of systematic efforts must be undertaken by the company to achieve ergonomic improvements. One effective way to obtain commitment and create an open attitude at all levels of the company is to provide opportunities for everyone to participate in the ergonomic improvement efforts. Such participation could take the form of a management/employee team to develop specific company procedures, a departmental ergonomic improvement committee, or monthly ergonomic meetings between supervisors and employees to solicit employee input. Of course, there are many other possibilities for encouraging participation. All these efforts will require that managers and employees receive training about WMSDs, ergonomics, the mechanics of the participatory process and specific job improvements (see chapter 8, dealing with training).

Organizational and job design experts have long proposed that employee involvement in work enhances motivation and generates production and product quality benefits (Lawler, 1986). Examples of this are the Scanlon Plans used in the United States during World War II, quality circles pioneered in Japan, high-involvement management, quality of working life committees and joint labour/management committees for solving production problems and ensuring product quality.

High-involvement management is a powerful tool for drawing on a broad range of employee expertise to achieve high quality solutions, gaining employee acceptance of decisions and motivating employee compliance with solutions. As indicated earlier, this approach does have some drawbacks, such as a lengthier decision-making process than traditional management styles and the sometimes difficult task of reaching a consensus across a wide array of opinions. Lawler (1986) has pioneered the use of the high-involvement management process and how this relates to implementing change in an organization. He proposes the following principles:

1. *Change needs a reason.* There must be justification for changing current organizational proprieties and processes. This justification must be articulated to all affected parties.
2. *There are many approaches to participatory management.* There is not just one correct way for participation to function as a management process, but rather many approaches to participation, each of which has positive benefits and drawbacks depending on the aims to be achieved. Steadfastly sticking to one approach as the only way to achieve success can undermine employee confidence in and the ultimate success of a management approach.
3. *Participatory management programmes should be evolutionary, not revolutionary.* Companies that have traditional hierarchical structures will have to start with limited participation and slowly work their way toward high-involvement management. This supports the concept that there is not one

best approach. It also suggests that change is a constant process, and organizational management and structure should not be viewed as static.

4. *Resurrection is more difficult than creation.* Participatory management is much easier to initiate and continue in new operations and plants than in established organizational structures. There is much inertia in existing organizational structures and procedures that has to be overcome, and this dilutes and diminishes the resources devoted to participatory management.

5. *Participatory islands often end up underwater.* Pilot or experimental efforts at participatory management are often undertaken to limit the expense of change, provide greater control over the change process and limit the extent of damage of a failure. However, even when such islands of experimentation are very successful, they rarely survive for long, primarily because they do not conform to the general organizational structure and process. If experimentation is the choice rather than wholesale change, then several islands of research need to be established to provide legitimacy and organizational support.

6. *It is hard to kill a good thing.* Once an organization decides to move toward high-involvement management, retreat becomes almost impossible. Employees will develop a strong commitment to and satisfaction with participatory approaches, and will be reluctant to revert to more traditional management structures and approaches. A retreat to the traditional management style may lead to dissatisfaction, unrest and poorer work performance.

7. *Vision is critical.* Leadership is a key element in successful management, whether it be traditional or participatory. Everyone in the organization needs to have a vision of the participatory management process and their roles in achieving successful performance. This vision starts at the top and radiates to the shop floor. It should make all employees feel important and essential for effective organizational success. Vision must be a 'shared' process by all employees.

8. *Values can energize change efforts.* All change efforts must create a sense of shared values in the participants. Management programmes that utilize existing values such as self-worth, democracy, equality and fairness have a much higher chance of success than those trying to create new values. In any case, values must be shared or of substantial importance to those involved in the process.

9. *Feeling successful leads to success.* Success is often a self-fulfilling prophecy. It is important to demonstrate some small success early in a programme. Talking about this success typically leads to greater acceptance of a programme that will lead to even greater long-term success. 'Success breeds success'.

10. *Planned change efforts rarely go as planned.* Problems and difficulties are to be expected when implementing change. Nothing ever goes as planned in organizational redesign efforts because of all of the uncontrolled events that can occur. Knowing that problems can and will occur helps to moderate

expectations, keeps people alert to potential problems and helps them to intervene quickly when problems do occur.

11. *Think long-term*. All change in organizational management has profound effects on institutional structures, historical patterns of behaviour and employees' knowledge and attitudes. These take time to evolve in the face of historical precedence and institutional inertia. Thus a long-term, evolutionary approach is less disruptive of these institutional processes. Evolutionary approaches tend to become institutionalized, while revolutionary approaches tend to be displaced by new revolutions.

7.4.2 Participatory ergonomics

Throughout this book, we have emphasized the need to involve workers in the change process in order not only to improve the nature and content of the change, but also to facilitate the management of change. Participatory ergonomics is one of several macroergonomic strategies for implementing ergonomic change. The fact that participatory ergonomics is anchored in the macroergonomic philosophy ensures adequate consideration for organizational design and management issues (Hendrick, 1991). The participatory ergonomics approach specifies that the 'end-users of ergonomics' (i.e. workers) should be heavily involved in planning and implementing the change, and therefore they cannot be passive. They have to take an active role in the identification and analysis of ergonomic risk factors as well as the design of ergonomic solutions. This is particularly important with regard to WMSDs because some of the risk factors may be hard to identify, quantify or change. Furthermore, worker behaviour may play an important role in the development of WMSDs. Therefore, solutions aimed at reducing WMSDs may involve in part that worker behaviours are modified. Participatory ergonomics can help in such situations because workers get a better understanding of the risk factors that can affect their own behaviour at work and their health and safety.

Participatory ergonomics is a technique that has proven successful in solving ergonomic problems and implementing ergonomic change (Noro and Imada, 1991). It can be particularly useful at the planning stage, by involving workers in the identification and analysis of ergonomic problems. Participatory ergonomics can take various forms, such as design decision groups, quality circles and worker-management committees. Some of the common characteristics of these various programmes are worker involvement in developing and implementing ergonomic solutions, dissemination and exchange of information, pushing down ergonomics expertise, co-operation between experts and non-experts (e.g. workers) and consideration for workers' opinions.

One of the characteristics of participatory ergonomics is the dissemination of information (Noro, 1991). Participatory ergonomics can be beneficial in reducing or preventing resistance to change not only because of worker involvement, but also because of the information provided to the various members of the organization concerned with WMSDs. Lack of information and uncertainty are two major causes of resistance to change. If employees are informed about ergonomics early on, they are less likely to actively resist the change.

The role of the expert varies considerably in the various forms of participatory ergonomics. At the beginning, the expert may play a very active role, then move to a coaching role, and finally become an advisor. The training and information received by the workers is important to the success of participatory ergonomics. Workers have a lot of knowledge and expertise about their jobs, but not necessarily about ergonomic risk factors, solutions and organizational workings (procedures, etc.). WMSDs are influenced by a great variety of factors, some of which are not fully understood by researchers. In this context, we can assume that workers will never become 'experts' in ergonomics and that well-trained ergonomists will still be needed. However, it is important to involve employees early in the change process. Early employee involvement can reduce resistance to change, and give employees a chance to start learning about WMSDs and ergonomic risk factors early on.

Imada (1991) lists several tools that have been successfully used in implementing participatory ergonomics: Pareto analysis, cause-and-effect diagrams, quantitative illustrations, five ergonomic viewpoints, link analysis, checklists, world maps, round-robin questionnaires, layout modelling and mock-ups and slides/videos. Many of these tools could be useful in planning for changes aimed at reducing WMSDs. For instance, a checklist could be used by workers to identify which of the WMSD risk factors are present in their job. Checklists are discussed in chapters 4 and 5 and examples are provided. Another useful tool for examining the potential benefits of an ergonomic solution is the use of models or mock-ups. Wilson (1991) uses 'design decision groups' to evaluate the potential advantages and weaknesses of new workstations or new environments. The design decision groups are comprised of employees who are likely to be users of the new workstations or the new environments. Workers actually construct cardboard models of potential work areas. The use of mock-ups helps workers visualize the changes in their workplace. Workers are less likely to fear or resist the change because they know more about it.

Worker involvement and participatory approaches have been very successful in improving productivity and quality in manufacturing and assembly and in dealing with health and safety problems. The application and success of these approaches has brought about a revolution in thinking about the specific responsibilities of management and labour. These programmes have succeeded where both sides have developed trust and respect for the contributions they can each make. But this shift from confrontation to co-operation does not occur overnight, or without some compelling force, such as the spectre of a serious hazard, that establishes the need to co-operate. Trust only comes later, after a track record of mutual benefit is demonstrated. As one long-time union member at a General Motors plant put it, 'It's a lot more difficult working together. When labour and management used to fight, hell, that was easy. You'd take a position and hold it – win or lose. Now it's compromising all the way through to solve problems' (Zino, 1988). Such compromise is difficult when you are not used to it.

7.4.3 Total Quality Management and ISO 9000

Major changes are occurring in the business world all over the planet. One of these changes is the focus on customer satisfaction and quality, so-called Total Quality Management (TQM). The concept of Total Quality Management has received tremendous attention from companies in many countries. It is a customer-driven approach that emphasizes the need for companies to design production processes and organizational structures that can produce high-quality products. ISO 9000 is an international standard series designed to encourage companies to set up these processes and structures. The principal concepts of ISO 9000 are outlined in the following quote (ISO, 1987):

> 'An organization should seek to accomplish the following three objectives with regard to quality:
> (a) The organization should achieve and sustain the quality of the products or service produced so as to meet continually the purchaser's stated or implied needs.
> (b) The organization should provide confidence to its own management that the intended quality is being achieved and sustained.
> (c) The organization should provide confidence to the purchaser that the intended quality is being, or will be, achieved in the delivered product or service provided. When contractually required, the provision of confidence may involve agreed demonstration requirements.'

The application of ISO 9000 and the implementation of Total Quality Management concepts have several implications for ergonomics and WMSDs. First, ISO 9000 emphasizes the need for TQM organizations to be concerned for the health and safety of their employees. Second, employee involvement is emphasized by ISO 9000 as a tool necessary for quality. Third, ISO 9000 specifies several criteria of product quality, among them product safety and reliability. Finally, ISO 9000 greatly emphasizes customer satisfaction. We will demonstrate how customer dissatisfaction could be a result of poor attention to ergonomics.

ISO 9000 emphasizes the need for TQM organizations to be concerned for the health, safety and well-being of their employees. Since 1993, companies that do (or want to do) business in the European Union have to meet the ISO 9000 standard. Given this economic implication, companies have to set up quality systems following ISO 9000 guidelines. They have to ensure that their employees' health, safety and well-being are ensured. They have to set up measurement and monitoring systems to collect data on health and safety. They have to design programmes to reduce health and safety problems, such as WMSDs. Ergonomics has a major role to play in this regard.

One of the very important characteristics of quality systems is employee involvement and the use of teams. Quality circles are an example of the application of these principles. Quality circles use participation, by organizing small groups of workers. Workers discuss quality problems, their sources and solutions. They understand the variety of factors with an impact on quality, some of them ergonomic factors. Quality circle methods can therefore be used to introduce ergonomics. Nagamachi (1991) and Zink (1991) describe the introduction of participatory ergonomics through quality circle activities in Japan and Germany, respectively. Noro (1991) reports that one-third of the improvements suggested by quality circles at a major Japanese steel company were about ergonomics. Workers understand that ergonomic deficiencies can affect quality both directly and indirectly. When workers are uncomfortable or feel pain, they are less likely to pay attention to their performance and the quality of their work. Ergonomic deficiencies can also make their jobs more difficult to perform, thereby affecting quality. Workers may pay even less attention to quality if they are under a lot of pressure to perform, and thus sacrifice quality for quantity.

ISO 9000 specifies various criteria for the evaluation of quality systems, such as the economics of the product and the ability to judge product quality and fitness for use on the basis of final product testing alone. Another important criterion that is very relevant to ergonomists and occupational safety and health professionals is the safety requirements of the product. A product is said to be of high quality if it is safe and reliable. Products, that is the tools and technologies used by workers, should be safe, i.e. they should not contribute to the development of WMSDs. They should be designed so that they can be used safely and do not put workers at high risk for WMSDs.

Companies whose products are of high quality and that follow the ISO 9000 guidelines are more likely to satisfy their customers than companies that are not able to manufacture high-quality products. Customer satisfaction is influenced not only by the objective characteristics of the product, but also by customer perceptions of the product and the company. The reputation of a company in producing high-quality products plays an important role in customer satisfaction, and can be badly damaged if the company is not concerned for the health and safety of its employees (Imada, 1990).

7.4.4 Successful safety programmes

Successful implementation methods used with other occupational health and safety problems, e.g. acute traumatic injuries, may also provide valuable insight into WMSDs. Research on successful safety programme performance in plants with high hazard potential has shown a number of factors that contribute to their success (Cohen, 1977; Smith *et al.*, 1978; Cleveland *et al.*, 1979). The primary factors are a formal, structured programme so that managers and workers know where to go for help; management commitment and involvement in the programme; good communications between supervisors and workers; and worker involvement in the safety and health activities.

Cleveland and co-workers (1979) found that in smaller companies, safety committee activities were more important in hazard control than having professional safety staff. In addition, having a committed workforce that participated in safety on a daily basis was a key element for success. To be successful, the workforce had to be trained, particularly the first-line supervisors together with the line workers. Finally, regular procedures for keeping up to date on the nature and extent of safety hazards was important for success. This research identified the importance of all levels of plant personnel playing an active role in hazard identification and control, and carrying these responsibilities out co-operatively. Research on successful safety programmes shows the need for an organisational approach that involves employees at different levels in the organisation.

7.5 Evaluating the change process

The need to collect data on WMSDs and risk factors was emphasized in chapters 4 and 5. Several surveillance methods have been proposed to identify WMSDs and risk factors. Evaluation of the change process should be based on data generated by these surveillance methods. The measurement process should provide feedback (i.e. data on the effects of the change) and direct efforts and behaviours toward the desired outcomes. Progress must be measured and followed up to assess the effectiveness of ergonomic redesign.

Theoretically, the change should produce a decrease in WMSDs. However, there is the possibility of increased WMSDs reporting either before or right after the change is implemented. Existing problems may become more visible: workers may be less afraid to speak out about their problems or they may realize that they have a WMSD problem. Another reason for the rise in WMSDs right after the change may be transient problems due to the change itself: workers having to get used to the new work procedures or new tools, or stress generated by the change process.

The change process should be evaluated over a lengthy period. Ergonomic changes may take time before positive results occur. Furthermore, according to the TQM philosophy and the ISO 9000 standards, evaluation should be ongoing so that improvements in working conditions are continuously implemented and WMSDs are reduced. If continuous improvements are expected, then employees at all levels must take part (Drury, 1991). Workers need ergonomic information and knowledge to ensure that changes are carried out effectively. Continuous improvement implies ongoing change and monitoring. Evaluating the impact of the change should be an integral part of the change process.

Earlier, we emphasized the complexity of the WMSD problem, which calls for multiple assessment methods and multiple solutions. The systems approach needed to examine WMSDs implies that multiple changes may be needed to achieve a 'balanced' work system. Hence the need to examine and evaluate work systems over time. Achieving a 'balanced' work system is likely to require many changes or adjustments in the way work is performed, structured and organized. It may take a long time before some of these changes are fully implemented.

8

WMSD-related training

8.1 Introduction

A survey of the literature suggests that training and education are widely seen as necessary complements to most other types of intervention. Beyond such widespread agreement, there are a number of problems. First, the terms 'training' and 'education' are never clearly defined. Second, it quickly becomes obvious that, having advocated training and education, most authors concentrate specifically on training. Stammers and Patrick (1975) provide the following definitions:

> *Education*: Activities which aim at developing the knowledge, moral values, and understanding required in all walks of life rather than knowledge and skill relating to only a limited field of activity. *Training*: The systematic development of the attitude/ knowledge/skill behaviour pattern required by an individual in order to perform adequately a given task or job.

From these definitions, it is fairly obvious that education is fairly general and nonspecific in nature, whereas training is much more specific and action-oriented. On the basis of these definitions, this chapter will focus primarily on 'training'.

Education does still warrant a mention. There are a number of changes that need to be implemented in, for instance, the education of physicians and engineers to ensure that the WMSD problem is dealt with effectively. Chapter 9 on medical management includes a general discussion on issues related to the education of physicians and other health practitioners in ergonomics. It is advocated that changes in values are needed, that the level of awareness must be raised, that the media should be used to reach the public, and that it might be useful to start at the primary school level, using an approach similar to that taken with automobile seatbelts. But all these educational changes are of a cultural and societal nature and will take a long time. This chapter will focus on training given at the company level to deal specifically with WMSD problems. The focus is on the training given by an organization to its members (workers, supervisors, managers/employers and specialists) to solve or prevent WMSD problems. In this chapter, we will not discuss the formal or academic education of specialists such as physicians and engineers, but instead will focus on the training of various members of an organization.

Training programmes designed to deal with WMSDs should be very diverse: they need to focus on different target audiences (e.g. workers, supervisors and managers), different training methods and techniques (e.g. lectures, on-the-job training and cognitive training), different approaches (e.g. train-the-trainer and workplace analysis schemes), different contents (e.g. working methods and WMSD risk factors) and different objectives (e.g. to modify working methods used by workers and convince managers of the validity of the WMSD problem). This phenomenon is not typical only of WMSD-related training, but can also be observed in safety training (Hale, 1984). This suggests that WMSD-related training must take a multifaceted approach.

Why do we need training? The impetus for WMSD-related training stems from various sources:

- standards, laws and regulations that require training as a part of any health and safety programme (see, for example, the ECE directive in the work of the Commission of the European Communities, 1992);
- the need to sensitize companies and employees to the WMSD problem (for example, the Musculoskeletal Injuries Prevention Programme 1992 of the Ontario, Canada, Workplace Health and Safety Agency);
- training of new employees;
- training necessary with changes in working conditions (changes in the work system: tasks, tools and technologies, organization, environment): for instance, the need to teach workers new work methods or to prepare the organization for change.

Training is not a panacea for reducing WMSDs. Training is an integral part of a larger work system with the individual at the centre. It is important that the entire system be taken into account (see the discussion on 'balance in the system' in section 6.4). The elements of this system are tasks, environment, organizational structure, technology and the individual. Training is one aspect of organizational design capable of affecting or being affected by the other elements of the system. Therefore, in the design of training programmes all the other elements of the work system need to be considered. For instance, teaching workers about neutral wrist postures may not actually be appropriate if the tools used in a task are not designed to allow such postures and the workers cannot actually keep their wrists in the neutral position.

8.2 Overview of the literature on WMSD-related training

Since very little research has been done on WMSD-related training, we will examine research on WMSD interventions having training as one component (subsection 8.2.1). Further, by way of example, training aimed at improving other musculoskeletal afflictions such as back problems (training in materials handling) (subsection 8.2.2) and training in VDT/office ergonomics (subsection 8.2.3) will also be considered.

8.2.1 Literature available on WMSD-related training

Worker training in WMSD-related issues needs to take the characteristics of the workers into account. For example, the training needs of new employees may be different from those of experienced employees. Parenmark *et al.* (1988) studied the effectiveness of ergonomic training on two groups of assembly workers. The first group was comprised of 33 newly hired workers (15 workers in the experimental group and 18 workers in the control group), whereas the second group included 60 assembly workers with more than one year's experience (30 workers in both the experimental and control groups). Workers in the experimental group were taught about potential adjustments in their workplace. They were equipped with biofeedback EMG units that recorded changes in the muscular load level in the upper extremities. Workers had to adjust their movement patterns to keep the muscular load below 10 per cent of the maximum voluntary contraction. Forty-eight weeks later, the amount of sick leave attributable to arm-neck-shoulder complaints had been reduced for the new workers (as compared with their control group).

Since it is very hard to change the work habits of experienced workers, initial training of new employees in ergonomics is very important for preventing WMSD problems. Training for new employees should be designed to avoid increases in muscle tension linked to learning new skills. A study by Sihvonen *et al.* (1989) shows that training in typing was accompanied by increased electrical activity in shoulder muscles. This initial skill training heightened the risk factors for musculoskeletal disorders. Training for new employees can be designed to prevent harmful working techniques that are risk factors for WMSDs by teaching such new workers 'healthy' work methods.

Some studies have examined the effectiveness of interventions to reduce WMSDs. Sometimes training is one part of the intervention. Chatterjee (1992) conducted a prospective longitudinal study at an electromechanical plant between 1980 and 1988. Several methods were used to identify the causes of WMSDs. A number of interventions were introduced simultaneously: (1) educational efforts targeting supervisors, engineers, workers, safety representatives and occupational health and safety personnel; (2) occupational health (a medical system was put in place to restrict work activities of workers with early signs of WMSDs); (3) ergonomic and engineering changes (e.g., new, adjustable workstations); and (4) organizational change – a steering committee, comprising management, engineers and occupational health specialists, was formed to ensure effective planning and implementation and adequate follow-up. The result of this multidisciplinary intervention programme was a dramatic reduction over a three-year period in the incidence of new cases of WMSDs after major engineering an ergonomic modifications were introduced. The results of the programme are positive. However, it is not possible to determine the contribution made by training alone to the overall effectiveness of the intervention programme.

Training is often complementary to other interventions. For instance, Van Velzer (1992) studied the training of supervisors to ensure that they considered the physical demands of the task in question when developing job rotation schedules. The application of this strategy may require that supervisors be trained in the design of rotation schedules so that job rotation is done in a 'healthy' manner (Van Velzer, 1992). The case study reported by McKenzie *et al.* (1985) describes a programme to control WMSDs in a telecommunications manufacturing facility with 6600 employees. The programme was managed by a task force that included managers, medical staff, industrial engineers, industrial hygienists and human factors engineers. There were several components in the programme: engineering controls, training, analysis of medical records and management of restricted workers. Training of all plant engineers and supervisors was used as a means of disseminating knowledge in the fields of ergonomics and health and safety. Engineers and supervisors were taught ergonomics, health and safety principles and techniques for job analysis and redesign, in addition to having WMSDs explained to them. The training sessions were attended by management and safety representatives to demonstrate to the engineers and supervisors the company's commitment to reducing WMSDs. During the training sessions, discussion was encouraged. The active participation of trainees was a positive factor in the training programme, which resulted in a reduction in the number of lost and restricted work days. The training of engineers and supervisors was seen as a factor facilitating the implementation of ergonomic change.

Training can also be the first step in a global strategy to reduce or prevent WMSD problems. Wands and Yassi (1992) described a programme aimed at improving ergonomics and reducing health and safety problems, with an emphasis on worker participation in ergonomic assessments and including an educational facet. By increasing the workers' knowledge of ergonomics and health and safety, the programme empowered them to approach management with their concerns and ideas for improvement. This case study showed that worker training in ergonomics and health and safety generated discussion between workers and management about ergonomic risk factors and ideas for improvement.

Some studies have examined the effectiveness of training programmes aimed at reducing musculoskeletal symptoms, including back problems and WMSDs.

Luopajärvi (1987) described three case studies performed in various environments with a view to reducing musculoskeletal symptoms by improving worker knowledge of health and safety and musculoskeletal symptoms. The objective of the first study was to reduce neck and upper limb disorders among data entry operators. One of the positive elements of this study was that training in health and safety was accompanied by changes in physical working conditions. The short-term effects were assessed through self-reports of WMSD at the beginning of the study and six months after the intervention. In the intervention group, tension neck syndrome decreased from 54 to 16 per cent, whereas the figures in the control group were 43 per cent at the beginning and 45 per cent at the final examination. However, the long-term benefits were reduced as a result of a lack of motivation among workers. The employees' working habits and attitudes remained the same. Involving workers in the planning and implementation of the programme would have fostered worker motivation. In the second study, forest workers with back pain were taught better work techniques, along with principles for preventing musculoskeletal strain. The proportion of awkward unhealthy postures of the back, upper and lower limbs decreased. Two positive features of this programme were: (1) the trainer was a forest worker; and (2) videotapes of actual work situations were used in the training. The third study involved the distribution of 'self-analysis' guidebooks to workers so that they could observe and analyze their own workplace. The guidebooks provided guidelines to the workers for the observation and analysis of their workstation, equipment and working methods. This method improved workers' knowledge, as well as their motivation to apply the ergonomic guidelines.

These three case studies were not aimed specifically at WMSDs. However, we can draw important conclusions from them regarding WMSD-related training. They all show the importance of:

- worker involvement in the design of training programs;
- an emphasis on active participation and learning by doing, and other newer methods such as cognitive training (see subsection 8.3.2 for definition of cognitive training);
- using workers as trainers;
- the use of audiovisual techniques;
- on-the-job training, as opposed to classroom teaching.

8.2.2 Literature available in similar areas: training in materials handling

In light of the scarcity of literature on WMSD-related training, we will review a number of studies on training in materials handling as a strategy for reducing or preventing low back problems. After reviewing this research evidence, we will draw lessons gleaned from the literature on training in materials handling and applied to WMSD-related training.

Kroemer (1992) reviewed the research evidence relating to training in safer materials handling. There are three types of training in materials handling: (1) training aimed at increasing knowledge of biomechanics, for instance; (2) training aimed at improving handling methods and skills; and (3) training the body via physical fitness to decrease its susceptibility to injury. Although a number of programmes have been developed to train workers in safer materials handling, very few have been systematically and scientifically evaluated (Kroemer, 1992).

There have been many studies in hospital facilities, most of them showing that 'counseling' on materials handling has very little impact on the number and severity of musculoskeletal symptoms (Dehlin *et al.*, 1981; Buckle, 1982; Stubbs *et al.*, 1983), and on objectively measured back loads (Scholey, 1983; Stubbs *et al.*, 1983). A study by Snook *et al.* (1978) showed that training in materials handling did not have any significant effect on the severity of self-reported back problems. With regard to the level of knowledge acquired, the evaluation of training in materials handling is positive. Subjects correctly assimilate the concepts taught (Hultman *et al.*, 1984).

In 1971, Brown suggested that, despite extensive information campaigns, very few workers used the 'straight back-bent knees' technique. Nowadays, training content and methods are more diversified and elaborate. Studies show, however, that most workers do not adopt the recommended positions when handling materials and people (Chaffin *et al.*, 1986; Hale and Mason, 1986; Wachs and Parker, 1987; St-Vincent *et al.*, 1989). This lack of success of training programmes has even been observed among trainers. For instance, Hale and Mason (1986) found that, following a 35-hour training course, only 10 per cent of the trainers-to-be could reproduce the 11 principles taught.

The reasons for this lack of success fall into four categories:

1. *The training period is too short.* The learning of appropriate behaviours requires more time than is normally allotted for training programmes. A study by Gagnon and Lortie (1987) showed that, after 120 trials, unexperienced subjects could not perform one simple task according to certain principles.
2. *It is difficult to transfer the training to the actual work context.* Many studies have demonstrated that the work context is very rarely compatible with the principles or behaviours taught in training (Park and Chaffin, 1974; St-Vincent *et al.*, 1989; Ayoub, 1982; Garg *et al.*, 1992). In addition, workers may tend to revert to old work habits if the principles are not reinforced or refreshed (Kroemer, 1992).
3. *The validity of the principles taught is questionable.* With the development of more-sophisticated biomechanical models, there is less and less consensus on 'appropriate' materials handling principles (Authier and Lortie, 1992; Kroemer, 1992).
4. *Training is not a panacea.* Encouraging safe and healthy behaviours may not eliminate the risk of musculoskeletal problems. As noted by Kroemer

(1992, p. 1131), 'designing a safe job is fundamentally better than training people to behave safely'.

Training in materials handling has not been successful thus far. From the literature on training in materials handling, we can draw a number of conclusions that could apply to WMSD-related training:

- training aimed at improving knowledge and attitudes does not necessarily lead to changes in behaviour;
- the work context must be taken into account when deciding on training content and methods;
- the broader organizational context must be considered: reinforcement, production pressures, support for proper behaviour;
- there should be a systematic, rigorous approach to developing training programmes: needs assessment, design of training, implementation and evaluation (see section 8.4);
- time factors must be taken into account: reinforcement, refresher training and adequate time to understand material and master new skills;
- training is not a panacea for addressing all risk factors, but rather a complement to other interventions.

8.2.3 Literature available in similar areas: VDT/office ergonomic training

Several guidelines and standards have been developed for VDT/office ergonomics, with a view to increasing employers' awareness of ergonomics and helping them in the procurement of VDT/office equipment and the design of computerized offices. Such guidelines are not sufficient to ensure that computerized offices are ergonomically designed and that workers do not suffer from working at a VDT workstation. Although experts in the field of VDT/office ergonomics have long recognized the need for training (Sauter *et al.*, 1985; Smith *et al.*, 1992), it is not included in the 1988 ANSI/HFS Standard for VDT Workstations.

Several training needs have been emphasized regarding VDT/office ergonomics. First, employees should be taught how to use the adjustable equipment. It is not sufficient to provide employees with ergonomic chairs and tables, for instance. They must be taught how to use them. Actually, some ergonomic chairs have many adjustments that employees may not know how (or why) to use. A study by Green and Briggs (1989) demonstrated that providing keyboard operators with adjustable workstations had very little effect on musculoskeletal disorders. The authors argued that the negative results could result from a lack of appropriate information given to the keyboard operators using the adjustable equipment. Training can be seen as a necessary component in ergonomic redesign. Second, employees should be 'sensitized' to ergonomics. This need is especially important if we expect employees to know how to use their VDT workstations in a 'healthy', efficient way. Third, it is important that supervisors learn about ergonomics too, so that they can encourage 'healthy' habits among their employees (e.g. encouraging employees to take mini-breaks) (Green and Briggs, 1989). Fourth, specialists may also need training, and purchasing department employees should be involved in the ergonomics training as well, since they require a knowledge of ergonomics in order to choose appropriate equipment.

These two examples of training (materials handling and VDT/office ergonomics) show the diversity of subject matter and trainee needs that must be addressed. They also illustrate that a holistic and systematic approach is necessary for effective training.

8.3 Special focus on various training methods discussed in the literature

Training methods may be defined as the way in which the training is actually conducted. The following are examples of training methods: cognitive training, the five-step approach, train-the-trainer, skills training, behaviour modification, training by doing and observational training.

We will examine two examples of training methods that can be used to train workers. The first method, the *train-the-trainer approach*, is particularly appropriate when a large number of workers need WMSD-related training (e.g. on risk factors, description of WMSDs and interventions). In the train-the-trainer approach, a group of workers is selected to train other workers. The members of the first group are given training and then go on to become trainers themselves. The active role of workers in providing training can be a good motivational factor. The second method described is the *five-step approach* developed by Vartiainen (1987). It is a particularly useful tool for achieving optimal learning because it is a combination of various training methods, drawing on their different strengths. It can be used to train workers on new jobs or tasks and can be applied whenever there is a change in working conditions requiring that workers be retrained to learn the new jobs. It is also useful for training new employees.

8.3.1 Train-the-trainer approach

The rationale of the train-the-trainer approach is that people can be trained to train other people. Knowledge can therefore be transferred at an 'exponential' rate. For instance, one person can train X people and these X people can then train Y people each, for a total of X×Y people trained. The number of people trained by the train-the-trainer approach is much greater than by the traditional approach where one (or several persons) must train every member of the group of employees requiring training.

The train-the-trainer approach was applied at GM under a UAW/GM national agreement. At each plant, ergonomic monitors were designated to conduct ergonomic evaluations and surveillance of their work areas and to work with their supervisors to implement simple job improvements. At each plant, an ergonomic co-ordinator was designated as the primary in-plant trainer. These ergonomic co-ordinators were taught about ergonomics and WMSDs. Then, they were assigned to train the ergonomic monitors and supervisors in ergonomics. The results showed that this train-the-trainer approach was effective for transferring introductory ergonomics knowledge to the plant (Silverstein, 1991).

The advantages of the train-the-trainer approach are as follows:

- Efficiency. The total number of people trained in a given period of time can be quite large, as compared with the single-trainer approach.
- Employee involvement. Selected employees are trained as trainers. Employee involvement can be a success factor for the training programme. Furthermore, employee involvement can increase their self-esteem.
- Job enrichment. The trainee-trainers, i.e. the employees trained to become trainers, have their jobs enriched and gain unique expertise in ergonomics.
- Increased problem-solving skills. The trainee-trainers can learn more by teaching ergonomics to their colleagues. They acquire knowledge about ergonomics and are able to solve ergonomic problems before they become too serious. This is especially important for WMSDs because of the time it takes for many risk factors to have an effect on workers. Prevention of problems is emphasized because workers know more about ergonomics.

The disadvantages of the train-the-trainer approach include:

- Loss of information in the transfer. Since the original source of knowledge (the original trainer) is remote from the employees who are trained, communication problems may arise and information may be lost or misinterpreted.
- Backfire. The trainee-trainers may feel overly confident and believe that they can solve ergonomic problems by themselves. They may not be able to recognize when they need expertise to help solve a problem.

The train-the-trainer approach seems effective for introductory ergonomics. It is an efficient means of training a large group of employees. Follow-up 'refresher' sessions on a regular basis are needed to increase the trainers' knowledge, clarify technical issues and provide support (Silverstein, 1991). The trainee-trainers need to have access to a resource person, i.e. an in-house or outside expert in ergonomics who can provide help and support when necessary.

8.3.2 Cognitive training and the five-step approach

Cognitive training is a training method in which the worker is trained by mentally analysing the tasks without actually performing the tasks. The rational is to help the worker develop a mental model of the tasks before its actual performance. The cognitive training method has been used in training of assembly tasks.

Vartiainen (1987) studied the benefits and drawbacks of different training methods by conducting both laboratory and field experiments. The results indicated that the traditional, broadly used method of learning-by-doing was not as effective as training methods with a cognitive emphasis. Laboratory experiments showed that groups participating in cognitively oriented training with various instruction methods learned the task in 20–36 per cent of the time needed by the group that learned by actually doing the task (Vartiainen, 1987).

The use of combined training methods has several advantages compared with the exclusive use of learning-by-doing. First, machine time, tools and materials are saved. Learning-by-doing requires workers to use machines, tools and materials, whereas other training methods (e.g. simulation) do not. Second, when a combination of training methods is used, training can be given to groups, as opposed to individuals, thereby lowering the cost of training. Third, mental strain during the learning process is significantly reduced when a combination of different training methods is used.

The five step training approach is a systematic training programme that takes into account the different stages in learning a skill and combines various training methods (Vartiainen, 1987). The five steps are as follows (see Figure 8.1):

Step 1. *Orientate the trainee and specify the goal.* In order to learn, the trainee must be motivated. The manner in which the goal and the entire goal structure of the task are demonstrated is of crucial importance. The goal can be communicated verbally, but also with drawings, videos or slides. This stage of the training process aims at encouraging the trainee to develop a preliminary mental model of the task and do some initial mental work.

Step 2. *Teach.* The trainer demonstrates how the task should be done and, beforehand, afterward or simultaneously, gives general or specific rules concerning the execution of the task. Heuristic rules are given concerning the most progressive phases of the work, the essential tricks of working, problems or facts hindering working and factors affecting workload and safety. The trainee observes and analyses the way in which the task is executed.

Step 3. *Have the trainee rehearse mentally.* This stage of learning a skill is the one that deviates the most from the traditional learning-by-doing method. In this phase of the learning process, the trainee is first asked to verbalize the goals and main phases of the task (verbal training). The trainer monitors and corrects, if necessary. Second, the trainee verbalizes the rules given and, again, the trainer makes any corrections necessary. Third, the trainee repeats the task mentally by internal speech (cognitive training).

In this part of the learning process, the aim is to develop a higher intellectual level of regulation. The trainee's skill level is estimated before he or she actually performs the task, by analysing the content of the trainee's externalization of the task. This is considered especially important in work processes involving dangerous, valuable or fragile machines and tools.

Step 4. *Try out the skill.* The trainee does the work and the trainer monitors his or her performance, points out any weaknesses in the trainee's task execution and gives positive feedback as appropriate. The actual work task can be replaced by simulations when there are risks to the worker or to the production process. Thorough repetition of the task is recommended. A large task entity can be divided into small parts to be learned in sequence. Only at this stage of the learning process does the trainee become physically active, even if the task is a mainly physical one.

Step 5. *Inspect.* The trainee works independently. The trainer observes his or her performance from time to time, offering advice and positive feedback. It is assumed that the trainee will continue to gain self-confidence, autonomy and responsibility.

Steps 3 to 5 are repeated until the trainee achieves a previously determined level of skill.

There are, however, some problems in the application of this training method. If some phases of the training process are neglected, learning is hindered. Especially when learning simple tasks, trainees may consider mental training unnecessary. It has also been noted that relatively inexperienced trainers may be embarrassed to ask trainees to analyze the task verbally (Vartiainen *et al.*, 1989).

With regard to worker health, steps 2 and 3 of the training process are the most important. Trainers should be able to give trainees the heuristic rules concerning workload, e.g. optimal working movements and postures. During the mental training, images of these postures and movements, as well as their rules, should be verbalized.

The five-step approach emphasizes the development of job skills based on the worker's mental representation of the job. The trainer develops images of tasks, activities, postures and movements before actually performing the job.

8.4 Designing a training programme

Previous experience with safety training programmes (Hale, 1984) and ergonomics training programmes (e.g. Kroemer, 1992) shows that the success of training programmes is largely dependent on a systematic and rigorous

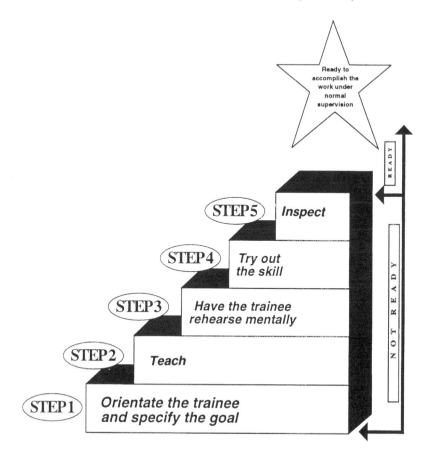

Figure 8.1 The five-step method for training (adapted from Vartiainen, 1987).

approach to their development and implementation. Training development and implementation is an elaborate process in which all steps are essential. Failure to take any of the steps into account in a rigorous way may compromise results and waste resources. This section describes a systematic approach to training development and implementation.

The model for training development and implementation consists of four phases: needs assessment, training development, training implementation and evaluation/follow-up (Gagné and Briggs, 1978; Goldstein, 1993; Larouche, 1984). Figure 8.2 shows the four phases of the training development model, featuring the following elements (Goldstein, 1993):

- assessment of training needs;
- specification of instructional objectives;
- training programme to achieve these objectives;

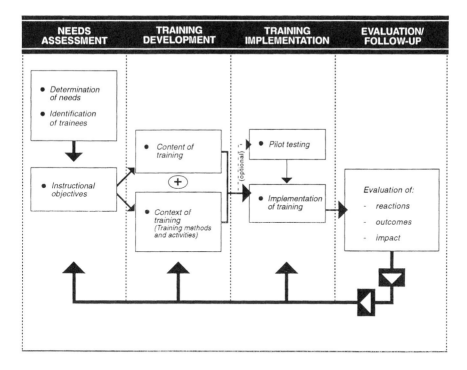

Figure 8.2 The four phases of the development of a training programme (adapted from Goldstein, 1993).

- development of criteria to measure performance (e.g. achievement of objectives);
- systematic evaluation;
- feedback to continually modify training programmes;
- complex interactions among the elements of the training system (e.g. one particular medium may be effective in achieving one objective, one specific method might be adapted to one type of trainees).

The first essential step of the approach involves identifying training needs. The following questions need to be asked: 'Who needs training? What do they need to be taught? Why do they need to be trained? Is training the appropriate intervention, or would some other type of intervention be more appropriate? What objectives does the training effort pursue?'. The answers to these questions are needed in order to design and evaluate training adequately. During the second phase, careful attention should be given to such topics as detailed definition of content, adequate schedule programming, careful choice of training methods and selection of resources (e.g. training materials, trainers). During the third phase, implementation, particularly for a new type of training, a pilot experiment is recommended and monitoring should be done in order to make the necessary changes. Last but not least, evaluation and follow-up are two essential ingredients that are too often overlooked. There are many ways to evaluate a training programme. The right method should be chosen according to the initial training objectives. Follow-up involves taking the appropriate actions to implement the results of the evaluation.

8.4.1 Needs assessment

8.4.1.1 Identifying trainees

The first step in the training development model is an accurate definition of training needs. 'Who needs training?' The very first step in needs assessment is a clear definition of the target audience. The content, methods and schedule will depend upon who the potential trainees are. The identification of trainees is related to the global objective of the training. For instance, if the training is required because of a change in tools, then the target audience will be the workers who use the tools and the supervisors who make sure that the workers are using the tools properly. If the training is motivated by a lack of awareness of the WMSDs problem in a company, then the audience might include not only the workers and the supervisors, but also managers and specialists, such as engineers and health professionals. WMSD-related training is required by several groups in an organization:

- workers,
- supervisors,
- managers/employers,
- staff with specific functions: e.g. procurement officers, and,
- specialists: e.g. engineers, health and safety professionals and medical staff.

Workers are an important target group for training. Supervisors should also receive training to reinforce that given to the workers, to recognize early signs of WMSDs and to help workers get medical help. Managers/employers are also an important audience for WMSD-related training because their commitment is necessary for the success of any intervention aimed at reducing or preventing WMSDs. Procurement officers buy the equipment, tools and technologies used by workers. They should be trained to ensure that the equipment they buy is compatible with the needs of workers to do the work and can be used safely. Engineers design production systems and therefore should be trained to ensure that their designs are in keeping with ergonomic and health and safety principles. Health and safety professionals may be involved in designing health and safety programmes aimed at reducing WMSDs, or may be asked to identify risk factors for WMSDs and design interventions. Training will help them in these tasks. Health and safety professionals may also be involved in the training of other employees. Medical staff should be trained to recognize early signs of WMSDs. They also should know about the rehabilitation and return to work of injured workers. Chapter 9 on medical management discusses general education and training issues for medical staff.

8.4.1.2 Identifying needs

Once the target audience has been identified, we need to measure the difference between what trainees know and/or do and what they should know and/or do, or between the current and desired skills and behaviours. This difference will define the training needs. First, we need to assess what the trainees should know and be able to do or what attitude they should have. The answer to this assessment depends on the global objective of the training, for example, increased productivity, reduced WMSDs, better workstation comfort, reduced worker compensation costs, fewer accidents, fewer job errors, improved motivation and so on. Second, we need to measure the trainee's current levels of knowledge, skills and existing attitudes. A common error in designing training is to assume that potential trainees know more than they actually do. This leaves gaps in content which may render the training useless, or at the very least greatly hamper the results.

In assessing current levels of knowledge or skills, a variety of techniques can be used. In some instances, a job analysis is needed to evaluate different facets of the job. The results of the job analysis can then be compared with a 'standard' job. The discrepancies between the job analysis and the 'standard' job will define the training needs. In other cases, the results of the job analysis are compared with the results of worker analysis. Again, the discrepancies between the job requirements and the worker's skills and knowledge will define the training needs. The results of the job analysis can also be used to identify WMSD risk factors. When part of the analysis is performed by the workers, this can serve two purposes: (1) identifying some WMSD risk factors that need to be reduced or eliminated with different strategies (e.g. ergonomic redesign or training); and (2) training workers to examine their own jobs and define their training needs.

Various job evaluation schemes (e.g. checklists) are very popular for collecting rapid, structured information on workplaces. Checklists were discussed in chapters 4 and 5. However for these tools to be most useful their users need to be trained thoroughly.

Job evaluation schemes are also important WMSD-related training tools. They have several advantages over qualitative methods:

- The organized structure of job evaluation schemes is useful for a non-trained observer. It cuts the job into observable chunks and provides guidance for further analysis.
- The structure of a job evaluation scheme can help organize the content of the training. The job evaluation scheme, along with some complementary material, can provide the basis for the training.

- The structured form of the schemes makes them appealing to technically oriented individuals. Division of the job/workplace into items and quantification of observations is similar to other tools and methodologies used by industrial engineers. Accordingly, these schemes are more readily accepted by engineers, who can use them to identify WMSD-related training needs.

Some of the newest job evaluation schemes focus on changes to the workplace (Thurman *et al.*, 1988). Several studies have shown the effectiveness of this approach (Kogi, 1985): with minimal training, participants can identify areas for improvement in the workplace and propose feasible solutions.

Other methods of assessing needs include holding interviews and meetings. Interviewing or meeting with the persons targeted by the training can be an effective way of identifying training needs.

8.4.2 Training development

Whether or not the training will produce the desired results will depend on a large number of factors. These factors fall into two categories: contextual factors and design factors.

8.4.2.1 Contextual factors

Contextual factors include mainly environmental and social factors. Some of them can be conducive to success in training, while others may present major obstacles. For example, the training needs of and methods used for newly hired and experienced workers may be quite different. Some studies have shown that the work habits of experienced workers are difficult to change (Parenmark *et al.*, 1988). Besides experience on the job, other individual characteristics need to be considered in the design of training programmes.

The organization can be seen as a social environment in which the training takes place. A preliminary analysis of the values regarding the training of the organizational members affected by or having an influence on it can shed light on the potential success (or failure) of the training. The following questions should be answered: 'Does the organization value training?' 'Is training seen as "rewarding"?' Negative answers to these questions may mean that education with regard to the value of training is necessary before any training programme is implemented.

8.4.2.2 Design factors

Many training design factors must be taken into account if the training effort is to produce the expected results. The first and most obvious is the identification and development of training content, but special attention must also be paid to choosing the right training methods and activities. Once the layout of the training is planned, the following design factors must be considered: human and material resources, location of training, trainers, training planning and schedule. The training programme should be designed so that it facilitates and encourages learning. The design features should be in keeping with the learning principles presented in Box 8.1 (Goldstein, 1975).

THE PLACE OF TRAINING WITHIN THE GLOBAL STRATEGY
It is important to realize that training is not done in a vacuum and to keep this in mind while developing the training programme. It is part of a more global strategy for reducing or preventing WMSDs. Efforts should be made to ensure that the training developed is compatible with the other components of the global strategy. WMSD-related training may also need to be integrated with other types of training, for instance occupational health and safety training or job skill training. The purpose of such integration is to ensure that there is no contradiction between the WMSD-related training and the other types of training. Commitment of the entire organization (management, supervisors, workers, workers' representatives and specialists) to the training is important.

TRAINING CONTENT
An important step in designing a training programme is to determine the exact content of the programme as a whole and/or the various training sessions. First, the content is devised and then validated. Validation is twofold: the content is tested against the training objectives, and the parties concerned are consulted to ensure that all essential elements are covered. Training content for some of the various groups of individuals in an organization is summarized below.

Workers can be trained in a variety of topics, ranging from general training (or education) about the importance of WMSD problems, to more specific issues such as workstation design.

Workers:
- basic instruction in the structure and functions of the human body at work;
- workplace biomechanics and physiology and the identification of occupational risk factors;
- work methods for control/prevention of occupational WMSDs;
- early recognition of WMSDs;
- information on obtaining medical care and reporting musculoskeletal disorders to supervisors or management;
- roles and responsibilities in co-operating with a recommended medical treatment programme and returning to work;
- understanding and managing the change process, including their role.

Recommended topics for training managers and supervisors include many of the same topics as for workers, along with the repercussions of WMSDs on the organization and ways of introducing ergonomic changes in the organization.

Box 8.1:   Learning principles (Goldstein, 1975) to be considered in the design of a training programme

- *Transfer and testing similarity.* To facilitate maximum positive transfer, the training environment should be designed to provide similar stimulus conditions and appropriate responses necessary for on-the-job performance.
- *Transfer and understanding.* Training programmes should be designed to increase positive transfer by presenting the underlying principles governing the solution of problems.
- *Transfer and practice on the original task.* In order for positive transfer to occur, there must be sufficient practice on the original task. If only limited practice is permitted, there is a possibility that negative transfer will result. This principle may be especially true for complex procedures, where it appears necessary to provide for extensive practice on the first part of a series of tasks.
- *Motivation.* Motivation of trainees to perform safely can be increased by making rewards explicit and contingent upon acceptable performance.
- *Feedback.* Knowledge of results serves a motivational function and a guidance function. As a general rule, training aids should be designed to provide maximum and immediate feedback to the trainees concerning the adequacy of their performance. It is also important to determine that the feedback is being used by the trainees.
- *Guidance toward correct responses.* Guidance should be provided in the learning process by physical, verbal or visual cues that are as similar as possible to operational cues. This is particularly useful in the early stages of the learning process, where it may be necessary to amplify critical cues not easily distinguished by trainees.
- *Opportunity for correct response.* Training time is best used by allowing practice of the correct response with feedback to inform the trainees about their performance.
- *Mental practice.* Mental practice, where the employees mentally rehearse procedures, can serve as an addition to the training process.
- *Learning to learn.* Practice on a variety of tasks can be expected to lead to increased facility in solving similar problems.
- *Massed vs spaced practice.* In general, the use of spaced (over time) rather than massed (all at once) practice is preferable.
- *Part vs whole task practice.* The following guidelines govern the determination of part versus whole practice: (a) if the size or complexity of the task is great, the part method of training is preferred; (b) if the task can be easily divided into discrete steps, the part method is preferred; (c) if parts of the task are highly interdependent, the whole method is preferred; (d) eventually the part tasks must be combined into the whole task.
- *Stages of training.* At each stage of the training, trainees should not be required to perform tasks that are beyond their capabilities. Trainees should be required to perform increasingly difficult tasks only after mastering the necessary simpler tasks. Prompting and feedback should be provided at each stage of the training.

Box 8.1: Learning principles (Cont.)

- *Verbal labelling.* Early in the training programme, trainees, who are being presented with unfamiliar environments, should be given verbal pre-training where they are taught the names of unfamiliar objects, practices and procedures.
- *Use of learning time.* Most training time should be spent on those tasks that are difficult to learn.
- *Over-learning.* Tasks or procedures that may have to be performed in emergencies or stressful situations should be practised until they become over-learned. Such over-learning increases the resistance of task performance to disruption or interference under stress.
- *Learning and individual differences.* Training programmes should be designed to take into account individual differences such as initial performance levels, age, gender and abilities.

Supervisors:
- ergonomics training;
- basic instruction in the structure and functions of the human body at work;
- workplace biomechanics and physiology and the identification of occupational risk factors;
- work methods for control/prevention of occupational WMSDs;
- early recognition of WMSDs;
- information on medical referral and obtaining workplace analysis;
- roles and responsibilities in return-to-work procedures, restricted duties and accommodating injured workers;
- understanding and managing the change process, including their role.

Managers:
- need for management commitment: commitment to reducing or preventing WMSDs and supporting injured workers; for instance, supporting training of supervisors and workers, set-up of WMSD prevention programme; recognition of the work relatedness of WMSDs; reinforcement of health and safety rules and principles;
- overview of ergonomics and WMSDs within the context of company overall health and safety policies, company work related injury compensation experience and costs, and professionally accepted standards, guidelines and regulations;
- macroergonomics and systems issues;
- management of change.

Procurement officers:
- ergonomic principles in the design of tools and technologies.

Engineers:
- ergonomic principles applied to the design of production systems;
- techniques for identifying WMSD risk factors and high-risk jobs;
- job redesign methods;
- process for involving workers in participatory engineering design and review.

Medical and health and safety professionals:
- work methods that contribute to the development of WMSDs;
- early detection of WMSDs;
- patient follow-up;
- returning injured workers to work (placement techniques);
- training for engineers, supervisors and managers.

More information on the educational and training content for medical management professionals is discussed in general terms in chapter 9 on medical management.

Training is also needed for those employees responsible for a surveillance system. This training should include a review of the criteria for interpreting surveillance results, such as:

- size of differences in surveillance results (magnitude), e.g. between departments;
- stability of estimate calculated (statistical testing);
- plausibility or reasonableness of conclusion;
- consideration of other factors that may explain results;
- utility of having the analyses and their interpretation reviewed by an experienced analyst.

TRAINING METHODS/TECHNIQUES
Various training methods can be used, such as:

- simulation,
- role playing,
- group discussion,
- on-the-job versus classroom,
- 'hands on' versus cognitive,
- orientation versus skill training,
- practical versus conceptual.

Goldstein (1993) describes the strengths and weaknesses of these various training methods. The method selected will depend on:

- time available for the training,
- number of participants,
- training objectives,
- effectiveness of the method,
- budget.

Training can be administered and scheduled in different ways:

- mass (all at once) or spaced over time,
- pre-employment training; teach new employees proper work methods and habits instead of letting them pick up incorrect methods on the job.

TRAINER
Several types of trainers can be used:

- use of selected workers as trainers (Smith *et al.*, 1978; Spokes, 1986),
- transfer of expertise/in-house trainers/trainee–trainers (Silverstein, 1991).

MATERIAL RESOURCES
The training materials given to trainees should include:

- subjects to be covered, summaries,
- description of methods,
- detailed timetable and names of participants,
- visual and audiovisual aids to be used,
- description of assessment techniques,
- description of follow-up.

8.4.3 Training implementation

Implementation of training normally requires a simple application of what has been prepared at the design stage. But two specific considerations merit attention here. The first involves pilot testing or pre-experiments. Even with extreme care at the design stage, there is no absolute guarantee that a training session will be adequate right from the start. Therefore, it is recommended that training be initiated on a small scale and monitored closely. This can help in making any necessary adjustments before the training programme is launched on a large scale. Pilot testing is especially important when a new or revised training programme is implemented. Second, training needs to be active/interactive. One way of making training active is to have trainees discuss the present situation in their organization. A training session can be an ideal opportunity for management, supervisors and workers to discuss certain problems together. The trainer will need to lead and control the discussion. Because of the potential benefits of such discussions, mixed audiences with representatives from different hierarchical levels are recommended. For instance, supervisors and workers can be trained together so that they know what each other learned. This can decrease mistrust and foster a climate of open discussion.

Involving workers in the training is another way of making training active. Workers can be involved in identifying training needs. They can define their training needs by examining their own jobs using workplace analysis schemes, for instance (as discussed in section 8.4.1.2). Workers can also be involved in running the training. The train-the-trainer approach described earlier is an example of how workers can be used as trainers. Worker involvement in training can take different forms and be very effective for the following reasons:

- increased motivation to learn and to apply the concepts taught;
- improved identification of training needs and training content based on 'local' experience and knowledge of the job and the work context (Kilbom, 1988);
- improved satisfaction and self-esteem, and reduced stress.

8.4.4 Evaluation/follow-up

Training evaluation is the systematic collection of information regarding the success of training programmes (Goldstein, 1993). There are three types of evaluation:

1. *Evaluating reactions* involves assessing trainees' attitudes towards the training they have received. Typically, it aims to obtain answers to such questions as: Did the trainees like the training? Would they recommend it to their colleagues? Do they feel the training will be useful in their jobs?
2. *Evaluation of outcomes* is designed to find out whether the training actually changed something in the trainees: Did it increase the trainees' knowledge? Did it modify their attitudes towards the object of training? Did it develop their skills, improve their abilities, change the way they do their jobs?
3. *Evaluation of impact* is concerned with overall results. The relevant questions are: Did it change something in the workplace? Can results be seen in the organization? Has the training produced improvements in the company? Are there fewer WMSD problems?

Measuring reactions is the type of evaluation that can be performed the soonest after the training. Some outcomes (e.g. knowledge) can also be measured shortly after the training is over. Other outcomes may not be noticeable right away. It will take some time before impacts on the organization are seen. Measuring the impact should be done last, and only after sufficient time has elapsed to allow effects to set in. Evaluation of impact can be tied to the surveillance system (see chapter 5). Since training may be only one element of a global strategy, evaluation of the specific impact of training (as distinct from the impact of other interventions) may not be possible. For instance, if training is offered to safety professionals to teach them how to set up a surveillance system, we cannot expect the training to have any direct effect on the rate of WMSDs.

Training can have an indirect impact on the organization. It is nonetheless important that these effects be measured.

Training must be followed up and may require changes or adjustments depending on the results of the evaluation. Current training programmes need to be re-evaluated whenever some change occurs, such as a change in tools or technologies or in the demographics of entry-level workers. Refresher training may be necessary to ensure the long-term success of the training. Evaluations should be repeated from time to time to examine the long-term effects of the training. If the effects disappear over time, refresher training is needed.

A detailed discussion on training evaluation is beyond the scope of this book. The reader is referred to various texts on this topic (e.g. House, 1980).

8.5 Summary

Training can be seen as part of a larger strategy to reduce or prevent WMSDs. This chapter discussed WMSD-related training programmes. There is little literature available on WMSD-related training programmes as such and that which does exist usually looks at training as only one of the possible interventions. Accordingly, there is too little evidence to evaluate the specific impact of WMSD-related training. However, information, data and lessons can be drawn from other areas which are relevant to WMSDs: training in materials handling and VDT/office ergonomic training. The literature on two methods of WMSD-related training was also examined: the train-the-trainer and five-step approaches. The pros and cons of these approaches were discussed.

Overall, training should be multifaceted: multiple needs and objectives, multiple audiences, multiple methods, diversity in subject matter, etc. To be effective, a training programme should be designed in a systematic and holistic way. The four-phase approach to developing training briefly presented here should be used. The entire organization should be involved in the training to ensure consistency. For instance, jobs should be designed so as to allow workers to apply and use the principles and concepts taught in the training. Training is not a panacea for WMSD problems. It should be seen as a complement to other strategies aimed at dealing directly with risk factors.

9

Medical management: an overview

9.1 Introduction

Medical management of WMSDs involves many aspects, some more specifically medical than others (e.g. diagnosis, treatment, rehabilitation). This chapter will not treat all aspects of the subject in depth.

'Medical management' can mean different things to different professionals in different countries. However, all will agree that in order to correctly identify and control problems related to WMSDs, access to the following expertise is thought to be essential:

- expertise in WMSD diagnosis, treatment, rehabilitation and return to work;
- expertise in WMSD surveillance;
- expertise in the surveillance and control of risk factors (at both the micro- and macroergonomic levels);
- expertise in training and education;
- expertise in management and leadership, with regard to WMSD-related aspects.

Whether these types of expertise all come under the umbrella of 'medical management' will depend on the firm's resources, training of personnel and the point of view of the various professionals involved. Although it is clear that some of this expertise will come from medical doctors, other expertise could very well be the responsibility of nurses, industrial hygienists with special training, ergonomists, engineers with special training, management personnel with special training, physical therapists, trainers, etc. Because various professionals are likely to be called upon to help identify and solve WMSDs, management and leadership skills will also be required; one individual could be responsible for pulling all this together.

Each firm will want to organize its expertise according to availability of personnel and other resources. For example, in some firms the medical management personnel may take on most of the responsibility, including the more medical aspects and the surveillance of both WMSDs and risk factors, leaving only the responsibility for the control of risk factors to, say, the ergonomists. In another firm, the medical management team may restrict its activities to the more medical aspects and the surveillance of WMSDs, while engineers and management with special training (for micro- and macroergonomic factors, respectively) will look into the surveillance of risk factors and their control. Again, irrespective of who is responsible for what, all those in the group must mesh together to succeed, perhaps using the forum of an 'ergonomic task force' (see subsection 9.3.5). In a small firm, for example, where most of this expertise would not be on hand, the employer may decide to train its personnel to gain the required expertise in certain areas, and to tap into services or organize access to appropriate professionals for the other missing elements. Again, the employer must make sure that some sort of a forum or ergonomic task force exists to bring all these experts (and their information) together.

Obviously, therefore, in any given firm, the team assembled will vary greatly depending on the role of the medical management team, the type of professionals on the team and the training they require. The following text outlines a programme that may be adapted to local resources and needs. Only one approach to medical management is described here among many possibilities.

9.2 Objectives of medical management

The primary objective of medical management in occupational health and safety programmes is the prevention of work related disorders and injuries. The American Medical Association has published the objectives, scope and functions of occupational health programmes and has listed the following as important aspects in the prevention of musculoskeletal disorders (American Medical Association, 1972):

1. Protecting employees against health and safety hazards in their work situation.
2. Evaluating workers' physical, mental and emotional capacity before job placement. Ensuring that employees can perform the work with an acceptable degree of efficiency and without endangering their own health and safety or that of others.
3. Ensuring adequate medical care and rehabilitation for the occupationally ill or injured.
4. Encouraging and assisting with measures for personal health maintenance, including the acquisition of a personal physician whenever possible.

9.3 Some basic activities in medical management: a brief overview

There are many activities besides diagnosis, treatment, rehabilitation and return to work which can be part of medical management. For example, surveillance and other ergonomic activities are essential for the detection and control of WMSDs. In a large firm these activities, with or without others, could be the responsibility of the medical management team; in a small firm, they may be allocated to some outside services, medical or other. This section briefly describes these various tasks. In the approach shown here (one of many possibilities), it is assumed that the medical management team is responsible for these surveillance and ergonomic-related duties.

9.3.1 Passive health surveillance

The medical management team should continuously perform medical surveillance for WMSDs (surveillance is defined in chapter 5). Surveillance may take many forms (under the categories of passive or active) and may vary in depth (level 1 or 2). The collection of WMSD data should be done at the same time as assessment of information concerning exposure at the different workplaces. For more information on health surveillance and how it links to risk factor surveillance, see chapter 5. Also, strategies for early recognition of WMSDs have to be developed by medical management and are discussed in chapter 5.

The use of injury reports for health surveillance is considered a form of passive health surveillance (described in chapter 5) since it is based on records currently in use. 'Injury reports' in this context are any recorded notification of accident or disease in the workplace, including WMSDs. For effective passive health (WMSD) surveillance, the medical management team needs sources of data which have a high sensitivity for WMSDs. Besides workers' compensation claims, any patient admittance to occupational medical care should be regarded as a possible injury report.

Injury reports (WMSDs, for our purposes) should be followed up by workplace visits and an evaluation (checklists are available for this purpose). In a study by Fontus *et al.* (1989) it was pointed out that both patients and physicians were reluctant to request workplace evaluations even if the job history and diagnosis suggested a work-related disease. The advantages of providing feedback to workers and management from a workplace evaluation of an injury report are twofold. It can not only initiate job redesign for the risk factors which caused the problem originally (to prevent the problem from recurring), but also initiate an examination of the whole work process, addressing a wide variety of occupational risk factors. A follow-up programme of injury (WMSD) reports usually initiates ergonomic changes. These changes require ergonomic expertise which can be provided by the medical management team. Thus, an injury follow-up programme will help to secure and develop the ergonomic competence of the medical management team.

9.3.2 Periodic medical surveillance (or active health surveillance)

Surveillance may take many forms, as already mentioned: passive or active, and may vary in depth (level 1 or 2) (see chapter 5). The medical management team should know when and how to perform periodic medical surveillance [synonymous with 'active health (WMSD) surveillance' used in chapter 5].

In a population of workers, or in a specific job category where there is a high risk of WMSDs, it may be appropriate to provide periodic medical evaluation to identify those individuals in the early stages of a progressive disease. These individuals can be the target of early secondary prevention, i.e. treatment. One example of medical surveillance can be found in the recommendations for periodic medical surveillance of vibration-exposed workers, as proposed by NIOSH, to identify individuals with vibration syndrome (NIOSH, 1989). For WMSDs and their associated risk factors, one approach to surveillance, including periodic health surveillance for WMSDs, is suggested in chapter 5.

9.3.3 Plant walk-throughs and risk factor surveillance

Working to achieve prevention of WMSDs requires that the medical management team be thoroughly familiar with the various job requirements and the workers' capacity. It is essential to remember that the strategy for successful ergonomics is defined as adjustment of the work to humans, not the opposite. Plant walk-throughs should be performed by the medical management team on a regular basis. Basic procedures and checklists for plant walk-throughs are available (Keyserling *et al.*, 1992; Kornberg, 1992). This type of surveillance is discussed in detail in chapter 5 under 'level 1 active risk factor surveillance', i.e. a quick assessment of the risk factors in the work situation. The medical management team can also conduct a more in-depth analysis of the work situation, i.e. 'level 2 active risk factor surveillance' in chapter 5. Videotaping is one simple method of performing this type of surveillance. Videotaping of potentially hazardous jobs should be arranged by management to provide information and documentation as a basis for selective job placement, job analysis and job redesign.

9.3.4 Job-skill training programmes

The medical management team should be involved in job-skill training and ensure that ergonomic concepts are understood and used. Production engineers are usually interested in working with the medical management team to help design, implement, evaluate and follow up on job-skill training and, in the process, learn about biomechanics and ergonomics. The medical management team also has much to learn from the production engineers. Outside expertise may be required on occasion: e.g. more in-depth knowledge of ergonomics, specific expertise in designing, implementing and evaluating training courses, etc. Results from job analysis, evaluation of the worker's capacity and optimization of work procedures are essential to design appropriate job-skill training programmes. Parenmark *et al.* (1987) examined a job-skill training programme at an assembly plant. The programme had been designed by the production engineers and the medical management team and used biofeedback. It was shown to be effective in reducing both overall absenteeism and absenteeism caused by musculoskeletal disorders in an assembling industry (Parenmark *et al.*, 1987).

9.3.5 Ergonomics task force

The medical management team can take the initiative to form a multidisciplinary ergonomics task force (ETF), the main objective being to promote a healthy and safe work environment along with efficient productivity. The ETF cannot neglect the concern for efficient productivity, since generating increased profit is the primary motive behind job and organizational redesign. Productivity and a safe work environment need not be opposing objectives. For the ETF to accomplish its interdisciplinary work, the team would ideally comprise representatives from the medical, health and safety, engineering and managerial departments as well as labour representatives. Lutz and Hansford (1987) present an ergonomics programme at one company that illustrates an example of an ETF, its possible composition and role.

9.3.6 Ergonomics project management

The medical management team should take an active role in promoting and initiating ergonomics projects, obviously within the mandate it has received and in conjunction with the ergonomics task force (if such a task force exists). As well, if required, it should have the competence to be able to manage these projects. Being active in initiating and counselling participatory ergonomics projects is seen as a vital role for the medical management team and requires that the members have management and leadership training. The success of medical management in a firm ultimately depends on the team's creativeness, ability and initiative for improving the work environment.

9.4 Medical management and some principles of diagnosis and treatment for WMSDs

Diagnosing and treating WMSDs are responsibilities of medical management, regardless of whether this work is performed 'in house' in a large firm with its own medical management team or by outside services in the case of a small firm. A very brief overview of these responsibilities is presented in this section, including some aspects that are sometimes overlooked or which may require more attention. A separate report is also being prepared by the Institut de recherche en santé et en sécurité du travail du Québec, (IRSST – Québec Research Institute on Occupational Health and Safety, Canada). It is intended as a practical physician's guide for diagnosing WMSDs of the upper extremity. The guide reviews the clinical and work relatedness aspects of the diagnoses, and explores the medical differential diagnosis in parallel with the contribution of activity-related risk factors. Some of the ideas presented in this chapter are discussed in greater detail in the guide.

9.4.1 Diagnostic aspects

There is a need for uniformity in the diagnostic terminology, locally as well as nationally and internationally. Case definitions used in clinical and paraclinical examinations (i.e. for clinical purposes) are not necessarily the same as those used in questionnaires and paramedical examinations (i.e. in studies). Even clinically, there may be discrepancies among clinicians regarding the exact definition of certain diagnoses (e.g. in shoulder tendinitis). Also, between studies there are differences in case definitions used. Standardization of diagnoses is required at many levels.

Examples of various disorder classifications which could help establish consensus standards for the diagnosis of WMSDs are found in the literature. The Québec Task Force Spitzer committee (Spitzer *et al.*, 1987) proposed a classification of activity-related spinal disorders, but it is uncertain whether it is now widely used. In review articles, single authors have proposed criteria for WMSDs, mostly for epidemiologic studies or medical textbooks (Brooke, 1977; Waris *et al.*, 1979; Hagberg, 1987; Hagberg and Wegman, 1987; Kuorinka and Viikari-Juntura, 1982; McRae, 1983; Neer, 1983; Silverstein, 1985). The American Association of Hand Surgeons has published the book *The Hand – Examination and Diagnosis* (American Society for Surgery of the Hand, 1990), but specific diagnostic criteria for hand disorders were not presented. The American College of Rheumatology has proposed standards for the diagnosis of primary fibromyalgia (Wolfe *et al.*, 1990); principles for the classification of work related neck and upper-limb disorders have been presented by Kuorinka and Viikari-Juntura (1982); epidemiologic screening criteria for diagnoses of neck-shoulder and upper-limb disorders were proposed by Waris *et al.* (1979); and Silverstein (1985) proposed criteria for upper-extremity cumulative trauma disorders in industry.

Before national and international consensus standards are established for musculoskeletal diagnoses, we recommend that criteria proposed in standard orthopedic and occupational medicine textbooks also be reviewed and considered, for example *Physical Examination of the Spine and Extremities*

(Hoppenfeld, 1976); *Orthopedic Neurology* (Hoppenfeld, 1977); *Orthopaedic Examinations* (McRae, 1983); and Textbook of *Occupational and Environmental Medicine* (Rosenstock, Cullen, 1994).

9.4.2 Therapeutic aspects of WMSDs

In the acute phase, the medical treatments for WMSDs are usually similar to those used for non-work related disorders. The following general therapeutic objectives may be of some guidance:

- promote rest for the affected anatomical structures;
- diminish spasms and inflammation;
- reduce pain;
- increase strength and endurance;
- increase range of motion;
- alter mechanical and neurological structures;
- increase functional and physical work capacity; and
- modify the work and social environment.

Standards of medical care change over time. It is the responsibility of the treating clinician to select treatments consistent with current science and practice at the time care is provided. Recently, a few treatment algorithms have been published; these could be of use for many practitioners (Spitzer *et al.*, 1987; Ambrose *et al.*, 1990; Uhthoff and Sarkar, 1990).

9.5 Medical management, rehabilitation, return to work and work hardening

Medical management includes the responsibilities of rehabilitation, return to work and work hardening, irrespective of whether these are carried out in house (in a large firm with a medical management team) or by outside services (in a smaller firm). The following section is a very brief overview of these responsibilities and emphasizes aspects which are sometimes overlooked.

9.5.1 Rehabilitation and return to work

Occupational rehabilitation refers to the process in which an injured worker follows a specific programme that promotes healing and helps him or her return to work. The injury may not always be just physical; in fact, psychosocial (at work and outside work) and psychic disability may constitute the main problem for the rehabilitation process. Many countries have legislation that establishes standards to encourage the employer to find ways to accommodate the disabled worker rather than to throw obstacles in the way of his or her employment, for example the *Americans with Disabilities Act* (St. Clair and Shults, 1992).

Part of the medical management team's role in occupational rehabilitation programmes is as job-placement experts. Since the team should be well aware of the different job tasks and work requirements at their firm, this gives them the knowledge to set up an early rehabilitation programme for workers with WMSDs. Rehabilitation for WMSDs could follow the outlines of programmes usually employed in sports medicine. After a WMSD occurs, the worker should keep active and in constant contact with the workplace. It is important that the worker be encouraged to adhere to fitness programmes that do not involve the injured anatomical region. Under the guidance of the appropriate experts, only controlled movements of the injured part should be performed; this should be done without an external load. The load on the injured part should be increased in increments, and the worker should return to work gradually. If the demands at work are a possible cause of the development of the WMSD, then a job redesign programme must be developed. It is unethical to send a worker back to work after a work related injury if a workplace redesign has not occurred. Engineering modifications are ultimately a necessary tool in effective medical case management (Lutz and Hansford, 1987).

Selective job placement requires that the medical staff be thoroughly familiar with job requirements and with the worker's capacity. This selectivity may generate an ethical controversy since it implies that the worker is fitted to the job. However, it is essential to remember that the strategy for good ergonomics is defined as adjustment of the work to humans, not the opposite. As a general rule, all jobs should be designed to fit at least 95 per cent of the working population, both men and women, and be adjustable to fit all workers. In a very small number of occupational tasks, specific, selective job placement requirements may be placed on the worker.

The medical management team has to instil confidence. The injured worker should believe in both the medical competence of the health team, and the team's ability to set realistic expectations for return to work (Peters, 1990).

9.5.2 Work hardening

Work hardening has been defined by the Commission on Accreditation of Rehabilitation Facilities (Commission on Accreditation of Rehabilitation Facilities, 1989) as 'a highly structured, goal-oriented, individualized treatment program designed to maximize the individual's ability to return to work. The programs use the conditioning tasks that are graded progressively to improve the biomechanical, neuromuscular, cardiovascular, and psychosocial functions with real or simulated work activities'.

Isernhagen (1992) described the main principles of work hardening as: (1) acceptance of the patient at his or her level of function; (2) establishment of the goal of raising the level of function for the purpose of increased abilities in life;

and (3) performance of the rehabilitation process by the patient, but facilitated and assisted by a therapist. The components of work hardening are exercise, aerobic training, education to teach the worker a safer, more productive way to work, and work simulation (Isernhagen, 1992). Other work related behaviour may also be fostered, such as attendance, punctuality and positive work attitude. The success of a work hardening programme depends on an accurate job analysis, and most often job redesign. Work hardening cannot solve WMSD problems due to inadequate job design and poor work organization.

9.6 Information on pre-employment and preplacement screening for medical management

In some companies the medical management team may have a role to play in pre-employment and preplacement screening. In a small firm, this may be done by outside medical consultants. In this section, information is discussed which may help the medical management team (or outside consultant) make decisions about the use of pre-employment and preplacement screening for WMSDs. Overall, it will become clear after reading this brief overview that there is no scientific evidence that screening can predict the development of WMSDs.

9.6.1 Definitions

Screening refers to the application of at least one test (or examination) to individuals with the intent of distinguishing apparently well persons who are likely developing a specific disorder (WMSD in our case) from those who are not. Screening tests are not diagnostic; persons with suspicious screening findings are referred to their physicians for diagnosis (Last, 1988). Screening can be seen to have many roles. The following definitions of pre-employment and preplacement screening are based on those of the American College of Occupational and Environmental Medicine (ACOEM) Committee on Occupational Medical Practice (ACOM, 1990). Pre-employment screening and examination are done before any offer of employment is made. If the applicant is found to have physical limitations that prevent safe and effective performance of the job, then no offer of employment is extended. The result of this examination by the medical team is simply a recommendation to management, not a decision. In the preplacement screening process an employee is examined who has already received an offer of employment. This examination will decide whether he or she should be placed in a specific job. If limitations related to the job are found in the preplacement examination, the company usually agrees to consider the person for other open positions. Again, the result of preplacement screening is a recommendation by the medical team to management, not a decision.

9.6.2 An overview of the literature available

A literature search revealed about 25 articles dealing with preplacement or pre-employment examination with some reference to WMSDs and low-back problems. Most of the authors are physicians, ergonomic consultants or osteopaths.

Some of the articles promote the use of pre-employment/preplacement for these disorders; however, they do not provide any scientific evidence. One article, 'Preplacement low back screening for high risk areas' (Steele and Hoefner, 1988), presents detailed criteria for limits regarding strength and flexibility tests which could be used in preplacement screening. However, there is no scientific basis to support the criteria presented. In another article, 'Revealing a true profile on musculoskeletal abilities', Althouse (1980) presents extensive tests to assess musculoskeletal work functions for pre-employment screening. However, in this article no criteria are presented, nor is there any justification for the different tests presented. Editorials have also been published in recent years proposing criteria for medical limitations; for example, for individuals with prior disk-related back disease there should be no heavy lifting, and for employees with chronic tendinitis there should be no frequent repetitive motions of the wrist and arm (Strasser, 1988). Again, there is no scientific evaluation of the proposed criteria. Even the question of organization of preplacement screening has been addressed by some authors. One article suggests that nurses perform preplacement health assessments (Flight and Schussler, 1976), while another recommends performing preplacement examinations at the same time as periodic monitoring examinations (Goldman, 1986). It has also been pointed out that the pre-employment and preplacement screening can be used by the insurance industry (Guthier, 1986).

Mention should also be made of a special Task Force which was convened by the Health Services and Promotion Branch of Health and Welfare Canada. This Task Force aimed to review the health examinations carried out in the occupational field (Task Force on the Health Surveillance of Workers, 1986). With regard to the examinations of relevance for our purposes, the Task Force did not recommend electromyography, nerve conduction velocity or back X-rays as screening tools.

One test also warrants a special mention: the vibration (Hettinger) test. This test, which is based on hand temperature changes, has been claimed to predict the individual susceptiblity to WMSDs. Originally proposed by Hettinger (1957), the test has been subject to a few studies concerning its reliability and discriminatory power with respect to diseased and non-diseased. No information is available on the predictive value of the test (in a prospective study). Taking into account its low reliability (Kuorinka *et al.*, 1981; Thompson *et al.*, 1987), it is not probable that the test is a useful tool for prediction of WMSDs. Some authors have been more optimistic in this respect (Brown *et al.*, 1985). For the time being, there is no prospective data available on the predictive value of this vibration test.

Overall, it is important to remember that there is no scientific evidence to show that pre-employment and preplacement screening can predict the risk of developing a WMSD. Results from scientific evaluation of pre-employment and preplacement screening have often evoked protests from practising physicians, as seen in 'letters to the editor', for example Bogni (1977) and Lewy (1986). Lastly, mention should be made of the legal status of pre-employment screening. In most countries with anti-discrimination laws the practice of pre-employment screening could come under legal scrutiny.

9.6.3 Screening and medical history

Prospective cohort studies have revealed that a previous history of back pain is a predictor of a new episode of back pain (Biering-Sørensen, 1983; Bigos *et al.*, 1991). It is doubtful, however, whether the criterion 'previous history of back pain' is effective in pre-employment and preplacement screening. In a prospective study of 5649 nurses, Venning *et al.* (1987) found that having a previously reported back injury resulted in a slightly higher risk of a back injury (odds ratio of 1.73). However, in the multivariate analysis, work related factors such as being in the service area attributed a risk of 4.26; the exposure factor of being a daily lifter constituted a risk of 2.19; and the job category of nursing aide compared to registered nurse constituted a risk of 1.77. All the above were work related risks, and these had a greater contribution to back injury than the variable 'having a previously reported back injury'. In a scientific review, Himmelstein and Andersson (1988) pointed out that the indicator 'history of low-back pain' lacks reliability and specificity. Thus, history of a back injury is not an effective approach to pre-employment and preplacement screening for future musculoskeletal problems.

Currently, no study has investigated the use of a previous medical history of WMSDs for pre-employment or preplacement screening. Therefore, no results are available that focus on the relationship between a previous medical history of WMSDs and a relapse, when the worker is exposed to job risk factors associated with these disorders. A worker's medical history of general health issues such as smoking and specific diseases, such as diabetes mellitus or seizures, may be important before placing him or her in a job. When employing smokers, smoking cessation programmes can be started. Employees with a history of seizures should be placed in positions where they are safe, employees with diabetes should be placed in non-shift work, etc.

9.6.4 Screening and strength testing

The assumption behind using strength testing as a pre-employment and preplacement screening instrument is that low strength is a predictor for the development of musculoskeletal disorders. In a prospective cohort study of neck-shoulder disorders (non-specified diagnosis), it was found that among women with jobs that required great strength (engine assemblers), low muscle strength was a predictor of deterioration of neck-shoulder disorders (Kilbom, 1988). In contrast, among female electronics assembly workers with tasks not requiring great strength, muscle strength was not a predictor of neck-shoulder disorders (Kilbom, 1988). In a prospective population study (one-year follow-up), Biering-Sørensen (1984) found that back endurance, but not back strength, was a predictor of the incidence of back pain reports. In the studies quoted above, the correlations between back pain and back endurance and shoulder muscle strength and neck-shoulder disorders were poor. These results do not suggest that muscle strength can be used as an effective screening tool in preplacement or pre-employment screening for musculoskeletal disorders.

If muscle strength testing is to be used as a tool in pre-employment and preplacement screening, it has been claimed that the strength testing has to be designed according to the demands of the task (Snook, 1987). In a study of 20 jobs in a tyre and rubber plant, it was found that the incidence of musculoskeletal injuries (back, shoulder, neck, etc.) in employees who were selected using strength testing was approximately one-third that of employees selected using traditional medical criteria (Keyserling *et al.*, 1980). The isometric strength tests used involved the back, arm and shoulder. However, this investigation was criticized because the observation time in the experimental group was only 4000 hours compared with 31 000 hours in the control group; furthermore, statistically significant testing was done using a significance level of 0.1 rather than the more commonly used 0.05. In a two-year prospective study of 171 nurses, isokinetic lifting strength was studied as a preplacement test (Mostardi *et al.*, 1992). Lifting strength was a poor predictor of back injury. Himmelstein and Andersson (1988) claimed that muscle strength testing may be a predictor of future musculoskeletal injury as a part of a well-designed programme that considers specific job demands related to the workers' capabilities, but unfortunately the conditions necessary for such a programme are rarely met. Lastly, it is important to note that the use of muscle strength testing will systematically discriminate against women and certain ethnic groups, and this may not be acceptable in countries with anti-discrimination laws.

9.6.5 Screening and ultrasonic and radiographic examination of the back

It is possible to measure by ultrasound the diameter of the lumbar spinal canal. In a case-control study (16 cases and 23 controls) it was found that individuals with a canal diameter of less than 14 mm (representing the lowest 10th percentile in this population) had an odds ratio of 10.7 for time missed from work because of low-back pain (Anderson *et al.*, 1988). The authors suggested that ultrasound of the low back was a possible screening tool to be used in industry. However, further studies are needed to establish the sensitivity, specificity and positive and negative predictive values in a cohort of workers before the method can be recommended.

A scientific review by Gibson (1988) addressed the value of preplacement screening using radiographic examination of the low back. He concluded that there was reasonable evidence, from case-control and cohort studies, that the only non-inflammatory, non-neoplastic conditions identifiable by standard low-back X-rays, that were associated with excess risk of low-back pain, were spondylolisthesis, degenerative disk and previous back surgery. Empirical evidence from workplace studies was fairly consistent: the use of low-back X-rays as a preplacement screening tool was poorly predictive of future back pain (Gibson, 1988). The conclusion was that preplacement screening using radiography of the low back had little value in industry (Gibson, 1988).

9.6.6 Screening and fitness

In an examination of 514 police officers it was found that, of those 35 years or older, 7 per cent of absenteeism could be explained by age, sex and physical fitness (Boyce *et al.*, 1991). It was concluded that the extent to which physical fitness capacity could predict absenteeism was low. In studies by Cady *et al.* (1979; 1985) it was reported that physical capacity was related to musculo-skeletal health. In a prospective longitudinal study at a Boeing plant in the USA, low cardiovascular fitness level was a risk factor for disabling low-back pain of at least three months' duration (Battié *et al.*, 1989). However, in all these studies the relationship between physical fitness and musculoskeletal health is not considered strong enough to be used as a screening tool. If pre-employment and preplacement testing should be done, present studies indicate that the best way of doing it is simply to allow the worker to attempt the job in question, with a subjective assessment (Rodgers, 1988).

9.6.7 Screening and biological markers

In a review by Mastin *et al.* (1992) it was suggested that biomarkers have potential use as pre-employment and preplacement screening tools. One report mentions that substance P, which is related to back pain of certain etiologies, might be one such useful biomarker. In general, the types of biological markers suggested for use are proteins that are constituents of musculoskeletal tissue (e.g. cartilage proteins), proteins associated with the inflammatory response (e.g. acute phase proteins) or pre-inflammatory response biomolecules, indicators or immune-mediated inflammatory processes. Primary proteoglycans and glycosaminoglycans, especially keratan sulphate, have been used as biomarkers of osteoarthritis. There is also a non-collagenous glycoprotein in articular cartilage that appears to be released from degenerating cartilage and could possibly be used as a biomarker. To date, however, there is no evidence that the use of biological markers to predict the development of WMSDs is an effective pre-employment or preplacement screening tool.

9.6.8 Cost-effectiveness of screening

In an evaluation of over 100 000 pre-employment medical examinations of applicants for governmental functions it was found that the medical diagnoses frequently encountered among rejected applicants were as common among accepted applicants (De Kort *et al.*, 1991). If selection aiming at reducing absenteeism or work disablement was considered to be the only goal of the pre-employment medical examinations then their efficiency appeared to be low for many job categories. Cost-benefit analysis in other studies also showed no effectiveness. Pre-employment screening in a prospective cohort study of 6125 job applicants for light-duty jobs at a telephone company showed that between groups of employees with or without employment risk factors (i.e. as determined by screening), there was no significant difference relative to sickness, accidents or absence, or to work performance during the first 12 months after hiring (Alexander *et al.*, 1977). A study by Lowenthal (1986) on 200 consecutive new employees with extensive preplacement health evaluation compared with 200 employees given minimal evaluation showed that, over four years of follow-up, there were no significant differences between the two groups. It was concluded that comprehensive preplacement evaluation for medical employees is not a cost-effective activity.

9.6.9 Conclusion

At present there is no scientific evidence to support the use of pre-employment or preplacement screening tests to predict the development of WMSDs, nor are they justified from an ethical and, in some countries, legal point of view. However, pre-employment and preplacement testing can be important tools to provide a safe work environment for aspects other than WMSDs, for example when dealing with infectious hazards among hospital workers (Lewy, 1985).

9.7 Information on health promotion programmes for medical management

Medical management involves initiating and setting up different health promotion programmes. Smoking cessation, exercise and other fitness programmes are common in the prevention of WMSDs. Health promotion programmes may reduce musculoskeletal symptoms and medical claims (Wheat *et al.*, 1992). However, care should be used in selecting and applying these programmes since they often do not reach all segments of the work force equally (Stange *et al.*, 1991).

9.7.1 Smoking cessation programmes

Smoking cessation programmes may help reduce certain disorders. For example, there is a strong consensus that smoking is associated with back disorders. It has been suggested that smoking accounts for up to 50 per cent of disablement pensions of back disorders (Heliövaara, 1988). Smoking has also been associated with symptoms of the neck-shoulder area and hand (Hagberg *et al.*, 1990; Mäkelä *et al.*, 1991). To date, there is no intervention study which has examined the effectiveness of smoking cessation programmes in reducing WMSDs.

9.7.2 Exercise programmes

It is possible that exercise programmes may prevent WMSDs (Cady *et al.*, 1979; Cady *et al.*, 1985). The Cady study of firefighters is the only one so far that points towards beneficial effects of exercise in terms of lowering the incidence of musculoskeletal injuries. In jobs where a high level of muscle strength is utilized, muscle strength training may prevent neck-shoulder disorders (Kilbom, 1988). However, the results of intervention studies showing that exercise programmes prevent musculoskeletal disorders are not consistent. It should also be mentioned that fitness programmes may have many benefits other than improved health, for example better corporate image, lower employee turnover, gains in productivity, and improved employee lifestyle, etc. (Shepard, 1992).

9.7.3 Weight control programmes

Being overweight is a risk factor for osteoarthrosis of the hip and knee (see section 9.8 on personal susceptibilities). Weight reduction programmes may improve musculoskeletal health, but to date there is no scientific evaluation available concerning the effects of weight reduction programmes on musculo-skeletal health.

9.8 Information on individual susceptibility for medical management: are personal characteristics important risk factors in WMSDs?

This book's mandate is to examine the evidence for the contribution of work to musculoskeletal disorders, set the foundations for establishing the importance of eliminating work related risk factors and help prevent WMSDs in the workplace. Yet it would not be complete without touching upon an issue which is often raised and is important in medical management: how might workers' personal attributes contribute to the development of musculoskeletal disorders? It should be remembered that workers' personal susceptibility to these disorders can be increased or decreased (salutary buffering effect), although the tendency has been to focus on the issue that the risk could be increased.

9.8.1 Brief definition

Individual susceptibility is defined as an increased vulnerability to musculoskeletal disorders in a person caused by disease, genetic code or lack of fitness. Increased susceptibility may create a situation where an exposure, at a lower than normal threshold, can cause an effect resulting in a work related musculoskeletal disorder. Furthermore, exposure may trigger symptoms early, and at an unusual location under local strain, in an individual with a preclinical systemic disease (Hagberg *et al.*, 1986). The influence on workers' susceptibility to developing musculoskeletal disorders is discussed for the following variables: age, gender, anthropometric measurements, anatomical differences, tissue type, alcohol and smoking habits, personality and psychiatric disorders, general inflammatory disorders, neuromuscular diseases, metabolic diseases and neoplasms.

9.8.2 Age

The ability to tolerate external stress on different tissues decreases with age, causing a shift of the stress-strain curve. The normal reparative and wound-healing process is slowed with age. The normal age degeneration of tissue is a result of changes in the chemical content of the tissue cells and the extracellular fluid (Freeman and Meachim, 1979). Degeneration is a predisposing factor or a prerequisite in some work related musculoskeletal disorders. Exertion may trigger an inflammatory 'foreign body' response to the debris of dead cells in the degenerated shoulder tendon, resulting in active tendinitis. Also, an infection (viral, urogenital) or systemic inflammation may predispose a subject to reactive tendinitis in degenerated shoulder tendons. One hypothesis proposes that an infection activates the immune system and increases the possibility of a 'foreign response' to the degenerative structures in the tendon (Hagberg, 1987). It is likely that the outcome to a specific exposure is greater among older workers than younger workers. The outcome of non-specified musculoskeletal pain (finger-ache and tingling in the hands at night) was related to increased age even after controlling for duration and level of exposure (Hagberg *et al.*, 1990).

9.8.3 Gender

The incidence of common musculoskeletal disorders such as carpal tunnel syndrome (CTS) has been reported to be greater in females. A male-to-female ratio of 1:3 was described in a population study (Stevens *et al.*, 1988). Yet there is no evidence of increased female susceptibility to work related CTS after controlling for exposure. In a population-based incidence study the ratio was 1.2:1 for occupational CTS when females were compared with males (Franklin *et al.*, 1991). In the study by Silverstein (1985) of CTS among industrial workers, no difference could be seen between genders after controlling for work exposure factors. Neck-shoulder muscular pain is more commonly reported among females than males, both in the general population and among industrial workers (Hagberg and Wegman, 1987). It is not clear whether this skewed distribution for neck-shoulder pain in gender is due to genetic differences between males and females or to gender difference in exposures, both at work and at home. One possibility, for example, could be that females may have an increased risk of work related myofascial pains in the trapezius muscle since females have more type I muscle fibres in the trapezius muscle compared with males (Lindman *et al.*, 1991a,b). As discussed in subsection 3.4.4.6, type I muscle fibres may be at the origin of some of the myofascial pain.

9.8.4 Anthropometry

Anthropometry may add to an individual's susceptibility by modifying exposure. A short person may have to elevate the arm more than a tall person; a tall person may have to bend the back more than a short person to pick up materials from floor level. Most furniture and work stations are designed for the average Caucasian male. Thus, a large proportion of the female population may be at a disadvantage due to generally smaller anthropometric dimensions. Weight and body mass index are related to arthrosis (Anderson and Felson, 1988; Vingård, 1991). Additional weight will cause additional trauma to the joints, thus increasing exposure. Height and weight are also related to certain strength and endurance measures, probably due to the correlation between muscle mass, height and weight. Carpal canal size is a controversial risk factor for CTS. There are different studies linking CTS with both small and large canal areas (Bleecker *et al.*, 1985; Winn and Habes, 1990).

9.8.5 Anatomical differences

A common type of neurogenic thoracic outlet syndrome (TOS) is caused by a cervical rib or a fibrous band that may appear as an anatomical difference or malformation in some individuals. The prevalence of cervical ribs in the population is claimed to be 0.1 per cent (Bateman, 1978); fibrous bands have even higher prevalences. These anatomical differences usually impinge the lowest part of the brachial plexus, causing pain down the ulnar side of the arm and forearm, and sometimes in the hand. This pain may be exacerbated by strenuous use of the hand (Hall, 1987). The intertubercular sulcus on the humeral head may have a different configuration from one individual to another (Bateman, 1978). If the sulcus surface is rough and has a sharp edge, the friction of the tendon to the long head of the biceps muscle increases. This may make a subject more prone to biceps tendinitis. There are also case histories of an aberrant lumbrical muscle appearing in the carpal canal that predisposed the individual to CTS.

9.8.6 Tissue type

A person's tissue type is determined genetically. HLA-B27 is one tissue type linked to ankylosing spondylitis (inflammatory rheumatic disease). Ankylosing spondylitis could render an individual more susceptible to various types of inflammations, for example tendinitis. There are almost no studies of the relationship between tissue type and work related musculoskeletal disorders. In a case-control study of chronic shoulder tendinitis among industrial workers (20 cases), none had tissue type HLA-B27, although its prevalence in the general population is approximately 15 per cent (Bjelle *et al.*, 1979). Thus there is no support at present for the idea that the tissue type HLA-B27 is an important confounder in work related shoulder tendinitis.

9.8.7 Alcohol and smoking

Alcohol-induced neuropathy and myopathy are well-known entities (Brooke, 1977). Patients with these neuropathies or myopathies are likely to have increased susceptibility to strain. Entrapment neuropathies may develop more easily if the axonal transport is impaired (Lundborg, 1988). For example, a higher median consumption of alcohol occurred among engine drivers reporting neck pain compared with those with no pain (Gerdle and Hedberg, 1988). However, it is not clear whether this relationship was confounded by smoking since the authors did not control for it. It should be noted that the work relatedness of CTS cannot be explained by alcohol-induced neuropathy or by other causes of polyneuropathies that afflict many nerves (Hagberg *et al.*, 1992). Although exposure to physical workload factors may result in CTS as an early manifestation of a neurologic disorder in some workers (Hagberg *et al.*, 1986), the frequency of such disorders is probably very low in the worker populations (Mühlau *et al.*, 1984; Hagberg *et al.*, 1991). Further, it is more likely that the most susceptible workers (i.e. workers with these neurologic disorders) would seek jobs with less physical workload exposure as soon as they realize this susceptibility.

Low-back pain has been associated with smoking. There is great consistency between different studies to support this statement. Among exposed workers with a disability pension due to low-back pain in Finland, it was claimed that smoking was an exposure factor with an attributable fraction of 51 per cent (Heliövaara *et al.*, 1987). To explain these findings, the suggested pathogenic mechanisms are based on accelerated degeneration of the low back due to nicotine-related reduced blood flow and smoke-induced coughing causing mechanical strain. It is likely that smoking is also related to neck-shoulder disorders and CTS. In a study of chronic neck pain in a population sample representing the Finnish population, Mäkelä *et al.* (1991) found that smoking represented a risk of 1.3 (95 per cent confidence interval of 1.03–1.61) for this problem (Mäkelä *et al.*, 1991). Smoking was an important determinant for both tingling in the hands at night and wrist ache in a study of manual workers (Hagberg *et al.*, 1990).

9.8.8 Personality and psychiatric disorders

Hysteria, hypochondria and depression (which in combination are referred to as asthenia) are personality variables in the MMPI test that were found to be related to musculoskeletal symptoms in cross-sectional studies (Åstrand and Isacsson, 1988). It is not completely clear whether this association between asthenia and musculoskeletal symptoms is due to the asthenic personality being at greater risk of contracting musculoskeletal symptoms or to the possibility that a patient with pain may develop an asthenic personality. In a 22-year longitudinal study among pulp and paper industrial workers, the asthenic personality variables had no predictive value for musculoskeletal disorders, although in a cross-sectional study there was an association between the asthenic personality type and symptoms (Åstrand and Isacsson, 1988). In a study of type A behaviour among 158 metal workers, type A personality was not related to neck-shoulder pain, but it was to tenderness (Salminen *et al.*, 1991). This finding supports the idea that type A personality is associated with increased muscle activity leading to muscle fatigue and tenderness.

Factors outside work (e.g. social confidence, hobbies, intelligence, alexithymia) were found to have minor roles as predictors of neck-shoulder and low-back pain in a lifelong follow-up study of 154 subjects (Viikari-Juntura *et al.*, 1991). Intelligence and alexithymia in adolescence had no consistent association with neck-shoulder or low-back symptoms in adulthood (Viikari-Juntura *et al.*, 1991). Psychotic and non-psychotic depression are frequently associated with musculoskeletal pain. There is no information regarding depressive disorders and the association with work related musculo-skeletal disorders.

9.8.9 General inflammatory disorders

Local musculoskeletal symptoms may be the first sign of a systemic inflammatory disorder (Bywaters, 1979). Examples of such disorders are rheumatoid arthritis, ankylosing spondylitis, systemic sclerosis and polymyositis. Colitis, respiratory or urinary tract infections and other non-rheumatic inflammatory diseases may also result in reactive inflammatory symptoms of the musculoskeletal system (Dumonde and Steward, 1978; Ford, 1979; Olhagen, 1980). In a case-control study of male industrial workers with chronic severe shoulder pain, 7 out of 20 patients had a concurrent systemic rheumatic disease (Bjelle *et al.*, 1979). Most of these were of the reactive tendinitis type, due to urinary tract or prostate infection. They were detected by careful history-taking, and by extensive physical and laboratory examination. It is likely that an individual with a manifest or subclinical systemic rheumatic disease is prone to developing symptoms in musculoskeletal sites that are subject to local loads. Workers with ankylosing spondylitis may have susceptibility to muscular strain during static loading since they have evidence of general subclinical myopathy (Hagberg *et al.*, 1987). Muscular strength and endurance may be reduced not only during an acute respiratory infection but also months after the signs of infection have subsided (Friman, 1978).

9.8.10 Neuromuscular diseases

Most neuromuscular diseases are rare in the working population, e.g. muscular dystrophies and idiopathic polyneuropathy. Exposure to job risk factors may result in CTS as an early manifestation of a neurologic disease in some workers (Hagberg *et al.*, 1986); however, the frequency of such neurologic disease is probably very low in the worker population (Hagberg *et al.*, 1992).

Juntunen *et al.* (1983) have questioned the association between vibration exposure and CTS, claiming that patients with neuropathic susceptibility (thus being more likely to develop CTS) tend to be selected into groups of patients with vibration syndrome. This could then lead to a detection problem, since workers with vibration exposure are likely to get more medical attention because of the risk of vibration syndrome than those not exposed (i.e. therefore increasing the likelihood of having CTS detected). However, a literature review of studies showing work relatedness ruled out this possibility of a detection bias (Hagberg *et al.*, 1992). It is not likely that workers with neuropathic susceptibility would be exposed to ergonomic factors to a greater extent than those without this susceptibility. Instead, it is more likely that susceptible workers would seek jobs with less potential hazardous exposure as soon as they became aware of this susceptibility (Hagberg *et al.*, 1992).

Primary fibromyalgia could perhaps be regarded as a neuromuscular disease and possibly as another example of a 'personal susceptibility' to WMSDs. This disease is characterized by general muscle pain localized in the four quadrants of the body. In primary fibromyalgia, a local load on the musculoskeletal system may trigger pain and the patient may perceive the symptoms as work related, although they are a result of the disease. It should be remembered, however, that primary fibromyalgia is not considered, by definition, work related. Specific diagnostic criteria for primary fibromyalgia have been set by the American College of Rheumatology (Wolfe *et al.*, 1990). Primary fibromyalgia is not work related because, by definition, trauma-induced myalgia is excluded by the diagnostic criteria.

9.8.11 Metabolic diseases

Diabetes mellitus is often associated with peripheral neuropathy. The diabetic patient is at greater risk of CTS; it is likely that a diabetic individual has an increased susceptibility to nerve entrapment, e.g. CTS due to the neuropathic process (Lundborg, 1988). Hypothyroid patients may have general aches and pain in the muscles. There is no information available on whether hypothyroid patients have a susceptibility to work related musculoskeletal disorders. Megaloblastic anemias may cause symptoms of numbness and tingling. There is no information available on whether anemic patients have a susceptibility to work related musculoskeletal disorders. The above three examples of metabolic diseases are regarded as those diseases which could, theoretically, increase an individual susceptibility to work related musculoskeletal disorders. Besides diabetes, at this stage there is little scientific evidence to support this postulate of increased susceptibility.

9.8.12 Neoplasm

Primary and secondary neoplasms may cause musculoskeletal symptoms. Cancer patients in advanced stages may suffer musculoskeletal pain due to the catabolic metabolism.

9.8.13 Conclusion

Overall, the following summarizes the impact of personal characteristics on work related musculoskeletal disorders:

1. Increased susceptibility may result in a lower than normal threshold for an exposure-effect relationship in a work related musculoskeletal disorder. Furthermore, exposure may trigger symptoms of a systemic disease early and at an unusual location.
2. However, to date, evidence exists only for a few factors. Important individual

factors that increase the risk of developing a work related musculoskeletal disorder are age, smoking, inflammatory disorders and diabetes.

It is worth noting, however, that the importance of the work exposure contributions to these disorders is not diminished by the above findings; these findings merely serve to remind us that these disorders can be the result of many factors, with, in some cases, the contribution of personal attributes. Depending on the intensity, frequency and duration of workplace exposure, the individual/personal factors may take on a more or less important role. In many situations discussed, where there has been extensive work exposure, individual/personal factors are likely to be less important (although they should still be controlled for in epidemiologic studies). Examples of the relative contributions of various risk factors follow. Wisseman and Badger (1977) reported, for repetitive trauma disorders, an odds ratio (OR) of 4.6 for females compared with males, but an OR of 95 for being in a specific department in a film manufacturing plant. Silverstein *et al.* (1986) noted an OR of 4.8 for females compared with males for hand-wrist disorders, but an OR of 29 for high-force/high-repetition jobs compared with low-force/low-repetition jobs, while controlling for gender. In an annual health checkup, Torell *et al.* (1988) found that there was no association between age and musculoskeletal symptoms or diagnosis among male shipyard workers; however, there was a strong association with increasing physical load (based on a study population of $n = 1561$ participants from an original sample of 1751).

9.9 *Training required for medical management: an overview*

The training required to appropriately accomplish medical management will correspond to the mandate delineated. To identify and control problems related to WMSDs overall, the following training is essential:

- training in WMSD diagnosis, treatment, rehabilitation and return to work;
- training in WMSD surveillance and risk factor surveillance (at both the micro- and macroergonomic levels), as well as epidemiology in general;
- training for the control of risk factors (at both the micro- and macroergonomic levels);
- training in personnel training and education;
- training in management and leadership.

The need for any particular medical management team to have or acquire all this training, or only part of it, will depend on the expertise of the individual team members and the role that the team was allocated within a given firm. For example, the small business, which may not have all the necessary expertise on hand, may decide to ensure that some of its personnel has some of this training while ensuring that the outside expertise it secures has the training for the other aspects described here.

The following section on training for medical management has been written keeping in mind the role of the medical management team as described in this chapter so far. Most of this section concentrates on training for the occupational physician. Rather than providing great detail on all the training required for occupational physicians, it highlights certain aspects that may need more emphasis in future training.

9.9.1 Training requirements for medical management: general comments

The occupational health medical management team must comprise professionals with special training in occupational medicine and health. Within the medical management team, occupational health physicians need special training and expertise in musculoskeletal health and disorders. Expertise in functional anatomy, work physiology and biomechanics is a requirement to correctly assess musculoskeletal exposure at the workplace. Furthermore, it is a requirement to evaluate the exposure-effect relationship in WMSDs. Of course, training in diagnostic procedures and treatment of musculoskeletal disorders is needed. In addition, appropriate training to achieve rehabilitation and return to work is also required. Note that this training should include either training to handle the more psychological aspects of the return to work or, at least, training to recognize/identify when and where to gain access to this expertise. Most medical schools do not have sufficient training programmes for common musculoskeletal disorders.

In addition to training in epidemiology, formal training in monitoring risk factors and health in the workplace should also be required; health personnel or other members of the medical management team could receive this training. A specialist with expertise in ergonomics is also an important member of the occupational health team, since ergonomics is essential to the duties of the medical management team, as defined here. This member could be someone whose specialty is ergonomics or a specially trained person with a background in, for example engineering, nursing, physical therapy or the behavioural sciences. Also, expertise in personnel training and education may be an essential requirement for the team if it is involved in these areas. Obviously, in order to mesh these different types of expertise together so that the team members are able to function effectively, at least one member of the team will need training in management and leadership skills.

9.9.2 Some aspects of the occupational physician's training: clinical evaluation

Occupational physicians need to be trained in the correct assessment of musculoskeletal symptoms. This is vital, since symptoms may be referred from disorders of pulmonary, cardiac or abdominal origin. Also, neck-shoulder disorders may cause symptoms referred to the arm and hand. Furthermore, compression of a nerve root and referred pain from musculoskeletal structures in the spine may cause radiation of symptoms from the spine and shoulder to the arm and hand, and the leg and foot.

9.9.2.1 History of the disorder

Occupational physicians need the appropriate training in history-taking for musculoskeletal disorders. The type and localization of symptoms must be explored in great detail. For example, the following questions could be used in obtaining the history of the disorder: Does the pain radiate? Where to?

Diffuse symptoms may indicate musculoskeletal referred pain, whereas radiating pain, based on affected dermatomes, points to a nerve root disorder (e.g. radiculopathy). Details for each symptom should be recorded: the character, quality, distribution, intensity, frequency and duration. The relationship between symptoms and the work activity has to be established. In addition, information should be obtained about symptoms and postures, about movements and loading during occupational activity, and about the relationship of symptoms to recreational activities and to rest (see chapters 3, 4 and 5).

Work related symptoms could also be the first sign of a systemic inflammatory disorder. For example, the family and individual medical history, along with questions about morning stiffness and signs of inflammatory activity (joint swellings) may uncover a rheumatologic disease. For more information see section 9.8 on personal susceptibilities.

9.9.2.2 Work and exposure history

The occupational physician should be trained in correctly assessing work and exposure history. Usually, the worker's job title offers insufficient information on work exposure for determining whether the disorder is work related and whether the patient can return to his/her job. Questions on the patient's work tasks should be detailed, including what the patient produces, work posture, movements, materials handling, psychosocial factors and work organization.

WORK POSTURES AND MOVEMENTS
Special attention should be devoted to the frequency and duration of extreme positions of the head, trunk and limbs. Enquiries about the duration and frequency of work in different positions should also be made.

MATERIALS HANDLING
The ratio of each lift to the individual capacity should be determined. This may provide more information than the total amount of kilograms lifted per day or week. For example, lifting a one-kilogram machine by the hand with a straight

arm at 90 degrees of forward flexion requires 20 per cent of maximal muscular capacity in an ordinary man, and 30–40 per cent of the capacity of a woman.

PSYCHOSOCIAL FACTORS AND WORK ORGANIZATION

A self-chosen work pace may impose a different exposure compared with a preset machine-paced work rate. One should ask the patient about the frequency, distribution and duration of pauses and breaks. Working in the same position for a long time, even when the work posture is not extreme, may result in a static loading of muscle and an increased risk of muscle pain syndromes. Psychosocial work factors such as job demands, decision latitude, social support and job satisfaction are all variables that have a strong influence on absenteeism. These variables may also influence muscle tension, causing discomfort and pain, and should be considered in the history-taking.

More information on work exposures, i.e. what they are and how to measure and evaluate them, can be found in chapters 4 and 5.

9.9.2.3 Physical examination

The occupational physician should be trained to conduct the appropriate physical examination for these disorders. The physical examination of the musculoskeletal system includes the following steps: (1) inspection; (2) testing of range of movement; (3) testing for muscle contraction pain and muscle strength; (4) palpating muscle tendons and insertions; (5) conducting specific tests pertinent to the WMSD in question, including neurological tests.

9.9.2.4 Laboratory examination

Occupational physicians should be trained to appropriately select laboratory examinations for WMSDs and correctly interpret their results. For most musculoskeletal disorders only a few laboratory examinations are necessary. Blood tests for sedimentation rate and rheumatoid factor are taken to rule out general inflammatory disorders. More importantly, radiographs of the affected anatomical parts and spine should be taken when appropriate.

9.10 Ethical considerations

An important principle of the medical management team in industry is the maintenance of confidentiality of information entrusted to them by employees. This must be done with the same consideration as that given by any other physicians, nurses, etc. However, in practice, there are often pressures to violate this principle, especially with respect to disease-detection programmes. Ultimately, industry, management and labour are served best when the medical management team can establish a reputation for trustworthiness, concern for individuals and ability to convey information that management legitimately needs without compromising private information (Beard, 1989).

References

Aarås, A., 1991, What is an acceptable load on the neck and shoulder regions during prolonged working periods? In: Kumashiro, M., Megaw, E.D. eds. *Towards Human Work: Solutions to Problems in Occupational Health and Safety: UOEH International Symposium, Pan-Pacific Conference on Occupational Ergonomics, Kitakyushu, Japan*, London: Taylor & Francis, pp. 115–25.

Aarås, A. and Westgaard, R.H., 1987, Further studies of postural load and musculoskeletal injuries of workers at an electro-mechanical assembly plant, *Appl Ergon* **18**(3), 211–9.

Aarås, A. and Stranden, E., 1988, Measurement of postural angles during work, *Ergonomics*, **31**, 935–44.

Aarås, A., Westgaard, R.H. and Stranden, E., 1987, Work load on local body structures assessed by postural angles measurements. In: Corlett, N., Wilson, J. and Manenica, I. eds. *New Methods in Applied Ergonomics: Proceedings of the Second International Occupational Ergonomics Symposium, Zadar, Yugoslavia, 14–16 April*, Philadelphia, PA: Taylor & Francis, pp. 273–8.

Aarås, A., Westgaard, R.H. and Stranden, E., 1988, Postural angles as an indicator of postural load and muscular injury in occupational work situations, *Ergonomics*, **31**(6), 915–33.

Abrahams, M., 1967, Mechanical behaviour of tendon in vitro, *Med Bio Eng*, **5**, 433–43.

ACOM, 1990, Committee on Occupational Medical Practice: preplacement/preemployment physical examinations, *JOM*, **32**, 295–9.

Adams, J.C., 1971, *Outline of Orthopaedics*, 7th Edn, Edinburgh: Churchill Livingstone, pp. 304–5.

Adams, M.A., Hutton, W.C. and Stott, J.R.R., 1980, The resistance to flexion of the lumbar intervertebral joint, *Spine*, **5**(3), 245–53.

Aguayo, A.J., 1975, Neuropathy due to compression and entrapment, in Dyck, P.J., Thomas, P.K. and Lambert, E.H. (Eds.) *Peripheral Neuropathy*, pp. 688–713, Philadelphia, PA: W.B. Saunders Co.

Ahlbäck, S., 1968, Osteoarthrosis of the knee: a radiographic investigation, *Acta Radiol*, (suppl 277).

Alexander, R.W., Brennan, J.C., Maida, A.S. and Walker, R.J., 1977, The value of preplacement medical examinations for nonhazardous light duty work, *JOM*, **19**(2), 107–12.

Althouse, H.L., 1980, Revealing a true profile of musculoskeletal abilities, *Occ Health & Sfty*, **49**, 25–30.

Amano, M., Umeda, G., Nakajima, H. and Yatsuki, K., 1988, Characteristics of work actions of shoe manufacturing assembly line workers and a cross sectional factor control study on occupational cervicobrachial disorders, *Jpn J Ind Health*, **30**(1), 3–12.

Ambrose, R.F., Kendall, L.G., Alarcon, G.S. et al., 1990, Rheumatology algorithms for primary care physicians, *Arthritis Care Res*, **3**, 71–7.

American Academy of Orthopaedic Surgeons, 1965, *Joint Motion: Method of Measuring and Recording*, Edinburgh: Churchill & Livingstone.

American Medical Association, 1972, *Scope, Objectives, and Functions of Occupational Health Programs*, Chicago: American Medical Association.

American Medical Association, 1990, *Guides to the Evaluation of Permanent Impairment*, 3rd Edn, Chicago: American Medical Association.

American Society for Surgery of the Hand, 1990, *The Hand Examination and Diagnosis*, New York: Churchill Livingstone.

American National Standards Institute, 1993, *Ergonomics: General Principles and Techniques for Employee Training*, draft paper, New York: ANSI. (Document No. Z-365).

Amick, B.C. and Smith, M.J., 1992, Stress, computer based work monitoring and measurement systems: a conceptual overview, *Appl Ergon*, **23**(1), 6–16.

Amis, A.A., 1987, Variation of finger forces in maximal isometric grasp tests on a range of cylinder diameters, *J Biomed Eng*, **9**(10), 313–20.

Anderson, C.T., 1965, *Wrist Joint Position Influences Normal Hand Function*, unpublished Master's thesis, Iowa City: University of Iowa.

Anderson, J.J. and Felson, D.T., 1988, Factors associated with osteoarthritis of the knee in the first national Health and Nutrition Examination Survey (HANES 1): evidence for an association with overweight, race and physical demands of work, *Am J Epidemiol*, **128**, 179–89.

Anderson, D.J., Adcock, D.F., Chovil, A.C. and Farrell, J.J., 1988, Ultrasound lumbar canal measurement in hospital employees with back pain, *Brit J Ind Med*, **45**, 552–5.

Andersson, S., Nilsson, B., Hessel, T., Saraste, M., Noren, A., Stevens-Andersson, S. and Rydhold, D., 1989, Degenerative joint disease in ballet dancers, *Clin Orthop Relat Res*, **238**, 233–6.

Argyris, C., 1971, *Management and Organizational Development*, New York: McGraw-Hill.

Armstrong, T.J., 1987, *The Neuromuscular Basis of Manual Work*, Conference held within the context of the Occupational ergonomics engineering summer conferences, Ann Arbor: University of Michigan.

Armstrong, T.J. and Chaffin, D.B., 1978, An investigation of the relationship between displacements of the finger and wrist joints and the extrinsic finger flexor tendons, *J Biomech*, **11**, 119–28.

Armstrong, T.J. and Chaffin, D.B., 1979, Some biomechanical aspects of the carpal tunnel, *J Biomech*, **12**, 567–70.

Armstrong, T.J. and Silverstein, B.A., 1987, Upper-extremity pain in the workplace: role of usage in causality, in Hadler, N.H. (Ed.) *Clinical Concepts in Regional Musculoskeletal Illness*, pp. 333–54, Orlando, Fl.: Grune and Stratton.

Armstrong, R.B., Warren, G.L. and Warren, J.A., 1991a, Mechanisms of exercize-induced muscle fibre injury, *J Sports Med*, **12**(3), 184–207.

Armstrong, T.J., Foulke, J.A., Joseph, B.S. and Goldstein, S.A., 1982, Investigation of cumulative trauma disorders in a poultry processing plant, *Am Ind Hyg Assoc J*, **43**(2), 103–16.

Armstrong, T.J., Castelli, W.A., Evans, F.G. and Diaz-Perez, R., 1984, Some histological changes in the carpal tunnel contents and their biomechanical implications, *JOM*, **26**(3), 197–201.

Armstrong, T.J., Werner, R.A., Waring, W.P. and Foulke, J.A., 1991b, Intra-carpal canal pressure in selected hand tasks: a pilot study. In: Queinnec, Y. and Daniellou, F. eds. *Designing for Everyone: Proceedings of the Eleventh Congress of the International Ergonomics Association held in Paris*, vol. 1, New York: Taylor and Francis, pp. 156–158.

Armstrong, T.J., Bir, C., Finsen, L., Foulk, J., Martin, B., Sjøgaard, G. and Tseng, K., 1993, Muscle responses to torques to hand held power tools. In: *Proceedings of the International Society of Biomechanics*, pp. 114–5.

Armstrong, T.J. Buckle, P., Fine, L.J., Hagberg, M., Jonsson, B., Kilbom, Å., Kuorinka, I.A.A., Silverstein, B.A., Sjøgaard, G. and Viikari-Juntura E.R.A., 1993, A conceptual model for work related neck and upper-limb musculoskeletal disorders, *Scand J Work Environ Health*, **19**(2), 73–84.

Arndt, R., 1987, Work pace, stress and cumulative trauma disorders, *J Hand Surg*, **12A**(5), 866–9.

Aronsson, G., Åborg, C. and Örelius, M., 1988, *Datoriseringens Vinnare och Förlorare: En Studie av Arbetsförhållanden och Hälsa Inom Statliga Myndigheter och Verk = Winners and Losers of Computerization: a Study of the Working Conditions and Health of Swedish State Employees*, text in Swedish with English summary, Solna: Arbetsmiljöinstitutet och författarna. (Arbete och Hälsa, 27).

Arthur, R.J. and Gunderson, E.K., 1965, Promotion and mental illness in the Navy, *J Occup Med*, **7**, 452–6.

ASF, 1986, *Industrial Welfare Board Statutory Code: Work at Checkout Counters*, Stockholm: ASF.

Åstrand, N.-E. and Isacsson, S.-O., 1988, Back pain, back abnormalities and competing medical, psychological and social factors as predictors of sick leave, early retirement, unemployment, labor turnover and mortality: a 22 year follow up of male employees in a Swedish pulp and paper industry, *Br J Ind Med*, **45**, 387–95.

Australian National Occupational Health & Safety Commission, 1992, *Draft National Code of Practice for Manual Handling (Occupational Overuse Syndrome): a Public Discussion Paper*, Canberra: Australian Government Publishing Service.

Authier, M. and Lortie, M., 1992, Analysis of criteria retained by handlingmen in the choice of a handling method. In: Kumar, S. ed. *Advances in Industrial Ergonomics and Safety IV: Proceedings of the Annual International Industrial Ergonomics and Safety Conference held in Denver, Colorado*, 10–14 June, London: Taylor & Francis, pp. 289–96.

Ayoub, M.A., 1982, Control of manual lifting hazards 1: training in safe handling, *J Occup Med*, **24**(8), 573–77.

Ayoub, M.M., 1988, *Psychophysical Basis for Manual Lifting Guidelines*, Cincinnati, OH: NIOSH. (Report no. 88–79313).

Ayoub, M.M. and Lo Presti, P., 1971, The determination of an optimum size cylindrical handle by use of electromyography, *Ergonomics*, **14**, 509–18.

Ayoub, M.A. and Wittels, N., 1989, Cumulative trauma disorders, *International Reviews of Ergonomics*, **2**, 217–72.

Bäcklund, L. and Nordgren, L., 1968, A new method for testing isometric muscle strength under standardized conditions, *Scand J Clin Lab Invest*, **21**, 33–41.

Baidya, K.N. and Stevenson, M.G., 1988, Local muscle fatigue in repetitive work, *Ergonomics*, **31**(2), 227–39.

Baleshta, M.M. and Fraser, T.M., 1986, An arm movement notation system in investigation of cumulative trauma disorder. In: Karwowski, W. ed. *Trends in Ergonomics/Human Factors III: Proceedings of the Annual International Industrial Ergonomics and Safety Conference held in Louiseville, Kentucky*, 12–14 June, New York: North-Holland, pp. 613–20.

Bammer, G., 1993, Work related neck and upper limb disorders: social, organisational, biomechanical and medical aspects. In: Gontijo, L.A. and de Souza, J. eds. *Segundo*

Congresso Latino Americano e Sexto Seminario Brasileiro de Ergonomia, Floria-nopolis: Ministerio do Trabalho Fundacentro/SC, pp. 23-38.

Bammer, G. and Martin, B., 1992, Repetition strain injury in Australia: medical knowledge, social movement, and de facto partisanship, *Soc Probl*, **39**(3), 219–37.

Bard, C.C., Sylvestre, J. and Dussault, R., 1984, Hand osteoarthropathy in pianists, *J Can Assoc Radiol*, **35**, 154–8.

Barnes, R.M., 1958, *Motion and Time Study*, New York: John Wiley & Sons.

Barnhart, S., Demers, P.A., Miller, M., Longstreth, W.T. and Rosenstock, L., 1991, Carpal tunnel syndrome among ski manufacturing workers, *Scand J Work Environ Health*, **17**(1), 46–52.

Basmajian, J.V., 1961, Weight-bearing by ligaments and muscles, *Can J Surg*, **4**, 166–70.

Bateman, J.E., 1978, *The Shoulder and Neck*, 2nd Edn, New York: W. Saunder.

Bateman, J.E., 1983, Neurologic painful conditions affecting the shoulder, *Clin Orthop Relat Res*, **173**(March), 44–54.

Bates, M.S., 1987. Ethnicity and pain: a biocultural model, *Soc Sci Med*, **24**(1), 47–50.

Battié, M., Bigos, S., Fisher, L. et al., 1989, A prospective study of the role of cardiovascular risk factors and fitness in industrial back pain complaints, *Spine*, **14**, 141–7.

Baty, D., Buckle, P.W. and Stubbs, D., 1986, Posture recording by direct observation, questionnaire assessment and instrumentation: a comparison based on a recent field study. In: Corlett, N., Wilson, J. and Manenica, I. eds. *The Ergonomics of Working Postures: Models, Methods and Cases: Proceedings of the First International Occupational Ergonomics Symposium, Zadar, Yugoslavia, 18–20 April 1985*, London: Taylor and Francis, pp. 283–92.

Beard, R.R., 1989, Industrial physician's responsibilities to his patients and to manage-ment: confidential records, *Sangyo Ika Daigaku Zasshi*, **11**, 83–9.

Bennett, E., 1982, Dupuytren's contracture in manual workers, *Br J Ind Med*, **39**, 98–100.

Berg, M., Sanden, A., Torell, G. and Järvholm, B., 1988, Persistence of musculoskeletal symptoms: a longitudinal study, *Ergonomics*, **31**(9), 1281–85.

Beyer, J.A. and Wright, I.S., 1951, The hyperabduction syndrome: with special reference to its relationship to Raynaud's syndrome, *Circulation: The Journal of the American Heart Association*, **IV**(2), 161–72.

Biering-Sørensen, F., 1983, A prospective study of low back pain in a general population I: occurrence, recurrence and aetiology, *Scand J Rehab Med*, **15**, 71–9.

Biering-Sørensen, F., 1984, Physical measurements as risk indicators for low back trouble over a one-year period, *Spine*, **9**(2), 106–19.

Bigos, S.J., Battié, M.C., Spengler D.M., Fisher, L.D., Fordyce, W.E., Hansson, T.H., Nachemson, A.L. and Wortley, M.D., 1991, A prospective study of work percep-tions and psychosocial factors affecting the report of back injury, *Spine*, **16**(1), 1–6.

Bishop, P.J., Norman, R.W., Wells, R., Ranney, D. and Skleryk, B., 1983, Changes in the centre of mass and moment of inertia of a headform induced by a hockey helmet and face shield, *Can J Appl Sport Sci*, **8**(1), 19–25.

Bishu, R.R., Manjunath, S.G. and Hallbeck, M.S., 1990, A fatigue mechanics approach to cumulative trauma disorders. In: Das, B. ed. *Advances in Industrial Ergonomics and Safety II: Proceedings of the Annual International Industrial Ergonomics and Safety Conference held in Montréal, Québec, Canada, 10–13 June*, New York: Taylor & Francis, pp. 215–222.

Bjelle, A., Hagberg, M. and Michaelsson, G., 1979, Clinical and ergonomic factors in prolonged shoulder pain among industrial workers, *Scand J Work Environ Health*, **5**, 205–10.

Bjelle, A., Hagberg, M. and Michaelson, G., 1981, Occupational and individual factors in acute shoulder-neck disorders among industrial workers, *Brit J Ind Med*, **38**, 356–63.

Bleecker, M.L., Bohlman, M., Moreland, R. and Tipton A., 1985, Carpal tunnel syndrome: role of carpal canal size, *Neurology*, **35**, 1599–604.

Bogni, J.A., 1977, Preplacement medical examinations, *JOM*, **19**, 517.

Bolle de Bal, M., 1982, Changement et intervention: pour une ergonomie stratégique, in Gaussin, J. and Van Laethem, A. (Eds.) *L'Ergonomie des Activités Mentales*, pp. 7–29, Louvain-la-Neuve: Cabay.

Bongers, P.M. and de Winter, C.R., 1992, *Psychosocial Factors and Musculoskeletal Disease: a Review of the Literature*, Leiden, Netherlands: Nederlands Instituut voor Praeventieve Gezondheidszorg TNO. (NIPG Publication number 92.028).

Boussenna, M., Corlett, E.N. and Pheasant, S.T., 1982, The relation between discomfort and postural loading at the joints, *Ergonomics*, **25**(4), 315–22.

Boyce, R.W., Jones, G.R. and Hiatt, A.R., 1991, Physical fitness capacity and absenteeism of police officers, *JOM*, **33**(11), 1137–43.

Breslow, L. and Buell, P., 1960, Mortality and coronary heart disease and physical activity on work in California, *J Chron Dis*, **11**, 615–26.

Brisson, C., Vinet, A. and Vezina, M., 1989, Disability among female garment workers: a comparison with a national sample, *Scand J Work Environ Health*, **15**, 323–28.

Brisson, C., Vinet, A., Vezina, M. and Gingras, S., 1989, Effect of duration of employment in piecework on severe disability among female garment workers, *Scand J Work Environ Health*, **15**, 329–34.

Brooke, M.H., 1977, *A Clinician's View of Neuromuscular Diseases*, Baltimore: Williams & Wilkins.

Brown, J.R., 1971, *Lifting as an Industrial Hazard*, Toronto: Labour Safety Council of Ontario, Ontario Department of Labour.

Brown, D.A., Coyle, I.R. and Beaumont, P.E., 1985, The automated Hettinger test in the diagnosis and prevention of repetition strain injuries, *Appl Ergon*, **16**(2), 113–8.

Browne, C.D., Nolan, B.M. and Faithfull, D.K., 1984, Occupational repetitive strain injuries, *Medical J Aust*, **140**, 329–32.

Buchanan, D.A. and Boddy, D., 1982, Advanced technology and the quality of working life: the effects of wordprocessing on video typists, *J Occup Psychol*, **55**, 1–11.

Buchholz, B., Frederick, L. and Armstrong, T.J., 1988a, An investigation of human palmar skin friction and the effect of materials, pinch force and moisture, *Ergonomics*, **31**(3), 317–25.

Buchholz, B., Wells, R.P. and Armstrong, T.J., 1988b, The Influence of object size on grasp strength: results of a computer simulation of cylindrical grasp. In: Hubbard, M. ed. *Proceedings of the 12th Annual Meeting of the American Society for Biomechanics held in Urbana, Illinois, 28–30 Sept*, Oxford: Pergamon, pp. 851–885.

Buckelew, S.P., Shutty, M.S., Jr, Hewett, J., Landon, T., Morrow, K. and Frank, R.G., 1990, Health locus of control, gender differences and adjustment to persistent pain, *Pain*, **42**, 287–94.

Buckhout, B.C. and Warner, M.A., 1980, Digital perfusion of handball players, *Am J Sports Med*, **8**(3), 206–7.

Buckle, P., 1982, *A Multidisciplinary Investigation of Factors Associated with Low Back Pain*, PhD thesis, Cranfield, United Kingdom: Institute of Technology.

Buckle, P.W., Stubbs, D.A. and Baty, D., 1986, Musculo-skeletal disorders (and discomfort) and associated work factors. In: Corlett, N., Wilson, J. and Manenica, I. eds. *The Ergonomics Wof Working Postures: Models, Methods and Cases: Proceedings of the First International Occupational Ergonomics Symposium, Zadar, Yugoslavia, 18–20 April 1985*, London: Taylor and Francis, pp. 19–30.

Bunnell, S., 1942, Surgery of the intrinsic muscles of the hand other than those producing opposition of the thumb, *J Bone Joint Surg*, **24**(1), 1–32.

Burdorf, A. and Monster, A., 1991, Exposure to vibration and self-reported health complaints of riveters in the aircraft industry, *Ann Occup Hyg*, **35**(3), 287–98.

Bureau of Labour Statistics, 1992, *Occupational Injuries and Ilnesses in the United States by Industry*, Washington, D.C.: US Department of Labor, BLS. (Bulletin 2399).

Burke, R.J., 1988, Sources of managerial and professional stress in large organizations, in Cooper, C.L. and Payne, R. (Eds.) *Causes, Coping and Consequences of Stress at Work*, pp. 77–114, New York: John Wiley & Sons.

Burns, T. and Stalker, G.M., 1961, *The Management of Innovation*, London: Tavistock.

Butsch, J.L. and Janes, J.M., 1963, Injuries of the superficial palmar arch, *J Trauma*, **3**, 505–15.

Byström, S., 1991, *Physiological Response and Acceptability of Isometric Intermittent Handgrip Contractions*, Stochholm: National Institute of Occupational Health. (Arbete och Hälsa, 38).

Byström, S. and Sjøgaard, G., 1992, Potassium homeostasis during and following exhaustive submaximal handgrip contractions, *Acta Physiol Scand*, **142**, 59–66.

Byström, S.E.G., Mathiassen, S.E. and Fransson-Hall, C., 1991, Physiological effects of micropauses in isometric handgrip exercise, *Eur J Appl Physiol*, **63**, 405–11.

Bywaters, E.G.L., 1979, Lesions of bursae, tendons and tendon sheats, *Clin Rheum Dis*, **5**, 883–925.

Cady, L.D., Thomas, P.C. and Karwasky, R.J., 1985, Program for increasing health and physical fitness of fire fighters, *J Occup Med*, **27**(2), 110–4.

Cady, L., Bischoff, D.P., O'Connell, E.R., Thomas, P.C. and Allan, J.H., 1979, Strength and fitness and subsequent back injuries in firefighters, *J Occup Med*, **21**(4), 269–72.

Cain, W.S., 1973, Nature of perceived effort and fatigue: roles of strength and blood flow in muscle contractions, *J Mot Behav*, **5**(1), 33–47.

Cannon, L.J., Bernacki, E.J. and Walter, S.D., 1981, Personal and occupational factors associated with carpal tunnel syndrome, *J Occup Med*, **23**(4), 255–8.

Caplan, R.D., Cobb, S., French, J.R.P., Jr, Harrison, R.V. and Pinneau, S.R., 1975, *Job Demands and Worker Health*, Washington, D.C.: US Department of Health, Education and Welfare. (NIOSH. Publication no. 75–169).

Carayon, P., Smith, M.J. and Miezio, K., 1987a, Comparing worker perceptions to engineering measurements of VDT workstations and environmental conditions. In: *Rising to New Heights with Technology: Proceedings of the Human Factors Society 31st Annual Meeting*, Santa Monica, CA: Human Factors Society, pp. 874–8.

Carayon, P., Swanson, N. and Smith, M.J., 1987b, Objective and subjective ergonomic evaluations of automated offices. In: Flach, J.M. ed. *Proceedings of the Fourth Midcentral Ergonomics/Human Factors Conference*, Urbana, IL: University of Illinois, pp. 358–66.

Castorina, J., Rempel, D., Jones, J., Osorio, A.M. and Harrison, R.J., 1990, *Carpal Tunnel Syndrome among Postal Machine Operators*, Berkeley, CA: California Department of Health Services. (Report 86–008).

Catovic, E., Catovic, A., Kraljevic, K. and Muftic, O., 1991, The influence of arm position on the pinch grip strength of female dentists in standing and sitting positions, *Appl Ergon*, **22**(3), 163–6.

CEN, 1990, *Ergonomic Principles of the Design of Work Systems (ISO 6385: 1981)*, Bruxelles: Comité Européen de Normalisation. (Standard no. ENV 26385:1990, CEN/TC 122).

CEN, 1991, *CEN Catalogue 1991: Catalogue of European Standards*, Bruxelles: Comité Européen de Normalisation.

Chaffin, D.B., 1973, Localized muscle fatigue: definition and measurement, *J Occup Med*, **15**, 346–54.

Chaffin, D.B. and Anderson, G.B.J., 1984, *Occupational Biomechanics*, New York: John Wiley & Sons.

Chaffin, D.B., Herrin, G.D. and Keyserling, W.M., 1978, Preemployment strength testing: an updated position, *J Occup Med*, **20**(6), 403–8.

Chaffin, D.B., Gallay, L.S., Woolley, C.B. and Kuciemba, S.R., 1986, An evaluation of the effect of a training program on worker lifting postures, *International Journal of Industrial Ergonomics*, **1**, 127–36.

Chao, E.Y., Opgrandi, F. and Axmere, M., 1976, Three dimensional force analysis of the finger joints in selected isometric hand functions, *J Biomech*, **9**, 387–96.

Chatterjee, D.S., 1992, Workplace upper limb disorders: a prospective study with intervention, *Occup Med*, **42**(3), 129–36.

Checkoway, H., Pearce, N.E. and Crawford-Brown, D.J., 1989, *Research Methods in Occupational Epidemiology*, New York: Oxford University Press.

Chiang, H.-C., Chen, S.-S., Yu, H.-S. and Ko, Y.-C., 1990, The occurence of carpal tunnel syndrome in frozen food factory employees, *Kaohsiung J Med Sci*, **6**, 73–80.

Christensen, H., 1986a, Muscle activity and fatigue in the shoulder muscles during repetitive work: an electromyographic study, *Eur J Appl Physiol*, **54**, 596–601.

Christensen, H., 1986b, Muscle activity and fatigue in the shoulder muscles of assembly plant employees, *Scand J Work Environ Health*, **12**, 582–7.

Clarke, R.G., 1961, The limiting hand skin temperature for unaffected manual performance in the cold, *J Appl Psychol*, **45**, 193–4.

Cleveland, R., Cohen, H.H. and Smith, M.J., 1979, *Safety Program Practices in Record-Holding Plants*, Morgantown, WV: DHEW, NIOSH, Division of safety research. (Publication No. 79–136).

Cobb, S. and Kasl, S.V., 1977, *Termination: the Consequences of Job Loss*, Washington, D.C.: US Government Printing Office.

Coburn, D., 1979, Job alienation and well-being, *Int J Health Serv*, **9**(1), 41–59.

Cochran, D.J., Albin,T.J., Bishu, R.R. and Riley, M.W., 1986, An analysis of grasp force degradation with commercially available gloves. In: *A Cradle for Human Factors: Proceedings of the 30th Human Factors Society 30th Annual Meeting, Dayton, Ohio, September 29–October 3*, vol. 2, Santa Monica, Ca.: Human Factors Society, pp. 852–5.

Cohen, A., 1977, Factors in successful safety programs, *J Safe Res*, **9**, 168–78.

Cohen, B.G.F., 1983, Organizational factors affecting stress in the clerical worker, *Occup Health Nurs*, **11**, 30–4.

Cohen, S. and Weinstein, N., 1981, Non-auditory effects of noise on behavior and health, *J Social Issues*, **37**, 36–70.

Cohen, M.L., Arroyo, J.F., Champion, G.D. and Browne, C.D., 1992, In search of the pathogenesis of refractory cervicobrachial pain syndrome, *Med J Aust*, **156**, 432–6.

Cole, D.C., Stock, S.R. and Gibson, E.S., 1992, Workplace based interventions to reduce overuse disorders of the neck and upper extremity: an epidemiologic review. In: Human Factors Association of Canada (Ed.) *The Economics of Ergonomics: Proceedings of the 25th Annual Conference of the Human Factors Association of Canada, October 25–28, Hamilton, Ont.*: Human Factors Association of Canada, pp. 263–9.

Colligan, M.J. and Murphy, L.R., 1979, Mass psychogenic illness in organizations: an overview, *J Occup Psychol*, **52**, 77–90.

Collins, D.H., 1950, *The Pathology of Articular and Spinal Diseases*, Liverpool: Edward Arnold and Co., 74–103.

Commission on Accreditation of Rehabilitation Facilities, 1989, *Standards Manual for Organizations Serving People with Disabilities*, report, Tucson, AZ: Commission on Accreditation of Rehabilitation Facilities.

Commission of the European Communities, 1992, *Training in Safety and Health at Work*, Luxembourg: Office for Official Publications of the European Communities.

Conn, J., Bergan, J.J. and Bell, J.L., 1970, Hypothenar hammer syndrome: post-traumatic digital ischemia, *Surgery*, **68**, 1122–8.

Cooper, C.L. and Marshall, J., 1976, Occupational sources of stress: a review of the literature relating to coronary heart disease and mental ill health, *J Occup Psychol*, **49**, 11–28.

Cooper, C.L. and Smith, M.J. (Eds.), 1985, *Job Stress and Blue Collar Work*, New York: John Wiley & Sons.

Copeman, W., 1940, The arthritic sequel of pneumatic drilling, *Ann Rheum Dis*, **2**, 141–6.

Corlett, E.N. and Bishop, R.P., 1976, A technique for assessing postural discomfort, *Ergonomics*, **19**(2), 175–82.

Corlett, E.N. and Bishop, R.P., 1978, The ergonomics of spot welders, *Appl Ergon*, **9**(1), 23–32.

Corlett, E.N., Madeley, S.J. and Manenica, I., 1979, Posture targetting: a technique for recording working postures, *Ergonomics*, **22**(3), 357–66.

Couture, L., 1986, *Health Problems Related to Standing*, Montréal: Commission de la Santé et de la Sécurité du Travail du Québec (CSST), CREF. (Report # 86082901).

Cox, T., 1985, Repetitive work: occupational stress and health, in Cooper, C.L. and Smith, M.J. (Eds.) *Job Stress and Blue Collar Work*, pp. 85–112, New York: John Wiley & Sons.

Croft, P., Coggon, D., Cruddas, M. and Cooper, C., 1992, Osteoarthritis of the hip: an occupational disease in farmers, *Br Med J*, **304**, 1269–73.

CSA, 1989, *A Guideline on Office Ergonomics: a National Standard of Canada*, Rexdale, Ont.: Canadian Standards Association. (Document no. CAN/CSA-Z412–m89).

Cuetter, A.C. and Bartoszek, D.M., 1989, The thoracic outlet syndrome: controversies, overdiagnosis, overtreatment, and recommendations for management, *Muscle & Nerve*, **12**, 410–9.

Dainoff, M.J., Hurrell, Jr, J.J. and Happ, A., 1981, A taxonomic framework for the description and evaluation of paced work, in Salvendy, G. and Smith, M.J. (Eds.) *Machine Pacing and Occupational Stress: Proceedings of the International Conference, Purdue University, March*, pp. 185–90, London: Taylor & Francis.

Dalton, G.D., 1969, *Influence and Organizational Change*, paper read at a conference on organizational behavior models, Kent, OH: Kent State University.

Danielsson, L., Lindberg, H. and Nilsson, B., 1984, Prevalence of coxarthrosis, *Clin Orthop*, **191**, 110–5.

Das, B., Kozey, J. and Tyson, J., 1989, Development of a computer-aided potentiometric system for anthropometric measurements. In: *Proceedings of the 22nd Annual Conference of the Human Factors Association of Canada, Toronto, Ontario, November 26–29*, Missisauga, Ont.: Human Factors Association of Canada, pp. 261–5.

Davis, P.R. and Stubbs, D.A., 1977, Safe levels of manual forces for young males (1), *Appl Ergon*, **8**(3), 141–50.

Dawson, D.M., Hallett, M. and Millender, L.H., 1983, *Entrapment Neuropathies*, Boston: Little, Brown and Company.

de Krom, M.C.T.F.M., Kester, A.D.M., Knipschild, P.G. and Spaans, F., 1990, Risk factors for carpal tunnel syndrome, *Am J Epidem*, **132**(6), 1102–10.

De Kort, W.L., Fransman, L.G. and Van Dijk, F.J.H., 1991, Preemployment medical examinations in large occupational health service, *Scand J Work Environ Health*, **17**, 392–7.

DeGreene, K., 1973, *Sociotechnical Systems*, Englewood Cliffs, New Jersey: Prentice-Hall.

Dehlin, O., Berg, S., Andersson, G.B.J. and Grimby, G., 1981, Effect of physical training and ergonomic counselling on the psychological perception of work and on the subjective assessment of low-back insufficiency, *Scand J Rehab Med*, **13**, 1–9.

Dennett, X. and Fry, H.J.H., 1988, Overuse syndrome: a muscle biopsy study, *Lancet*, **1**(8581), 905–8.

Diffrient, N., Tilley, A.R. and Bardagjy, J.C., 1974, *Humanscale 1/2/3: a Portfolio of Information: 1. Sizes of People, 2. Seating Considerations, 3. Requirements for the Handicapped and Elderly*, Cambridge, MA: MIT Press.

Diffrient, N., Tilley, A.R. and Harman, D., 1981a, *Humanscale 4/5/6: a Portfolio of*

Information: 4. Human Strength and Safety, 5. Controls and Displays, 6. Designing for People, Cambridge, MA: MIT Press.

Diffrient, N., Tilley, A.R. and Harman, D., 1981b, *Humanscale 7/8/9: a Portfolio of Information: 7. Standing and Sitting at Work, 8. Space Planning for the Individual and the Public, 9. Access for Maintenance, Stairs, Light and Color*, Cambridge, MA: MIT Press.

Dimberg, L., 1987, The prevalence and causation of tennis elbow (lateral humeral epicondylitis) in a population of workers in an engineering industry, *Ergonomics*, **30**(3), 573–80.

Dimberg, L., Olafsson, A., Stefansson, E., Aagaard, H., Andres, O., Andersson, G.B.J., Hagert, C.-G. and Hansson, T., 1989a, Sickness absenteeism in an engineering industry: an analysis with special reference to absence for neck and upper extremity symptoms, *Scand J Soc Med*, **17**, 77–84.

Dimberg, L., Olafsson, A., Stefansson, E., Aagaard, H., Oden, A., Andersson, G.B., Hansson, T. and Hagert, C., 1989b, The correlation between work environment and the occurrence of cervicobrachial symptoms, *J Occup Med*, **31**(5), 447–53.

Dobyns, J.H., O'Brien, E.T., Linscheid, R.L. and Farrow, G.M., 1972, Bowler's thumb: diagnosis and treatment: a review of seventeen cases, *J Bone Joint Surg*, **54A**(4), 751–5.

Doolittle, T.L. and Kaiyala, K., 1986, Strength and musculo-skeletal injuries of firefighters. In: *Proceedings of the Human Factors Association of Canada, 19th Annual Meeting, Richmond (Vancouver), British Columbia, August 22–23; Theme: Human factors on the move*, Rexdale, Ont.: Human Factors Association of Canada, pp. 49–52.

Doolittle, T.L. and Kaiyala, K., 1987, A generic performance test for screening firefighters. In: Asfour, S.S. ed. *Trends in Ergonomics/Human Factors IV: Proceedings of the Annual International Industrial Ergonomics and Safety Conference held in Miami, Florida, 9–12 June*, New York: North-Holland, pp. 603–10.

Dorland's Illustrated Medical Dictionary, 1974, 25th Edn, Philadelphia: W.B. Saunders.

Drury, C.G., 1987, A biomechanical evaluation of the repetitive motion injury potential of industrial jobs, *Seminars in Occupational Medicine*, **2**(1), 41–9.

Drury, C.G., 1991, Ergonomics practice in manufacturing, *Ergonomics*, **34**(6), 825–39.

Dul, J., Douwes, M. and Smitt, P., 1991, A work-rest-model for static postures. In: Queinnec, Y. and Daniellou, F. eds. *Designing for Everyone: Proceedings of the Eleventh Congress of the International Ergonomics Association held in Paris*, vol. 1, New York: Taylor and Francis, pp. 93–5.

Dumonde, D.C. and Steward, M.W., 1978, The role of microbial infection in rheumatic disease, in *Copemans Textbook of the Rheumatic Diseases*, London: Churchill Livingstone.

Duncan, R.B., 1972, Characteristics of organizational environments and perceived environmental uncertainty, *Adm Sci Q*, **17**, 313–27.

Easterby, R.S., 1967, Ergonomics checklists: an appraisal, *Ergonomics*, **10**(5), 549–56.

Eastman Kodak Company, Human Factors Section, 1983, *Ergonomic Design for People at Work: Workplace, Equipment and Environmental Design and Information Transfer*, Vol. 1, Belmont, CA: Lifetime Learning Publications.

Eastman Kodak Company, Ergonomics Group, 1986, *Ergonomic Design for People at Work*, Vol. 2, New York: Van Nostrand Reinhold.

Edwards, R.H.T., 1988, Hypotheses of peripheral and central mechanisms underlying occupational muscle pain and injury, *Eur J Appl Physiol*, **57**(3), 275–81.

Eisele, S.A. and Sammarco, G.J., 1993, Fatigue fractures of the foot and ankle in the athlete, *J Bone Joint Surg*, **75A**(2), 290–8.

Emery, F.E. and Trist, E.L., 1960, Sociotechnical systems, in Churchman, C.W. and Verhulst, M. (Eds.) *Management Science: Models and Techniques*, vol. 2, Oxford: Pergamon press.

Emery, F.E. and Trist, E.L., 1965, The causal texture of organizational environments, *Human Relat*, **18**, 21–3.

Falck, B. and Aarnio, P., 1983, Left-sided carpal tunnel syndrome in butchers, *Scand J Work Environ Health*, **9**(3), 291–7.

Federal Agency for Industrial Safety, 1985, *Evaluation Scheme for Cashier's Worksites*, Dortmund, Germany: Federal Agency for Industrial Safety.

Feindel, W. and Stratford, J., 1958, The role of the cubital tunnel in tardy ulnar palsy, *Can J Surg*, **1**, 87–300.

Feldman, R.G., Goldman, R. and Keyserling, W.M., 1983, Peripheral nerve entrapment syndromes and ergonomic factors, *Am J Ind Med*, **4**, 661–81.

Fellows, G.L. and Freivalds, A., 1991, Ergonomics evaluation of a foam rubber grip for tool handles, *Appl Ergon*, **22**(4), 225–30.

Fernandez-Palazzi, F., Rivas, S. and Mujica, P., 1990, Achilles tendinitis in ballet dancers, *Clin Orthop*, **257**, 257–61.

Finelli, P.F., 1975, Mononeuropathy of the deep palmar branch of ulnar nerve, *Arch Neurol*, **32**(August), 564–5.

Fisher, S., 1985, Control and blue collar work, in Cooper, C.L. and Smith, M.J. (Eds.) *Job Stress and Blue Collar Work*, pp. 19–48, New York: John Wiley and Sons.

Fisher, D.L., Andres, R.O., Airth, D. and Smith, S.S., 1993, Repetitive motion disorders: the design of optimal rate-rest profiles, *Hum Fact*, **35**(2), 283–304.

Fletcher, R.H., Fletcher, S.W. and Wagner, E.H., 1982, *Clinical Epidemiology: the Essentials*, Baltimore: Williams & Wilkins.

Flight, R.M. and Schussler, T., 1976, A post-hire evaluation of nurse-conducted preplacement health assessments, *JOM*, **18**(4), 231–4.

Fontus, H.M., Levy, B.S. and Davis, L.K., 1989, Physician-based surveillance of occupational disease, part II: experience with a broader range of diagnoses and physicians, *J Occup Med*, **31**(11), 929–32.

Ford, D.K., 1979, The clinical spectrum of Reiter's Syndrome and similar postenteric arthropathies, *Clin Orthop*, **143**, 59–65.

Forssberg, H., Eliasson, A.C., Kinoshita, H., Johansson, R.S. and Westling, G., 1991, Development of human precision grip, I: Basic coordination of force, *Exp Brain Res*, **85**(2), 451–7.

Foster, C.V.L., Harman, J., Harris, R.C. and Snow, D.H., 1986, ATP distribution in single muscle fibers before and after maximal exercise in the thoroughbred horse, *J Physiol*, **378**, 64.

Frankel, V.H. and Nordin, M., 1980, *Basic Biomechanics of the Skeletal System*, Philadelphia: Lea & Febiger, p. 303.

Frankenhaeuser, M. and Gardell, B., 1976, Underload and overload in working life: outline of a multidisciplinary approach, *J Human Stress*, **2**(sept), 35–46.

Franklin, G.M., Haug, J., Heyer, N., Checkoway, H. and Peck, N., 1991, Occupational carpal tunnel syndrome in Washington State, 1984–1988, *Am J Pub Health*, **81**(6), 741–6.

Fransson, C. and Kilbom, Å., 1991, Tools and hand function: the sensitivity of the hand to external surface pressure. In: Queinnec, Y. and Daniellou, F. eds. *Designing for Everyone: Proceedings of the Eleventh Congress of the International Ergonomics Association held in Paris*, vol. 1, New York: Taylor & Francis, pp. 188–90.

Fransson, C. and Winkel, J., 1991, Hand strength: the influence of grip and span and grip type. *Ergonomics*, **34**(7), 881–92.

Fransson, C., Gloria, R., Kilbom, Å., Karlqvist, L., Nygård, C.-H., Wiktorin, C., Winkel, J. and Stockholm-Music I Study Group, 1991, Presentation and evaluation of a portable ergonomic observation method (PEO). In: Queinnec, Y and Daniellou, F. eds. *Designing for Everyone: Proceedings of the Eleventh Congress of the International Ergonomics Association held in Paris*, vol. 1, New York: Taylor & Francis, pp. 242–4.

Freeman, M.A.R. and Meachim, G., 1979, Ageing and degeneration, in Freeman, M.A.R. (Ed.) *Adult Articular Carteladge*, pp. 487–543, London: Pitman.

French, J.R.P., Jr, 1963, The social environment and mental health, *J Social Issues*, **19**, 39–56.

Fridén, J., Sjöström, M. and Ekblom, B., 1981, A morphological study on delayed muscle soreness, *Exp*, **37**, 506–7.

Friedman, G.D., 1987, *Primer of Epidemiology*, New York: McGraw-Hill.

Friman, G., 1978, Effect of acute infections disease on human isometric muscle endurance, *Ups J Med Sci*, **83**, 105–8.

Gagné, R.M. and Briggs, L.J., 1978, *Principles of Instructional Design*, New York: Holt, Rinehart and Winston.

Gagnon, M. and Lortie, M., 1987, A biomechanical approach to low-back problems in nursing aides. In: Asfour, S.S. ed. *Trends in Ergonomics/Human Factors IV: Proceedings of the Annual International Industrial Ergonomics and Safety Conference held in Miami, Florida, 9–12 June*, part B, New York: North-Holland, pp. 795–802.

Gamberale, F., Ljungberg, A.-S., Annwall, G. and Kilbom, Å., 1987, An experimental evaluation of psychophysical criteria for repetitive lifting work, *Appl Ergon*, **18**(4), 311–21.

Gamsa, A. and Vikis-Freigbergs, V., 1991, Psychological events are both risk factors in and consequences of chronic pain, *Pain*, **44**, 271–7.

Garg, A., Hagglund, G. and Mericle, K., 1986, A physiological evaluation of time standards for warehouse operations as set by tradition work measurement techniques, *IIE Transactions*, **18**(3), 235–45.

Garg, A., Owen, B.D. and Carlson, B., 1992, An ergonomic evaluation of nursing assistants' job in a nursing home, *Ergonomics*, **35**(9), 979–95.

Gelberman, R.H., Herginroeder, P.T., Hargens, A.R., Lundborg, G.N. and Akeson, W.H., 1981, The carpal tunnel syndrome: a study of carpal canal pressures, *J Bone Joint Surg*, **63A**(3), 380–3.

Gemne, G., Pyykko, I., Taylor, W. and Pelmear, P.L., 1987, The Stockholm Workshop Scale for the classification of cold-induced Raynaud's phenomenon in the hand-arm vibration syndrome (revision of the Taylor-Pelmear scale), *Scand J Work Environ Health*, **13**(4), 275–8.

Gerdle, B.U. and Hedberg, G.E., 1988, Alcohol consumption and complaints from the musculoskeletal system among engine drivers: an epidemiological study, *Scand J Soc Med*, **16**, 105–9.

Gerr, F., Letz, R. and Landrigan, P.J., 1991, Upper-extremity musculoskeletal disorders of occupational origin, *Annu Rev Public Health*, **12**, 543–66.

Gibson, E.S., 1988, The value of preplacement screening radiography of the low back, *Occup Med*, **3**, 91–107.

Gilbreth, F.B., 1911, *Motion Study*, New York: Van Norstrand Co..

Glass, D.C. and Singer, J.E., 1972, *Urban Stress: Experiments on Noise and Social Stressors*, New York: Academic Press.

Goldman, R.H., 1986, General occupational health history and examination, *JOM*, **28**, 967–74.

Goldstein, I.L., 1975, Training, in Margolis, B.L. and Kroes, W.H. (Eds.) *The Human Side of Accident Prevention*, pp. 92–113, Springfield, Illinois: Charles C. Thomas Publisher.

Goldstein, S.A., 1981, *Biomechanical Aspects of Cumulative Trauma to Tendons and Tendon Sheaths*, Ph.D. thesis, Ann Arbor: University of Michigan.

Goldstein, J., 1988, A far-from-equilibrium systems approach to resistance to change, *Organ Dynam*, (Autumn), 16–26.

Goldstein, I.L., 1993, *Training in Organizations: Needs Assessment, Development and Evaluation*, 3rd Edn, Pacific Grove, CA: Brooks/Cole Publishing Company.

Goldstein, S.A., Armstrong, T.J., Chaffin, D.B. and Matthews, L.S., 1987, Analysis of cumulative strain in tendons and tendon sheaths, *J Biomech*, **20**(1), 1–6.

Gordon, C., Bowyer, B.L. and Johnson, E.W., 1987, Electrodiagnostic characteristics of acute carpal tunnel syndrome, *Arch Phys Med Rehab*, **68**, 545–8.

Gore, S., 1986, Perspectives on social support and research on stress moderating processes, *J Orgl Bhvr Mgt*, **8**(2), 85–101.

Grandjean, E., Hünting, W. and Pidermann, M., 1983, VDT workstation design: preferred settings and their effects, *Hum Fact*, **25**(2), 161–75.

Grant, K.A., Habes, D.J. and Stewart, L.L., 1992, An analysis of handle designs for reducing manual effort: the influence of grip diameter, *International Journal of Industrial Ergonomics*, **10**, 199–206.

Green, R.A. and Briggs, C.A., 1989, Effect of overuse injury and the importance of training on the use of adjustable workstations by keyboard operators, *J Occup Med*, **31**(6), 557–62.

Greiner, L., 1967, Patterns of organization change, *Harvard Bus Rev*, (May–June), 119–30.

Grieco, A., Occhipinti, E. and Colombini, D., 1989, Work postures and musculo-skeletal disorders in VDT operators, *Boll Ocul*, **68**(suppl 7), 99–111.

Grosshandler, S. and Burney, R., 1979, The myofascial syndrome, *North Car Med J*, **40**, 562–5.

Guthier, W.E., 1986, Medical screening and monitoring as noted by the insurance industry, *JOM*, **28**(8), 765–7.

Hackman, J.R., Oldham, G.R., Janson, R. and Purdy, K., 1975, A new strategy for job enrichment, *Calif Manage Rev*, **17**(4), 57–71.

Hadler, N.M., 1990, Cumulative trauma disorders: an iatrogenic concept, *J Occup Med*, **32**(1), 38–41.

Hadler, N., Gillings, D.B. and Imteus, H.R., 1978, Hand structure and function in an industrial setting: the influence of three patterns of stereotyped repetitive usage, *Arth Rheum*, **21**, 210–20.

Hagberg, M., 1981, Electromyographic signs of shoulder muscular fatigue in two elevated arm positions, *Am J Phys Med*, **60**(3), 111–21.

Hagberg, M., 1981, Work load and fatigue in repetitive arm elevations, *Ergonomics*, **24**, 543–55.

Hagberg, M., 1984, Occupational musculoskeletal stress and disorders of the neck and shoulder: a review of possible pathophysiology, *Int Arch Occup Environ Health*, **53**(3), 269–78.

Hagberg, M., 1987, Shoulder pain: pathogenesis, in Hadler, N.M. (Ed.) *Clinical Concepts of Regional Musculoskeletal Illness*, pp. 191–200, New York: Grune & Stratton.

Hagberg, M., 1988, *Occupational Musculoskeletal Disorders: A New Epidemiological Challenge?*, Amsterdam: Elsevier Science Publishers BV, pp. 15–26.

Hagberg, M., 1992, Exposure variables in ergonomic epidemiology, *Am J Indust Med*, **21**, 91–100.

Hagberg, M. and Kvarnström, S., 1984, Muscular endurance and electromyographic fatigue in myofascial shoulder pain, *Arch Phys Med Rehab*, **65**, 522–5.

Hagberg, M. and Wegman, D.H., 1987, Prevalence rates and odds ratios of shoulder-neck diseases in different occupational groups, *Br J Ind Med*, **44**, 602–10.

Hagberg, M., Michaelson, G. and Örtelius, A., 1982, Serum creatine kinase as and indicator of local muscular strain in experimental and occupational work, *Int Arch Occup Environ Health*, **50**, 377–85.

Hagberg, M., Hagner, I.M. and Bjelle, A., 1987, Shoulder muscular strength, endurance and electromyographic fatigue in ankylosing spondylitis, *Scand J Rheum Dis*, **16**, 161–5.

Hagberg, M., Nyström, Å. and Zetterlund, B., 1991, Recovery from symptoms after carpal tunnel syndrome surgery in males in relation to vibration exposure, *J Hand Surg*, **16A**(1), 66–71.

Hagberg, M., Morgenstern, H. and Kelsh, M., 1992, Impact of occupations and job tasks on the prevalence of carpal tunnel syndrome: a review, *Scand J Work Environ Health*, **18**, 337–45.

Hagberg, M., Almay, B., Kolmodin-Hedman, B. and Zetterlund, B., 1986, Vibration exposure: a modifier of the onset of amyloid polyneuropathy, *Scand J Work Environ Health*, **12**, 277–9.

Hagberg, M., Ängquist, K.A., Eriksson, N.E., Gerdle, B., Lindman, R. and Thornell, L.E., 1988, EMG-relationship in patients with occupational shoulder-neck myofascial pain, in de Groot, G.H.P., Huijing, P.A. and van Ingen Schenau, G.J., (Eds.) *Biomechanics XI-A*, pp. 450–4, Amsterdam: Free University Press.

Hagberg, M., Hansson-Risberg, E., Jorulf, L., Lindstrand, O., Milosevich, B., Norlin, D., Thomasson, L. and Widman, L., 1990, Höga risker för besvär i händerna hos vissa yrkesgrupper = High risk of problems with the hands in certain occupations, *Läkartidningen*, **87**(4), 201–5.

Hägg, G.M., Suurküla, J. and Kilbom, Å., 1990, *Prediktorer för Belastnings Besvär I Skuldra/Nacke: en Longitudinell Studie på Kvinnliga Montörer = Predictors for Work related Shoulder/Neck Disorders: a Longitudinal Study of Female Assembly Workers*, text in Swedish with English summary, Solna: Arbetsmiljöinstitutet och författarna. (Arbete och Hälsa, 10).

Hagner, I. and Hagberg, M., 1989, Evaluation of two floor-mopping work methods by measurement of work load, *Ergonomics*, **32**, 401–8.

Hale, A.R., 1984, Is safety training worthwhile?, *J Occup Accid*, **6**, 17–33.

Hale, A.R. and Mason, I.D., 1986, L'évaluation du rôle d'une formation kinétique dans la prévention des accidents de manutention, *Trav Humain*, **49**(3), 195–208.

Hales, T., Sauter, S., Petersen, M., Putz-Anderson, V., Fine, L., Ochs, T., Schleifer, L. and Bernard, B., 1992, *US West Communications (USWC): Phoenix, Arizona; Mineapolis, Minnesota; Denver, Colorado: Health Hazard Evaluation Report*, Cincinnati: NIOSH, Centers for Disease Control. (HETA Report, 89–299–2230).

Hall, C.D., 1987, Neurovascular syndromes at the thoracic outlet, in Hadler, N.M. (Ed.) *Clinical Concepts in Regional Musculoskeletal Illness*, pp. 227–44, New York: Grune & Stratton.

Hall, R.H., Haas, J.E. and Johnson, N.J., 1967, Organizational size, complexity and formalization, *Am Soc Rev*, (December), 905–12.

Hammarskjöld, E., Harms-Ringdahl, K. and Ekholm, J., 1992, Reproducibility of carpenters work after cold exposure, *International Journal of Industrial Ergonomics*, **9**, 195–204.

Hammer, A.W., 1934, Tenosynovitis, *Med Rec*, **140**, 353–5.

Hansen, N.S. and Jeune, B., 1982, Incidence of disability pensions among slaughterhouse workers in Denmark with special regard to diagnosis of the musculo-skeletal system, *Scand J Soc Med*, **10**, 81–5.

Hansford, T., Blood, H., Kent, B. and Lutz, G., 1986, Blood flow changes at the wrist in manual workers after preventive interventions, *J Hand Surg*, **11A**(4), 503–8.

Harms-Ringdahl, K. and Ekholm, J., 1986, Intensity and character of pain and muscular activity levels elicited by maintained extreme flexion position of the lower-cervical-upper-thoracic spine, *Scand J Rehab Med*, **18**, 117–26.

Hawkins, R.J. and Kennedy, J.C., 1980, Impingement syndrome in athletes, *Am J Sports Med*, **8**, 151–8.

Hazelton, F.T., Smidt, G.L., Flatt, A.E. and Stephens, R.I., 1975, The influence of wrist position on the force produced by the finger flexors, *J Biomech*, **8**, 301–6.

Heliövaara, M., 1988, *Epidemiology of Sciatica and Herniated Lumbar Intervertebral Disc*, thesis, Finland: The Social Insurance Institution.

Heliövaara, M., Knekt, P. and Aromaa, A., 1987, Incidence and risk factors of herniated lumbar intervertebral disc or sciatica leading to hospitalization, *J Chron Dis*, **1**, 251–8.

Hendrick, H.W., 1980, Human factors in management. In: Corrick, G.E., Haseltine, E.C. and Durst, R.T. eds. *Human Factors: Science for Working and Living hf'80 Proceedings of the Human Factors Society, 24th Annual Meeting, Los Angeles, October*, Santa Monica: Human Factors Society, pp. 25–50.

Hendrick, H.W., 1984, Wagging the tail with the dog: organizational design considerations in ergonomics. In: *Proceedings of the Human Factors Society 28th Annual Meeting*, Santa Monica: Human Factors Society, pp. 899–903.

Hendrick, H.W., 1986a, Macroergonomics: a concept whose time has come, *Human Factors Society Bulletin*, **30**, 1–3.

Hendrick, H.W., 1986b, Macroergonomics: a conceptual model for integrating human factors with organizational design. In: Brown, O. and Hendrick, H.W. eds. *Human Factors in Organizational Design and Management II: Proceedings of the Second Symposium held in Vancouver, B.C., Canada, 19–21 August*, Amsterdam: North-Holland, pp. 467–77.

Hendrick, H.W., 1987, Organizational design, in Salvendy, G. (Ed.) *Handbook of Human Factors*, pp. 470–94, New York: John Wiley and sons.

Hendrick, H.W., 1991, Ergonomics in organizational design and management, *Ergonomics*, **34**(6), 743–56.

Henriksson, K.G. and Bengtsson, A., 1991, Fibromyalgia: a clinical entity? *Can J Physiol Pharm*, **69**, 672–7.

Hensyl, W.R. (Ed.), 1990, *Stedman's Medical Dictionary*, 25th Edn, Baltimore: Williams & Wilkins.

Herberts, P., Kadefors, R. and Broman, H., 1980, Arm positioning in manual tasks: an electromyographic study of localized muscle fatigue, *Ergonomics*, **23**(7), 655–65.

Herberts, P., Kadefors, R., Andersson, G. and Petersén, I., 1981, Shoulder pain in industry: an epidemiological study on welders, *Acta Orthop Scand*, **52**, 299–306.

Herberts, P., Kadefors, R., Högfors, C. and Sigholm, G., 1984, Shoulder pain and heavy manual labor, *Clin Orthop Relat Res*, **191**, 166–78.

Hernberg, S., 1992 *Introduction to Occupational Epidemiology*, Chelesez, MI: Lewis Publishers.

Herrin, G.D., Jaraedi, M. and Anderson, C.K., 1986, Prediction of overexertion injuries using biomechanical and psychological models, *Am Ind Hyg Assoc J*, **47**(6), 322–30.

Hertzberg, H.T.E., 1955, Some contributions of applied physical anthropology to human engineering, *Ann NY Acad Sci*, **63**, 616–29.

Herzberg, F., 1966, *Work and the Nature of Man*, New York: Thomas Y. Crowell Co.

Hettinger, T., 1957, Ein Test zur Erkennung der Disposition zu Sehnenscheidenentzündungen, *Int Z Angew Physiol Einschl Arbeitsphysiol*, **16**, 472–9.

Hill, J.A., 1983, Epidemiologic perspective on shoulder injuries, *Clin Sports Med*, **2**(2), 241–6.

Himmelstein, J.S. and Andersson, G.B.J., 1988, Low back pain: risk evaluation and preplacement screening, *Occup Med*, **3**(2), 255–69.

Hochberg, F.H., Leffert, R.D., Heller, M.D. and Merriman L., 1983, Hand difficulties among musicians, *JAMA*, **249**(14), 1869–72.

Hocking, B., 1987, Epidemiological aspects of repetition strain injury in telecom Australia, *Med J Aust*, **147**(5), 218–22.

Holling, H.E. and Verel, D., 1957, Circulation in the elevated forearm, *Clin Sci*, **16**, 197–213.

Holzmann, P., 1982, ARBAN: a new method for analysis of ergonomic effort, *Appl Ergon*, **13**(2), 82–6.

Hoppenfeld, S., 1977, *Orthopaedic Neurology*, Philadelphia: J.B. Lippincott Company.

Hoppenfeld, S., 1976, *Physical Examination of the Spine and Extremities*, New York: Appleton Century Croft.

House, E.R., 1980, *Evaluating with Validity*, Beverly Hills, CA: Sage.

House, J.S. and Wells, J.A., 1978, Occupational stress, social support and health. In: *Reducing Occupational Stress: Proceedings of a Conference, May 10–12 1977 at Westchester Division, New York Medical Center, White Plains, New York, Division of Biomedical and Behavioral Science, NIOSH*, Cincinnati, OH: DHEW. (Publication no. 78-140).

Huang, J., Ono, Y., Shibata, E., Takeuchi, Y. and Hisanaga, N., 1988, Occupational musculoskeletal disorders in lunch centre workers, *Ergonomics*, **31**(1), 65–75.

Hueston, J.T., 1963, *Dupuytren's Contracture*, Edinburgh: E & S Livingstone, pp. 100.

Hultman, G., Nordin, M. and Örtengren, R., 1984, The influence of a preventive educational programme on trunk flexion in janitors, *Appl Ergon*, **15**(2), 127–33.

Hünting, W., Läubli, T. and Grandjean, E., 1981, Postural and visual loads at VDT workplace: 1. constrained postures, *Ergonomics*, **24**(12), 917–31.

Hurrell, J.J. and McLarey, M.A., 1988, Exposure to job stress: a new psychometric instrument, *Scand J Work Environ Health*, **14**(Suppl 1), 27–8.

Hviid-Andersen, J. and Gaardboe-Poulsen, O., 1990, *Helbredsprofil Bland Kvindelige Bekædningsindustriarbejdere*, in Danish with English summary, Copenhagen: The Work Environment Fund.

ILO, 1987, *Safety-health and Working Conditions: Training Manual*, Geneva: International Labour Office.

Imada, A.S., 1990, Ergonomics: influencing management behaviour, *Ergonomics*, **33**(5), 621–8.

Imada, A.S., 1991, The rationale and tools of participatory ergonomics, in Noro, K. and Imada, A. (Eds.) *Participatory Ergonomics*, pp. 30–50, London: Taylor & Francis.

Imrhan, S.N., 1989, Trends in finger pinch strength in children, adults and the elderly, *Human Factors*, **31**(6), 689–701.

Interdisciplinary task force on rehabilitation, 1988, *Set of Evaluative and Decision-Making Forms to Be Used by Rehabilitation Councellors in Adapting Work and Working Conditions: Musculoskeletal Impairments*, Montreal: Institut de recherche en santé et en sécurité du travail du Québec.

Isernhagen, S.J., 1992, Work hardening, in *Occupational Musculoskeletal Disorders: Occurrence, Prevention and Therapy*, pp. 119–22, Basel: EULAR Publishers.

ISO, 1981, *Ergonomic Principles of the Design of Work Systems*, Geneva: International Organisation for Standardization. (Reference no. ISO 6385:1981).

ISO, 1986, *Mechanical Vibration: Guidelines for the Measurement and the Assessment of Human Exposure to Hand-transmitted Vibration*, Geneva: International Organisation for Standardization. (Reference no. ISO 5349:1986).

ISO, 1987, *International Standard ISO 9000: Quality Management and Quality Assurance Standards: Guidelines for Selection and Use*, 1st Edn, Geneva: International Organisation for Standardization. (Reference no. ISO 9000: 1987 E).

ISO, 1992a, *Ergonomic Requirements for Office Work with Visual Display Terminals (VDTs): Part 1. General Introduction*, Geneva: International Organisation for Standardization. (Reference no. ISO 9241–1:1992).

ISO, 1992b, *Ergonomic Requirements for Office Work with Visual Display Terminals (VDTs): Part 2. Guidance on Task Requirements*, Geneva: International Organisation for Standardization. (Reference no. ISO 9241–2:1992).

Itani, T., Onishi, N., Sakai, K. and Shindo, H., 1979, Occupational hazard of female film rolling workers and effects of improved working conditions, *Arh Hig Rada Toksikol*, **30**, 1243–51.

Jackson, S.E. and Schuler, R.S., 1985, A meta-analysis and conceptual critique of research on role ambiguity and role conflict in work settings, *Organ Behav Hum Decis Process*, **36**, 16–78.

Jackson, M.J., Jones, D.A. and Edwards, R.H.T., 1985, Vitamin E and muscle diseases, *J Inherited Metab Dis*, **1**(suppl 8), 84–7.

James, F.R., Large, R.G., Bushnell, J.A. and Wells, J.E., 1991, Epidemiology of pain in New Zealand, *Pain*, **44**, 279–83.

Järvholm, U., Palmerud, G., Karlsson, D., Herberts, P. and Kadefors, R., 1990, Intramuscular pressure and electromyography in four shoulder muscles, *J Orthop Res*, **9**, 609–19.

Järvholm, U., Palmerud, G., Styf, J., Herberts, P. and Kadefors, R., 1988, Intramuscular pressure in the supraspinatus muscle, *J Orthop Res*, **6**(2), 230–8.

Jeyaratnam, J., Ong, C.N., Kee, W.C., Lee, J. and Koh, D., 1989, Musculoskeletal symptoms among VDU operators. In: Smith, M.J. and Salvendy, G. eds. *Work with Computers: Organizational, Management, Stress and Health Aspects*, Amsterdam: Elsevier Science Publishers, pp. 330–7.

Johansson, G. and Lindström, B.O., 1975, *Paced and Un-Paced Work under Salary and Piece-Rate Conditions*, Stockholm: University of Stockholm. Department of Psychology. (Report no. 459).

Johansson, G. and Aronsson, G., 1984, Stress reactions in computerized administrative work, *J Occup Behav*, **5**, 159–181.

Johansson, H. and Sojka, P., 1991, Pathophysiological mechanism involved in genesis and spread of muscular tension in occupational muscle pain and in chronic musculoskeletal pain syndromes: a hypothesis, *Med Hypotheses*, **35**, 196–203.

Johansson, G., Aronsson, G. and Lindström, B.O., 1978, Social psychological and neuroendocrine stress reactions in highly mechanized work, *Ergonomics*, **21**, 583–99.

Jonsson, B., 1982, Measurement and evaluation of local muscular strain in the shoulder during constrained work, *J Hum Ergol*, **11**, 73–88.

Jonsson, B., 1988, The static load component in muscle work, *Eur J Appl Physiol*, **57** , 305–10.

Jonsson, B.G., Persson, J. and Kilbom, Å., 1988, Disorders of the cervicobrachial region among female workers in the electronics industry: a two-year follow up, *International Journal of Industrial Ergonomics*, **3**(1), 1–12.

Juntunen, J., Matikainen, E., Seppäläinen, A.M. and Laine, A., 1983, Peripheral neuropathy and vibration syndrome, *Int Arch Occup Environ Health*, **52**, 14–24.

Kahn, H. and Cooper, C.L., 1986, Computing stress, *Curr Psychol Res Rev*, **5**(2), 148–62.

Kalimo, R. and Leppänen, A., 1985, Feedback from video display terminals, performance control and stress in text preparation in the printing industry, *J Occup Psychol*, **58**, 27–38.

Kamwendo, K., Linton, S.J. and Moritz, U., 1991, Neck and shoulder disorders in medical secretaries, *Scand J Rehab Med*, **23**, 127–33.

Karas, S.E., 1990, Thoracic outlet syndrome, *Clin Sports Med*, **9**, 297–310.

Karasek, R.A., 1979, Job demands, job decision latitude and mental strain: implications for job redesign, *Admin Sci Q*, **24**, 285–308.

Karasek, R.A., Baker, D., Marxer, F., Ahlbom, A. and Theorell, R., 1981, Job decision latitude, job demands, and cardiovascular disease, in Salvendy, G. and Smith, M.J. (Eds.) *Machine-Pacing and Occupational Stress*, pp. 694–705, London: Taylor & Francis.

Karhu, O., Kansi, P. and Kuorinka, I., 1977, Correcting working postures in industry: a practical method for analysis, *Appl Ergon*, **8**(4), 199–201.

Kasl, S.V., 1978, Epidemiological contributions to the study of work stress, in Cooper, C.L. and Payne, R. (Eds.) *Stress at Work*, pp. 3–48, New York: John Wiley and Sons.

Katz, J.N., Larson, M.G., Fossel, A.H. and Liang, M.H., 1991, Validation of a surveillance case definition of carpal tunnel syndrome, *Am J Pub Health*, **81**(2), 189–93.

Keikenwan Shōkōgun Iinkaï, 1973, Nihon sangyp-eisei gakkai keikenwan shōkōgun iinkai hōkokusho (Report of the Committee on Occupational Cervicobrachial

Disorder of the Japan Association of Industrial Health), *Jpn J Ind Health*, **15**, 304–311.

Keir, P.J. and Wells, R.P., 1992, MRI of the carpal tunnel: implications for carpal tunnel syndrome. In: Kumar, S. ed. *Advances in Industrial Ergonomics and Safety IV: Proceedings of the Annual International Industrial Ergonomics and Safety Conference held in Denver, Colorado, 10–14 June*, London: Taylor & Francis, pp. 753–60.

Kellgren, J.H. and Lawrence, J.S., 1952, Rheumatism in miners, part II: X-ray study, *Brit J Ind Med*, **9**, 197–207.

Kellgren, J.H. and Lawrence, J.S., 1957, Radiological assessment of osteoarthritis, *Ann Rheum Dis*, **16**, 494–502.

Kennedy, J.C., Hawkins, R. and Krissoff, W.B., 1978, Orthopaedic manifestations of swimming, *Am J Sports Med*, **6**, 309–22.

Kerguelen, A., 1986, *L'Observation systématique en Ergonomie: Élaboration d'un Logiciel d'Aide au Recueil et à l'Analyse des Données*, Thèse présentée en vue d'obtenir le diplôme d'ergonomiste, Paris: Conservatoire National des Arts et Métiers.

Keyserling, W.M., 1986, A computer-aided system to evaluate postural stress in the workplace, *Am Ind Hyg Assoc J*, **47**(10), 641–9.

Keyserling, W.M., Herrin, G.K. and Chaffin, D.B., 1980, Isometric strength testing as a means of controlling medical incidents on strenuous jobs, *J Occup Med*, **22**(5), 332–6.

Keyserling, W.M., Brouwer, M. and Silverstein, B.A., 1992, A checklist for evaluating ergonomic risk factors resulting from awkward postures of the legs, trunk and neck, *International Journal of Industrial Ergonomics*, **9**, 283–301.

Keyserling, W.M., Stetson, D.S., Silverstein, B.A. and Brouwer, M.L., 1993, A checklist for evaluating ergonomic risk factors associated with upper extremity cumulative trauma disorders, *Ergonomics*, **36**(7), 807–31.

Keyserling, W.M., Herrin, G.D., Chaffin, D.B., Armstrong, T.J. and Foss, M.L., 1980, Establishing an industrial strength testing program, *Am Ind Hyg Assoc J*, **41**(October), 730–6.

Kilbom, Å., 1988, Intervention programmes for work related neck and upper limb disorders: strategies and evaluation, *Ergonomics*, **31**(5), 735–47.

Kilbom, Å., 1988, Isometric strength and occupational muscle disorders, *Eur J Appl Physiol*, **57**, 322–6.

Kilbom, Å. and Persson, J., 1987, Work technique and its consequences for musculoskeletal disorders, *Ergonomics*, **30**(2), 273–9.

Kilbom, Å., Persson, J. and Jonsson, B.G., 1986, Disorders of the cervicobrachial region among female workers in the electronics industry, *International Journal of Industrial Ergonomics*, **1**(1), 37–47.

Kilbom, Å., Liew, M., Lagerlöf, E. and Broberg, E., 1984, *Ergonomisk Studie av Muskuloskeleta Sjukdomar Anmlda som Arbetsskador*, pp. 7–41, Stockholm: National Institute of Occupational Health. (Arbete och Hälsa, 45).

Kitai, E., Itay, S., Ruder, A., Engel, J. and Modan, M., 1986, An epidemiological study of lateral epicondylitis (tennis elbow) in amateur male players, *Ann Chir Main*, **5**(2), 113–21.

Kivi, P., 1984, Rheumatic disorders of the upper limbs associated with repetitive occupational tasks in Finland in 1975–1979, *Scand J Rheumatology*, **13**, 101–7.

Kivimäki, J., 1992, Occupational related ultrasonic findings in carpet and floor layer's knees, *Scand J Work Environ Health*, **18**, 400–2.

Kivimäki, J., Riihimäki, H. and Hänninen, K., 1992, Knee disorders in carpet and floor layers and painters, *Scand J Work Environ Health*, **18**, 310–6.

Kjellberg, A., 1990, Subjectives behavioral and psychophysiological effects of noise, *Scand J Work Environ Health*, **16**(suppl 1), 29–38.

Kjellberg, A., Sköldström, B. and Tesaiz, M., 1991, Equal EMG response levels to a 100 and 1000 Hz tone. In: *Proceedings of Internoise*, pp. 847–50.

Klaucke, D.N., Klaucke, D.N., Buehler, J.W., Thacker, S.B., Parrish, R.G., Trowbridge, F.L. and Berkelman, R.L., 1988, Guidelines for evaluating surveillance systems, *MMWR*, **37**(suppl 5), 1–18.

Kleinert, H.E. and Volianitis, G.J., 1965, Thrombosis of the palmar arterial arch and its tributaries: etiology and newer concepts in treatment, *J Trauma*, **5**, 447–57.

Knave, B. and Widebäck, P.-G. (Eds.), 1987, *Work with Display Units 86: Selected Papers from the International Scientific Conference on Work with Display Units, Stockholm, Sweden, May 12–15*, 1986, Amsterdam: North-Holland.

Knave, B.G., Wibom, R.I., Voss, M., Hedström, L.D. and Bergqvist, U.O.V., 1985, Work with video display terminals among office employees: subjective symptoms and discomfort, *Scand J Work Environ Health*, **11**, 457–66.

Kogi, K., 1982, Finding appropriate work-rest rhythm for occupational strain on the basis of electromyographic and behavioural changes. In: Buser, P.A., Cobb, W.A. and Okuma, T. eds. *Kyoto Symposia*, Amsterdam: Elsevier Biomedical Press, pp. 738–49.

Kogi, K., 1985, *Improving Working Conditions in Small Enterprises in Developing Asia*, Geneva: International Labour Office.

Kohatsu, N.D. and Schurman, D.J., 1991, Risk factor for the development of osteoarthrosis of the knee, *Clin Orthop Relat Res*, **261**, 242–6.

Komi, P.V., Salonen, M., Järvinen, M. and Kokko, O., 1987, In vivo registration of achilles tendon forces in man: I. Methodological development, *Int J Sports Med*, **8**(suppl), 3–8.

Kornberg, J.P., 1992, *The Workplace Walk-through*, Boca Raton: Lewis Publishers. (Kornberg's operational guideline series in occupational medicine, 1).

Koski, A.V. and McGill, S.M., in press, Shoulder flexion strength for use in occupational risk analysis, Accepted by *Clinical Biomechanics* in 1993.

Kraft, G. and Detels, P., 1972, Position of function of the wrist, *Arch Phys Med Rehab*, **53**, 272–5.

Kroemer, K.H., 1986, VDU workstation design: research, recommendations and standards. In: Knave, B. and Widebäck, G.P. eds. *Working with VDUs: Proceedings of the International Scientific Conference, Stockholm, May 12–15*, part II, Amsterdam: Elsevier Science Publishers B.V., pp. 882–6.

Kroemer, K.H.E., 1992, Personnel training for safer material handling, *Ergonomics*, **35**(9), 1119–34.

Kroemer, K.H.E. and Hill, S.G., 1986, Preferred line of sight angle, *Ergonomics*, **29**(9), 1129–34.

Kuhlman, J.R., Iannotti, J.P., Kelly, M.J., Riegler, F.X., Gevaert, M.L. and Ergin, T.M., 1992, Isokinetic and isometric measurement of strength of external rotation and abduction of the shoulder, *J Bone Joint Surg*, **74A**(9), 1320–33.

Kukkonen, R., Luopajärvi, T. and Riihimäki, V., 1983, Prevention of fatigue amongst data entry operators, in Kvalseth, T.O. (Ed.) *Ergonomics of Workstation Design*, pp. 28–34, London: Butterworths.

Kuorinka, I., 1981, Discomfort and motor skills in semi-paced tasks, in Salvendy, G. and Smith, M.J. (Eds.) *Machine Pacing and Occupational Stress: Proceedings of the International Conference, Purdue University, March*, pp. 295–301, London: Taylor & Francis.

Kuorinka, I. and Koskinen, P., 1979, Occupational rheumatic diseases and upper limb strain in manual jobs in a light mechanical industry, *Scand J Work Environ Health*, **5**(suppl 3), 39–47.

Kuorinka, I. and Viikari-Juntura, E., 1982, Prevalence of neck and upper limb disorders (NLD) and work load in different occupational groups: problems in classification and diagnosis, *J Hum Ergol*, **11**, 65–72.

Kuorinka, I., Videman, T. and Lepistö, M., 1981, Reliability of a vibration test in screening for predisposition to tenosynovitis, *Eur J Appl Physiol*, **47**, 365–76.

Kuorinka, I., Jonsson, B., Kilbom, Å., Vinterberg, H., Biering-Sørensen, F., Andersson, G. and Jørgensen, K., 1987, Standardised Nordic questionnaire for the analysis of musculoskeletal symptoms, *Appl Ergon*, **18**(3), 233–7.

Kurppa, K., Waris, P. and Rokkanen, P., 1979, Peritendinitis and tenosynovitis, *Scand J Work Environ Health*, **5**(suppl 3), 19–24.

Kurppa, K., Viikari-Juntura, E., Kuosma, E., Huuskonen, M. and Kivi, P., 1991, Incidence of tenosynovitis or peritendinitis and epicondylitis in a meat processing factory, *Scand J Work Environ Health*, **17**, 32–7.

Langenskjöld, A., Michelsson, J.E. and Videman, T., 1979, Osteoarthritis of the knee in the rabbit produced by immobilization: attempts to achieve a reproducible model for studies on pathogenesis and therapy, *Acta Orthop Scand*, **50**(1), 1–14.

Larouche, V., 1984, *Formation et Perfectionnement en Milieu Organisationnel*, Montréal: JCL.

Larsson, S.E., Bodegård, L., Henriksson, K.G. and Öberg, P.A., 1990, Chronic trapezius myalgia: morphology and blood flow studied in 17 patients, *Acta Orthop Scand*, **61**(5), 394–8.

Larsson, S.-E., Bengtsson, A., Bodegård, L., Henriksson, K.G. and Larsson, J., 1988, Muscle changes in work related chronic myalgia, *Acta Orthop Scand*, **59**(5), 552–6.

Last, J.M. (Ed.), 1986, *A Dictionary of Epidemiology*, New York: Oxford Press.

Last, J.M. (Ed.), 1988, *A Dictionary of Epidemiology*, 2nd Edn, New York: Oxford University Press.

Laville, A., 1982, Postural reactions related to activities on VDU. In: Grandjean, E. and Vigliani, E. eds. *Ergonomic Aspects of Visual Display Terminals: Prodeedings of the International Workshop, Milan, March 1980*, London: Taylor & Francis, pp. 167–74.

Lawler, E.E., 1986, *High Involvement Management*, San Francisco: Jossey-Bass Publishers.

Lawrence, J.S., 1961, Rheumatism in cotton operatives, *British Journal of Occupational Medicine*, **18**, 270–6.

Lawrence, P.R. and Lorsch, J.W., 1969, *Organization and Environment*, Homewood, Ill: Irwin.

Layzer, R.B. and Rowland, L.P., 1971, Muscular pain, *New Engl J Med*, **285**, 31.

Leach, R.E. and Miller, J.K., 1987, Lateral and medial epicondylitis of the elbow, *Clin Sports Med*, **6**(2), 259–72.

Leavell, H.R. and Clark, E.G., 1965, *Preventive Medicine for the Doctor in his Community: An Epidemiologic Approach*, 3rd Edn, New York: McGraw-Hill Book Company.

Lederman, R.J., 1986, Nerve entrapment syndromes in instrumental musicians, *Med Probl Performing Artists*, (June), 45–8.

LeGrande, D.E., 1993, VDT repetitive motion health concerns in the U.S. telecommunications industry. In: Luczak, H., Çakin, A.E. and Çakin, G. eds. *Work with Display Units WWDU '92, Berlin, 1–4 September 1992*, Amsterdam: Elsevier Science Publishers, pp. 240–7.

Leino, P., 1989, Symptoms of stress predict musculoskeletal disorders, *J Epidemiol Community Health*, **43**, 293–300.

Leplat, J., 1989, Error analysis, instrument and object of task analysis, *Ergonomics*, **32**(7), 813–22.

Leppilahti, J., Orava, S., Karpakka, J. and Takala, T., 1991, Overuse injuries of the achilles tendon, *Ann Chir Gynaecol*, **80**(2), 202–7.

Levi, L., 1972, *Stress and Distress in Response to Psychosocial Stimuli*, New York: Pergamon Press.

Lewin, K., 1947, Group decision and social change, in Maccoby, E.E., Newcomb, T.M. and Hartley, E.L. (Eds.) *Readings in Social Psychology*, 3rd Edn, pp. 197–211, New York: Holt, Rinehart & Winston.

Lewy, R., 1985, Preplacement examination of temporary hospital workers, *JOM*, **27**, 122–4.

Lewy, R., 1986, Preplacement examinations for medical center workers, *JOM*, **28**(11), 1189.

Lifshitz, Y. and Armstrong, T.J., 1986, A design checklist for control and prediction of cumulative trauma disorders in intensive manual jobs. In: *Proceedings of the Human Factors Society 30th Annual Meeting: A Cradle for Human Factors, held in Daytona, Ohio, September 29–October 3*, Volume 2, Ann Arbor, MI: Human Factors Society, pp. 837–41.

Lindberg, H. and Danielsson, L.G., 1984, The relationship between labour and coxarthrosis, in *Clin Orthop Relat Res*, **191**, 159–161.

Lindberg, H. and Montgomery, F., 1987, Heavy labor and the occurrence of gonarthrosis, *Clin Orthop Relat Res*, **214**, 235–6.

Lindman, R., Eriksson, A. and Thornell, L.-E., 1991a, Fiber type composition of the human female trapezius muscle, *Am J Anat*, **190**, 385–92.

Lindman, R., Hagberg, M., Ängqvist, K.-A., Söderlund, K., Hultman, E. and Thornell, L.-E., 1991b, Changes in muscle morphology in chronic trapezius myalgia, *Scand J Work Environ Health*, **17**, 347–55.

Linton, S.J., 1990, Risk factors for neck and back pain in a working population in Sweden, *Work & Stress*, **4**(1), 41–9.

Little, J.M. and Ferguson, D.A., 1972, The incidence of the hypothenar hammer syndrome, *Arch Surg*, **105**, 684–5.

Lowenthal, G., 1986, Medical center worker preplacement screening: a follow-up study, *JOM*, **28**(6), 451–2.

Lowrey, C.W., Chadwick, R.O. and Wattman, E.M., 1976, Digital vessel trauma from repetitive impacts in baseball catchers, *J Hand Surg*, **1**, 236–8.

Luchetti, R., Schoenhuber, R., Alfarano, M., Deluca, S., De Cicco, G. and Landi, A., 1990, Carpal tunnel syndrome: correlations between pressure measurement and intraoperative electrophysiological nerve study, *Muscle & Nerve*, **13**, 1164–8.

Lundborg, G., 1988, *Nerve Injury and Repair*, Edinburgh: Churchill Livingstone.

Lundborg, G., Dahlin, L.B., Danielsen, N. and Kanje, M., 1990, Vibration exposure and nerve fibre damage, *J Hand Surg*, **15A**, 346–51.

Lundborg, G., Gelberman, R.H., Minteer-Convery, M., Lee, Y.F. and Hargens, A.R., 1982, Median nerve compression in the carpal tunnel: functional response to experimentally induced controlled pressure, *J Hand Surg*, **7**(3), 252–9.

Lundström, R. and Johansson, R.S., 1986, Acute impairment of the sensitivity of skin mechanoreceptive units caused by vibration exposure of the hand, *Ergonomics*, **29**(5), 687–98.

Luopajärvi, T., 1987, Workers' education, *Ergonomics*, 30(2), 305–11.

Luopajärvi, T., Kuorinka, I. and Kukkonen, R., 1982, The effects of ergonomic measures on the health of the neck and the upper extremities of assembly-line packers: a four year follow-up study. In: *Proceedings of the 8th Congress of the International Ergonomics Association, Tokyo, Japan, August 23–27*, pp. 515–6.

Luopajärvi, R., Kuorinka, I., Virolainen, M. and Holmberg, M., 1979, Prevalence of tenosynovitis and other injuries of the upper extremities in repetitive work, *Scand J Work Environ Health*, **5**(suppl 3), 48–55.

Lutz, G. and Hansford, T., 1987, Cumulative trauma disorders controls: the ergonomics program at Ethicon, Inc., *J Hand Surg*, **12A**, 863–6.

Mairiaux, P., Bettonville, M.N., Mawet, M. and Malchaire, J., 1986, Serum creatine kinase relationship to postural constraints in manual work, *Int Arch Occup Environ Health*, **58**, 61–8.

Mäkelä, M., Heliövaara, M., Sievers, K., Impivaara, O., Knekt, P. and Aromaa, A., 1991, Prevalence, determinants, and consequences of chronic neck pain in Finland, *Am J Epidemiol*, **134**(11), 1356–67.

Margolis, B.L., Kroes, W.M. and Quinn, R.P., 1974, Job stress: an unlisted occupational hazard, *J Occup Med*, **16**(10), 659–61.

Markison, R.E., 1990, Treatment of musical hands: redesign of the interface, *Hand Clin*, **6**(3), 525–44.

Marras, W.S., 1990, Industrial electromyography (EMG), *International Journal of Industrial Ergonomics*, **6**, 89–93.

Marras, W.S. and Schoenmarklin, R.W., 1991, Wrist motions and CTD risk in industrial and service environments. In: Queinnec, Y. and Daniellou, F. eds. *Designing for Everyone: Proceedings of the Eleventh Congress of the International Ergonomics Association, Paris*, vol. 1, New York: Taylor & Francis, pp. 36–8.

Marras, W.S. and Schoenmarklin, R.W., 1993, Wrist motions in industry, *Ergonomics*, **36**(4), 341–51.

Marras, W.S., Lavender, S.A., Leurgans, S.E., Rajulu, S.L., Allread, W.G., Fathallah, F.A., Ferguson, S.A., 1993, The role of dynamic three-dimensional trunk motion in occupational-related low back disorders: the effects of workplace factors trunk position and trunk motion characteristics on risk of injury, *Spine*, **18**(5), 617–28.

Marriott, I.A. and Stuchly, M.A., 1986, Health aspects of work with visual display terminals, *J Occup Med*, **28**(9), 833–47.

Martin, D.K. and Dain, S.J., 1988, Postural modifications of VDU operators wearing bifocal spectacles, *Appl Ergon*, **19**(4), 293–300.

Maslow, A.H., 1970, *Motivation and Personality*, 2nd Edn, New York: Harper and Row.

Mastin, J.P., Henningsen, G.M. and Fine, L.J., 1992, Use of biomarkers of occupational musculoskeletal disorders in epidemiology and laboratory animal model development, *Scand J Work Environ Health*, **18**(suppl 2), 85–7.

Mathiassen, S.E., 1993, *Variation in Shoulder-Neck Activity: Physiological, Psychophysical and Methodological Studies of Isometric Exercise and Light Assembly Work*, Stockholm: National Institute of Occupational Health. (Arbete och Hälsa, 7).

Mathiassen, S.E. and Winkel, J., 1991, Quantifying variation in physical load using exposure-vs-time data, *Ergonomics*, **34**(12), 1455–68.

Mathiassen, S.E. and Winkel, J., 1992, Can occupational guidelines for work-rest schedules be based on endurance time data, *Ergonomics*, **35**(3), 253–9.

McAtamney, L. and Corlett, E.N., 1993, RULA: a survey method for the investigation of work related upper limb disorders, *Appl Ergon*, **24**(2), 91–9.

McCormack, R.R., Inman, R.D., Wells, A., Berntsen, C. and Imbus, H.R., 1990, Prevalence of tendinitis and related disorders of the upper extremity in a manufacturing workforce, *J Rheumatol*, **17**(7), 958–64.

McGill, S.M. and Norman, R.W., 1986, Partitioning of the L4/L5 dynamic moment into disc, ligamentous and muscular components during lifting, *Spine*, **11**(7), 666–78.

McKenzie, F., Storment, J., Van Hook, P. and Armstrong, T.J., 1985, A program for control of repetitive trauma disorders associated with hand tool operations in a telecommunications manufacturing facility, *Am Ind Hyg Assoc J*, **46**(11), 674–8.

McLellan, D.L. and Swash, M., 1976, Longitudinal sliding of the median nerve during movements of the upper limb, *J Neurol Neurosurg Psychiat*, **39**, 566–70.

McRae, R., 1983, *Clinical Orthopaedic Examination*, 2nd Edn, Edinburgh: Churchill Livingstone.

Mileti, D.S., Gillespie, D.F. and Haas, J.E., 1977, Size and structure in complex organizations, *Social Forces*, **56**(1), 208–17.

Miller, M. and Wells, R., 1988, The Influence of wrist flexion, wrist deviation and forearm pronation on pinch and power grasp strength, In: *Proceedings of the Fifth Biennial Conference of the Canadian Society for Biomechanics*, London, Ont.: Spodym Publishers, 112–3.

Mills, K.R. and Edwards, R.H.T., 1983, Investigative strategies for muscle pain, *J Neurol Sci*, **58**, 73–88.

Mital, A. and Sanghavi, N., 1986, Comparison of maximum volitional torque exertion capabilities of males and females using common hand tools, *Hum Fact*, **28**(3), 283–94.

Monk, T.H. and Tepas, D.I., 1985, Shift work, in Cooper, C.L. and Smith, M.J. (Eds.) *Job Stress and Blue Collar Work*, pp. 65–84, New York: John Wiley and Sons.

Moore, A.E. and Wells, R., 1992, Towards a definition of repetitiveness in manual tasks. In: Mattila, M. and Karwowski, W. eds. *Computer Applications in Ergonomics, Occupational Safety & Health: Proceedings of the International Conference on Computer-aided Ergonomics and Safety held in Tampere, Finland, 18–20 May*, Amsterdam: North-Holland, pp. 401–8.

Moore, A., Wells, R. and Ranney, D., 1991, Quantifying exposure in occupational manual tasks with cumulative trauma disorder potential, *Ergonomics*, **34**(12), 1433–53.

Morgan-Hughes, J.A., 1979, Painful disorders of muscle, *Brit J Hosp Med*, **22**, 360–5.

Morgenstern, H., Kilsh, M., Kraus, J. and Margolis, W., 1991, A cross-sectional study of hand/wrist symptoms in female grocery checkers, *Am J Ind Med*, **20**, 209–18.

Morris, D.B., 1991, *The Culture of Pain*, Berkeley: University of California Press.

Mostardi, R.A., Noe, D.A., Kovacik, M.W. and Porterfield, J.A., 1992, Isokinetic lifting strength and occupational injury: a prospective study, *Spine*, **17**(2), 189–93.

Mühlau, G., Both, R. and Kunath, H., 1984, Carpal tunnel syndrome: course and prognosis, *J Neurol*, **231**, 83–6.

Murphy, L. and Hurrell, J.J., Jr, 1980, Machine pacing and occupational stress, in Schwartz, R. (Ed.) *New Developments in Occupational Stress*, Washington, DC: DHHS (NIOSH). (Publication no. 81–102).

Nachemson, A.L., Andersson, G.B.J. and Schultz, A.B., 1986, Valsalva maneuvre biomechanics: effects on lumbar trunk loads of elevated intra abdominal pressure, *Spine*, **11**(5), 476–9.

Nagamachi, M., 1991, Application of participatory ergonomics through quality-circle activities, in Noro, K. and Imada, A. (Eds.) *Participatory Ergonomics*, pp. 139–64, London: Taylor & Francis.

Nagamachi, M. and Imada, A.S., 1992, A macroergonomic approach for improving safety and work design. In: *Proceedings of the Human Factors Society 36th Annual Meeting*, Santa Monica, CA: Human Factors Society, pp. 859–61.

Nathan, P.A., Meadows, K.D. and Doyle, L.S., 1988, Occupation as a risk factor for impaired sensory conduction of the median nerve at the carpal tunnel, *J Hand Surg*, **13B**(2), 167–70.

Nathan, P.A., Keniston, R.C. and Myers, L.D., 1992, Obesity as a risk factor for slowing of sensory conduction of the median nerve in industry: a cross-sectional and longitudinal study involving 429 workers, *JOM*, **34**(4), 379–83.

Neer, C.S., 1983, Impingement lesions, *Clin Orthop*, **173**, 70–7.

Negandhi, A.R., 1977, A model for analyzing organization in cross cultural settings: a conceptual scheme and some research findings, in Negandhi, A.R., England, G.W. and Wilbert, B. (Eds.) *Modern Organization Theory*, pp. 285–312, Kent State, OH: Kent State University Press.

Neviaser, J.S., 1980, Adhesive capsulitis and the stiff and painful shoulder, *Orthop Clin North Am*, **11**, 327–31.

Nielsen, R., 1986, Clothing and thermal environments: field studies on industrial work in cool conditions, *Appl Ergon*, **17**(1), 47–57.

Nilsson, T., Burström, L. and Hagberg, M., 1989, Risk assessment of vibration exposure and white fingers among platers, *Int Arch Occup Environ Health*, **61**, 473–81.

Nilsson, T., Hagberg, M., Burström, L, 1990, Prevalence and odds ratios of numbness and carpal tunnel syndrome in different exposure categories of platers, in Okada, A. and Dupuis, W.T.H. (Eds) *Hand-arm Vibration*, pp. 235–9, Kanazawa, Japan: Kyoei Press Co.

NIOSH, 1981, *Work Practices Guide for Manual Lifting*, Cincinnati, OH.: U.S. Departent of Health and Human Services. (Technical Report No. 81–122).

NIOSH, 1989, *Criteria for a Recommended Standard: Occupational Exposure to Hand-arm Vibration*, Washington D.C.: US Department of Health and Human Services.

NIOSH, 1990, *OSHA Instruction CPL 2.85: Directorate of Compliance Programs: Appendix C, Guidelines suggested by NIOSH for Videotape Evaluation of Work Station for Upper Extremities Cumulative Trauma Disorders*, Washington D.C.: US Department of Health and Human Services.

Nordgren, B., 1972, Anthropometric measures and muscle strength in young women, *Scand J Rehab Med*, **4**, 165–9.

Norkin, C.C. and Levangie, P.K., 1992, *Joint Structure and Function: a Comprehensive Analysis*, 2nd Edn, Philadelphia: F.A. Davis Company.

Norman, R.W. and McGill, S.M., 1984, *WATBAK: a Computer Software Package to Estimate Low Back Compressive and Shear Forces and NIOSH 'Action' and 'Maximum Permissible' Limits for the Assessment of Acute Strength Demands and Effects of Manually Lifting Loads: Final Report to Dr. W.S. Myles*, Toronto: Defence and Civil Institute of Environmental Medicine.

Noro, K., 1991, Concepts methods and people, in Noro, K. and Imada, A. (Eds.) *Participatory Ergonomics*, pp. 3–29, London: Taylor & Francis.

Noro, K. and Imada, A.S., 1991, *Participatory Ergonomics*, London: Taylor &9 Francis.

O'Conor, D.S., 1933, Early recognition of iliopectineal bursitis, *Surg Gynecol Obstet*, **57**, 674–84.

O'Driscoll, S.W., Horii, E., Ness, R. Cahalan, T.D., Richards, R.R. and An, K.-N., 1992, The relationship between wrist position, grasp size, and grip strength, *J Hand Surg*, **17A(1)**, 169–77.

O'Hanlon, J.F., 1981, Stress in short-cycle repetitive work: general theory and an empirical test, in Salvendy, G. and Smith, M.J. (Eds.) *Machine Pacing and Occupational Stress: Proceedings of the International Conference, Purdue University, March*, pp. 213–22, London: Taylor & Francis.

Odenrick, P., Eklund, J., Malmkvist, A.-K., Örtengren, R. and Parenmark, G., 1988, Influence of work pace on trapezius muscle activity in assembly-line work. In: *Ergonomics International 88: Proceedings of the 10th Congress of the International Ergonomics Association, Sydney, Australia, 1–5 August*, London: Taylor & Francis, pp. 418–20.

Ohara, H., Aoyama, H. and Itani, T., 1976, Health hazard among cash register operators and the effects of improved working conditions, *J Hum Ergol*, **5**, 31–40.

Ohara, H., Itani, T. and Aoyama, H., 1982, Prevalence of occupational cervicobrachial disorder among different occupational groups in Japan, *J Hum Ergol*, **11**, 55–63.

Ohlsson, K., Attewell, R. and Skerfving, S., 1989, Self-reported symptoms in the neck and upper limbs of female assembly workers, *Scand J Work Environ Health*, **15**, 75–80.

Olhagen, B., 1980, Postinfective or reactive arthritis, *Scand J Rheumatol*, **9**, 193–202.

Ong, C.N., Koh, D., Phoon, W.O. and Low, A., 1988, Anthropometrics and display station preferences of VDU operators, *Ergonomics*, **31(3)**, 337–47.

Onishi, N., Namura, H., Sakai, K., Yamamoto, T., Hirayama, K. and Itani, T., 1976, Shoulder muscle tenderness and physical features of female industrial workers, *J Hum Ergol*, **5**, 87–102.

Ontario Ministry of Labour, 1988, *Handicapped Employment Program: Physical Demands Analysis*.

OSHA, 1991, *Ergonomics Program Management Guidelines for Meatpacking Plants*, reprinted, US Department of Labor, Occupational Safety & Health Administration. (OSHA, no. 3123).

Östberg, O. and Nilsson, C., 1985, Emerging technology and stress, in Cooper, C.L. and

Smith, M.J. (Eds.) *Job Stress and Blue Collar Work*, pp. 149–69, New York: John Wiley and Sons.

Osterman, A.L., Moskow, L. and Low, D.W., 1988, Soft-tissue injuries of the hand and wrist in racquet sports, *Clin Sports Med*, **7**(2), 329–48.

Östlin, P., 1989, The 'health-related selection effect' in occupational morbidity rates, *Scand J Soc Med*, **17**, 265–70.

Oxenburgh, M., 1984, Musculoskeletal injuries occurring in word processor operators. In: Adams and Stevenson eds. *Ergonomics and Technological change: Proceedings of the 21st Annual Conference of the Ergonomics Society of Australia and New Zealand, Sydney, Nov 28–30*, Victoria, Australia: Ergonomics Society of Australia and New Zealand, pp. 137–43.

Oxenburgh, M., 1991, *Increasing Productivity and Profit through Health & Safety*, Sydney, Australia: CCH International.

Oxenburgh, M.S., Rowe, S.A. and Douglas, D.B., 1985, Repetitive strain injury in keyboard operators, successful management over a two year period, *J Occupational Health & Safety, Australia and New Zealand*, **1**(2), 106–12.

Parenmark, G., Engvall, B., et al., 1987, Största arbetsskadan i verkstadsindustri halverad genom ergonomisk inskolning, *LAKAA*, **84**, 2204–6.

Parenmark, G., Engvall, B. and Malmkvist, A.-K., 1988, Ergonomic on-the-job training of assembly workers: arm-neck-shoulder complaints drastically reduced amongst beginners, *Appl Ergon*, **19**(2), 143–6.

Park, K.S. and Chaffin, D.B., 1974, A biomechanical evaluation of two methods of manual load lifting, *AIIE Trans*, **6**(2), 105–13.

Parsons, C.A. and Thompson, D., 1990, Comparison of cervical flexion in shop assistants and data input VDT operators. In: Lovesey, E.J. ed. *Contemporary Ergonomics 1990: Proceedings of the Ergonomics Society's 1990 Annual Conference, Leeds, England, 3–6 April*, London: Taylor & Francis, pp. 299–304.

Partridge, R.E.H. and Duthie, J.J., 1968, Rheumatism in dockers and civil servants: a comparison of heavy manual and sedentary workers, *Ann Rheum Dis*, **27**, 559–67.

Patkin, M. and Gormley, J., 1991, Skill, excess effort, and strain. In: Kumashiro, M. and Megaw, E.D. eds. *Towards Human Work: Solutions to Problems in Occupational Health and Safety: UOEH Xth International Symposium, Kitakyushu, Japan*, Bristol, Pa.: Taylor & Francis, pp. 145–50.

Pelmear, P.L., Taylor, W., Wasserman, D.E. (Eds.), 1992, *Hand-arm Vibration: a Comprehensive Guide for Occupational Health Professionals*, New York: Van Nostrand Reinhold.

Perrow, C., 1967, A framework for the comparative analysis of organizations, *Am Sociol Rev*, **32**(2), 194–208.

Peters, P., 1990, Successful return to work following a musculoskeletal injury, *AAOHN J*, **38**, 264–70.

Peyron, J.G., 1986, Review of the main epidemiological: etiologic evidence that implies mechanical forces as factors in osteoarthritis, *Eng Med*, **15**, 77.

Pheasant, S., 1987, *Ergonomics: Standards and Guidelines for Designers*, Milton Keynes: British Standards Institution.

Pien, F.D., Ching, D. and Kim, E., 1991, Septic bursitis: experience in a community practice, *Orthopedics*, **14**(9), 981–4.

Pitner, M.A., 1990, Pathophysiology of overuse injuries in the hand and wrist, *Hand Clin*, **6**(3), 355–66.

Poole, C.J.M., 1993, Seamstress's finger, *Brit J Ind Med*, **50**, 668–9.

Priest, J.D. and Nagel, D.A., 1976, Tennis shoulder, Am J Sports Med, **4**, 28–42.

Priest, J.D., Braden, V. and Gerberich, S.G., 1980, The elbow and tennis, part 1: an analysis of players with and without pain, *Physician Sportsmed*, **8**, 81–91.

Pryce, J.C., 1980, The wrist position between neutral and ulnar deviation that facilitates the maximum power grip strength, *J Biomech*, **13**, 505–11.

Punnett, L. and Robins, J., 1985, *Adjusting for Selection Bias in Cross-sectional Studies: Soft Tissue Disorders of the Upper Limb*, thesis, pp. 119–46, Boston: Harvard School of Public Health.

Punnett, L. and Keyserling, W.M., 1987, Exposure to ergonomic stressors in the garnment industry: application and critique of job-site work analysis methods, *Ergonomics*, **30**(7), 1099–116.

Punnett, L., Robins, J.M., Wegman, D.H. and Keyserling, W.M., 1985, Soft tissue disorders in the upper limbs of female garment workers, *Scand J Work Environ Health*, **11**, 417–25.

Punnett, L., Fine, L.J., Keyserling, W.M., Herrin, G.D. and Chaffin, D.B., 1991, Backdisorders and nonneutral trunk postures of automobile assembly workers, *Scand J Work Environ Health*, **17**(5), 337–46.

Quintner, J. and Elvey, R., 1990, *The Neurogenic Hypothesis of RSI*, Canberra, Australia: The Australian National University, National Centre for Epidemiology and Population Health. (Working paper no. 24).

Raddatz, D.A., Hoffman, G.S. and Franck, W.A., 1987, Septic bursitis: presentation, treatment and prognosis, *J Rheumatol*, **14**, 1160–3.

Radin, E.L., Paul, I.L. and Rose, R.M., 1972, Role of mechanical factors in pathogenesis of primary osteoarthritis, *LANCA*, **1**(March), 519–21.

Radwin, R.G. and Lin, M.L., 1993, An analytical method for characterizing repetitive motion and postural stress using spectral analysis, *Ergonomics* **36**(4), 379–89

Radwin, R.G., Armstrong, T.J. and Chaffin, D.B., 1987, Power hand tool vibration effects on grip exertions, *Ergonomics*, **30**, 833–35.

Rais, O., 1961, Peritenomyositis (peritendinitis) crepitans acuta, *Acta Chir Scand*, (suppl 268), 1–101.

Ranney, D., Wells, R. and Moore, A., 1992, Forearm muscle pain and tenderness in manual workers. In: *Occupational Disorders of the Upper Extremities, Proceedings of a Seminar held at the University of Michigan, Sept. 30–Oct 1.*

Régie nationale des usines Renault, Services des conditions de travail, 1976, *Les Profils de Postes: Méthode d'Analyse des Conditions de Travail*, Paris: Masson.

Reilly, C.H. and Marras, W.S., 1989, Simulift: a simulation model of human trunk motion, *Spine*, **14**(1), 5–11.

Rempel, D., Bloom, T., Tal, R., Hargens, A. and Gordon, L., 1992, A method of measuring intracarpal pressure and elementary hand manoeuvers. In: Hagberg, M. and Kilbom, Å. eds. *International Scientific Conference on Prevention of Work related Musculoskeletal Disorders, PREMUS, Stockholm, May 12–14*, Stockholm: National Institute of Occupational Health, pp. 249–50. (Arbete och hälsa, 17).

Reneman, R.S., 1975, The anterior and the lateral compartmental syndrome of the leg due to intensive use of muscles, *Clin Orthop Relat Res*, **113**, 69–80.

Replogle, J.O., 1983, Hand torque strength with cylindrical handles. In: Pope, A.T. and Haugh, L.D. eds. *Proceedings of the Human Factors Society, 27th Annual Meeting, Norfolk, Virginia, October 10–14*, Santa Monica, Ca.: Human Factors Society, 1, pp. 412–6.

Richardson, W.J. and Pape, E.S., 1982, Work sampling, in Salvendy, G. (Ed.) *Handbook of Industrial Engineering*, pp. 4.6.1–21, New York: John Wiley & Sons.

Rissanen, A., Heliövaara, M., Knekt, P., Aromaa, A. and Maatela, J., 1990, Risk of disability and mortality due to overweight in a Finnish population, *Br Med J*, **301**, 835–7.

Robbins, S.R., 1983, *Organization Theory: the Structure and Design of Organizations*, Englewood Cliffs, NJ: Prentice-Hall.

Roberts, W.J., 1986, A hypothesis on the physiological basis for causalgia and related pains, *Pain*, **24**, 297–311.

Rodgers, S.H., 1987, Recovery time needs for repetitive work, *Seminars in Occupational Medicine*, **2**(1), 19–24.

Rodgers, S.H., 1988, Job evaluation in worker fitness determination, *Occup Med*, **3**, 219–39.

Roebuck, J.A., Jr, Kroemer, K.H.E. and Thomson, W.G., 1975, *Engineering Anthropometry Methods*, Ann Arbor, MI: University Microfilms International: Van Nostrand Reinhold.

Roethlisberger, F.J. and Dickson, W.J., 1939, *Management and the Worker*, Cambridge, MA: Harvard University.

Rohmert, W., 1973, Problems of determination of rest allowances, part 2: determining rest allowances in different human tasks, *Appl Ergon*, **4**(3), 158–62.

Rooney, E.F., Morency, R.R. and Herrick, D.R., 1993, Macroergonomics and total quality management: a case study. In: Nielsen, R. and Jorgensen, K. eds. *Advances in Industrial Ergonomics and Safety V*, London: Taylor & Francis, pp. 493–8.

Rose, M.J., 1991, Keyboard operating posture and actuation force: implications for muscle over-use, *Appl Ergon*, **22**(3), 198–203.

Rose, L., 1992, *Ergo-index: Development of a Model to Estimate Physical Load, Pause Need and Production Time in Different Working Situations*, thesis for Licentiate of Engineering, G?teborg, Sweden: Chalmers University of Technology, Department of Injury Prevention.

Rosenstock, L. and Cullen, M.R., 1994, *Textbook of Occupational and Environmental Medicine*, Philadelphia: W.B. Saunders Company.

Rossignol, A.M., Morse, E.P., Summers, V.M. and Pagnotto, L.D., 1987, Video display terminal use and reported health symptoms among Massachusetts clerical workers, *J Occup Med*, **29**(2), 112–8.

Roto, P. and Kivi, P., 1984, Prevalence of epicondylitis and tenosynovitis among meatcutters, *Scand J Work Environ Health*, **10**, 203–5.

Rutenfranz, J., Colquhoun, W.P., Knauth, P. and Ghata, J.N., 1977, Biomedical and psychosocial aspects of shift work, *Scand J Work Environ Health*, **3**, 165–82.

Ryan, G.A., 1989, The prevalence of musculo-skeletal symptoms in supermarket workers, *Ergonomics*, **32**(4), 359–71.

Rydevik, B. and Lundborg, G., 1977, Permeability of intraneural micro-vessels and perineurium following acute, graded experimental nerve compression, *Scand J Plast Reconstr Surg*, **11**, 179–87.

Rydevik, B., Lundborg, G. and Bagge, U., 1981, Effects of graded compression on intraneural blood flow: an in vivo study on rabbit tibial nerve, *J Hand Surg*, **6**, 3–12.

Rydholm, U., Werner, C. and Ohlin, P., 1983, Intracompartmental forearm pressure during rest and exercise, *Clin Orthop*, **175**, 213–5.

Sainfort, P.C., 1991, Stress, job control and other job elements: a study of office workers, *International Journal of Industrial Ergonomics*, **7**(1), 11–23.

Sakakibara, H., Miyao, M., Kondo, T., Yamada, S., Nakagawa, T. and Kobayashi, F., 1987, Relation between overhead work and complaints of pear and apple orchard workers, *Ergonomics*, **30**(5), 805–15.

Sällström, J. and Schmidt, H., 1984, Cervicobrachial disorders in certain occupations with special reference to compression in the thoracic outlet, *Am J Ind Med*, **6**, 45–52.

Salminen, J.J., Pentti, J., Wickström, G., 1991, Tenderness and pain in the neck and shoulders in relation to type A behaviour, *Scand J Rheumatol*, **20**, 344–50.

Salter, N. and Darcus, H.D., 1952, The effect of the degree of elbow flexion on the maximum torques developed in pronation and supination of the right hand, *J Anat*, **86**, 197–202.

Salter, R.B. and Field, P., 1960, The effects of continuous compression on living articular cartilage: an experimental investigation, *J Bone Joint Surg*, **42A**(1), 31–49.

Salvendy, G., 1981, Classification and characteristics of paced work, in Salvendy, G. and Smith, M.J. (Eds.) *Machine Pacing and Occupational Stress: Proceedings of the International Conference, Purdue University, March*, pp. 5–12, London: Taylor & Francis.

Salvendy, G. and Smith, M.J. (Eds.), 1981, *Machine Pacing and Occupational Stress: Proceedings of the International Conference, Purdue University, March*, London: Taylor & Francis.

Sauter, S.L., Chapman, L.J. and Knutson, S.J., 1985, *Improving VDT Work: Causes and Control of Health Concerns in VDT Use*, Lawrence, KS: The Report Store.

Sauter, S.L., Dainoff, M.J. and Smith, M.J., 1990, *Promoting Health and Productivity in the Computerized Office*, London: Taylor & Francis.

Sauter, S.L., Gottlieb, M.S., Rohrer, K.M. and Dodson, V.N., 1983, *The Well-Being of Video Display Terminal Users*, Madison, Wis.: University of Wisconsin, Department of Preventive Medicine.

Sauter, S.L., Chapman, L.J., Knutson, S.J. and Anderson, H.A., 1987, Case example of wrist trauma in keyboard use, *Appl Ergon*, **18**(3), 183–6.

Schmidt, R.T. and Toews, J.V., 1970, Grip strength as measured by the Jamar dynamometer, *Arch Phys Med Rehab*, 321–7.

Scholey, M., 1983, Back stress: the effects of training nurses to lift patients in a clinical situation, *Int J Nurs Stud*, **20**(1), 1–13.

Schottland, J.R., Kirschberg, G.J., Fillingim, R., Davis, V.P. and Hogg, F., 1991, Median nerve latencies in poultry processing workers: an approach to resolving the role of industrial 'cumulative trauma' in the development of carpal tunnel syndrome, *JOM*, **33**(5), 627–31.

Schuind, F. Garcia-Elias, M., Cooney, W.P. and An, K.-N., 1992, Flexor tendon forces: in vivo measurements, *J Hand Surg*, **17A**(2), 291–8.

Schüldt, K., Ekholm, J., Harms-Ringdahl, K., Németh, G. and Arborelius, U.P., 1987, Effects of arm support or suspension on neck and shoulder muscle activity during sedentary work, *Scand J Rehab Med*, **19**, 77–84.

Schultz, A.B., Andersson, G.B.J., Haderspeck, K., Örtengren, R., Nordin, M. and Björk, R., 1982, Analysis and measurement of lumbar trunk loads in tasks involving bends and twists, *J Biomech*, **15**(9), 669–75.

Shepard, R.J., 1992, A critical analysis of work-site fitness programs and their postulated economic benefits, *Med Sci Sports Exerc*, **24**, 354–70.

Shirazi-Adl, S.A., 1989, Strain in fibres of a lumbar disc: analysis of the role of lifting in producing disc prolapse, *Spine*, **14**(1), 96–103.

Sihvonen, T., Baskin, K. and Hänninen, O., 1989, Neck-shoulder loading in word-processor use, *International Archives of Occupational and Environmental Health*, **61**, 229–33.

Silverstein, B.A., 1985, *The Prevalence of Upper Extremity Cumulative Trauma Disorders in Industry*, Ph.D. thesis, Ann Arbor: University of Michigan, University Microfilms International.

Silverstein, B.A., 1987, Evaluation of interventions for control of cumulative trauma disorders, in ACGIH (Ed.) *Ergonomic Interventions to Prevent Musculoskeletal Injuries in Industry*, pp. 87–99, Chelsea, MI: Lewis Publishers. (Industrial Hygiene Science Series, 2).

Silverstein, B.A., 1991, Developing shop-floor ergonomic expertise using a train-the-trainer approach. In: *Healthy Work Environments, Healthy People: Participatory Approaches to Improving Workplace Health: International Conference, June 3–5, Ann Arbor, University of Michigan*, Ann Arbor: University of Michigan, pp. 74–?.

Silverstein, B.A. and Fine, L.J., 1984, *Evaluation of Upper Extremity and Low Back Cumulative Trauma Disorders: a Screening Manual*, Ann Arbor: University of Michigan, School of Public Health.

Silverstein, B.A., Fine, L.J. and Armstrong, T.J., 1986, Hand wrist cumulative trauma disorders in industry, *Br J Ind Med*, **43**, 779–84.

Silverstein, B.A., Fine, L.J. and Armstrong, T.J., 1987, Occupational factors and the carpal tunnel syndrome, *Am J Ind Med*, **11**, 343–58.

Silverstein, B., Fine, L. and Stetson, D., 1987, Hand-wrist disorders among investment casting plant workers, *J Hand Surg*, **12A**(5), 838–44.

Silverstein, B., Armstrong, T., Longmate, A. and Woody, D., 1988, Can in-plant exercise control musculoskeletal symptoms?, *J Occup Med*, **30**(12), 922–7.

Simard, M., Marchand, A., Couvrette, J. and Duquette, M.-J., 1993, *Étude des Stratégies de Développement de l'Implication Participative des Contremaîtres en Prévention des Accidents du Travail dans l'Industrie Manufacturière*, Montréal: Université de Montréal, GRASP.

Sjøgaard, G., 1986, Intramuscular changes during long-term contraction. In: Corlett, N., Wilson, J. and Manenica, I. eds. *The Ergonomics of Working Postures: Models, Methods and Cases: Proceedings of the First International Occupational Ergonomics Symposium, Zadar, Yugoslavia, 18–20 April 1985*, London: Taylor and Francis, pp. 136–43.

Sjøgaard, G., Savard, G. and Juel, C., 1988, Muscle blood flow during isometric activity and its relation to muscle fatigue, *Eur J Appl Physiol*, **57**, 327–35.

Sjøgaard, G., Ekner, D., Schibye, B., Simonsen, E.B., Jensen, B.R., Christiansen, J.U. and Pedersen, K.S., 1987, *Skulder/Nakke-besvær hos Syersker: En Epidemiologists og Arbejdsfysiologisk Undersögelse*, in Danish with English summary, Copenhagen: Arbejdsmiljøfondet.

Skie, M., Zeiss, J., Ebraheim, N.A. and Jackson, W.T., 1990, Carpal tunnel changes and median nerve compression during wrist flexion and extension seen by magnetic resonance imaging, *J Hand Surg*, **15A**(6), 934–9.

Smith, A., 1870, *The Wealth of Nations*, Harmondsworth: Penguin.

Smith, E.M., Sonstegard, D., Anderson, W., 1977, Carpal tunnel syndrome: contribution of flexor tendons, *Arch Phys Med Rehab*, **58**, 379–85.

Smith, M.J., 1985, Machine-paced work and stress, in Cooper, C.L. and Smith, M.J. (Eds) *Job Stress and Blue Collar Work*, pp. 51–64, New York: John Wiley & Sons.

Smith, M.J., 1986, Job stress and VDUs: Is the technology a problem ? In: *Work with Display Units: Proceedings of the International Scientific Conference, Stockholm, May 12–15*, Part 1, Amsterdam: Elsevier Science Publishers, pp. 189–95.

Smith, M.J. and Sainfort, P.C., 1989, A balance theory of job design for stress reduction, *International Journal of Industrial Ergonomics*, **4**, 67–79.

Smith, M.J. and Zehel, D., 1992, A stress reduction intervention programme for meat processors emphasizing job design and work organization (United States) *Conditions of Work Digest*, **11**(2), 204–13.

Smith, M.J., Cohen, B.G.F. and Stammerjohn, L.W., Jr, 1981, An investigation of health complaints and job stress in video display operations, *Hum Factors*, **23**(4), 387–400.

Smith, M.J., Carayon, P. and Miezio, K., 1987, VDT technology: psychosocial and stress concerns. In: Knave, B. and Widebäck, P.-G. eds. *Work with Display Units 86: Selected Papers from the International Scientific Conference on Work with Display Units, Stockholm, Sweden, May 12–15, 1986*, Amsterdam: North-Holland, pp. 695–712.

Smith, M.J., Cohen, H.H., Cohen, A. and Cleveland, R.J., 1978, Characteristics of successful safety programs, *J Safe Res*, **10**(1), 5–15.

Smith, M.J., Colligan, M.J., Frockt, I.L. and Tasto, D.L., 1979, Occupational injury rates among nurses as a function of shift schedule, *J Saf Res*, **11**(4), 181–7.

Smith, M.J., Carayon, P., Eberts, R. and Salvendy, G., 1992, Human-computer interaction, in Salvendy, G. (Ed.) *Handbook of Industrial Engineering*, 2nd Edn, pp. 1107–44, New York: John Wiley & Sons.

Smith, M.J., Carayon, P., Saunders, K.J., Lim, S.Y. and LeGrande, D., 1992, Employer stress and health complaints in jobs with and without electronic performance monitoring, *Appl Ergon*, **23**(1), 17–27.

Smutz, W.P., Bloswick, D.S. and France, E.P., 1992, An investigation into the effect of low force high frequency manual activities on the development of carpal tunnel syndrome. In: Kumar, S. ed. *Advances in Industrial Ergonomics and Safety IV: Proceedings of the Annual International Industrial Ergonomics and Safety Conference held in Denver, Colorado, 10–14 June*, London: Taylor & Francis, pp. 805–12.

Snook, S.H., 1978, The design of manual handling tasks, *Ergonomics*, **21**(12), 963–85.

Snook, S.H., 1985, Psychophysical considerations in permissible loads, *Ergonomics*, **28**(1), 327–30.

Snook, S.H., 1987, Approaches to preplacement testing and selection workers, *Ergonomics*, **30**, 241–7.

Snook, S.H., 1992, Psychophysical studies of hand work. In: *Occupational Disorders of The Upper Extremities, Proceedings of a Seminar held at the University of Michigan, Sept 30–Oct 1.*

Snook, S.H. and Ciriello, V.M., 1991, The design of manual handling tasks: revised tables of maximum acceptable weights and forces, *Ergonomics*, **34**(9), 1197–214.

Snook, S.H., Campanelli, R.A. and Hart, J.W., 1978, A study of three preventive approaches to low back injury, *J Occup Med*, **20**(7), 478–81.

Spilling, S., Eitrheim, J. and Aarås, A., 1986, Cost-benefit analysis of work environment investment at STK's telephone plant at Kongsvinger. In: Corlett, N., Wilson, J. and Manenica, I. eds. *The Ergonomics of Working Postures: Models, Methods and Cases: The Proceedings of the First International Occupational Ergonomics Symposium, Zadar, Yugoslavia, 15–17 April 1985*, London: Taylor & Francis, pp. 380–97.

Spitzer, W.O., LeBlanc, F.E., Dupuis, M., Abenhaim, L., B)langer, A.Y., Bloch, R., Bombardier, C., Cruess, R.L., Drouin, G., Duval-Hesler, N., Laflamme, J., Lamoureux, G., Nachemson, A., Pagé, J.J., Rossignol, M., Salmi, L.R., Salois-Arsenault, S., Suissa, S. and Wood-Dauphinée, S., 1987, Scientific approach to the assessment and management of activity-related spinal disorders: a monograph for clinicians: report of the Quebec Task Force on Spinal Disorders, *Spine*, **12**(7S), S1–59.

Spokes, E.M., 1986, New look at underground coal mine safety, *Mining Eng*, (April), 266–70.

St. Clair, S. and Shults, T., 1992, Americans with disabilities act: Considerations for the practice of occupational medicine, *J Occup Med*, **34**(5), 510–7.

St-Vincent, M., Tellier, C. and Lortie, M., 1989, Training in handling: an evaluative study, *Ergonomics*, **32**(2), 191–210.

Stammerjohn, L.W., Smith, M.S. and Cohen B.G.F., 1981, Evaluation of work station design factors in VDT operations, *Hum Fact*, **23**(4), 401–12.

Stammers, R. and Patrick, J., 1975, *The Psychology of Training*, London: Methuen.

Stange, K.C., Strogatz, D., Schoenbach, V.J., Shy, C., Dalton, B. and Cross, A.W., 1991, Demographic and health characteristics of participants and nonparticipants in a work site health-promotion program, *J Occup Med*, **33**, 474–8.

Steele, K.M. and Hoefner, V.C., 1988, Preplacement low-back screening for high-risk areas, *AOA*, **88**, 499–505.

Stellman, J.M., Klitzman, S., Gordon, G.C. and Snow B.R., 1985, Air quality and ergonomics in the office: survey results and methodologic issues, *Am Ind Hyg Assoc J*, **46**(5), 286–93.

Stenlund, B., Goldie, I., Hagberg, M., Hogstedt, C. and Marions, O., 1992, Radiographic osteoarthrosis in the acromioclavicular joint resulting from manual work or exposure to vibration, *Brit J Ind Med*, **49**, 588–93.

Stetson, D.S., Keyserling, W.M., Silverstein, B.A. and Leonard, J.A., 1991, Observational analysis of the hand and wrist: a pilot study, *Applied Occupational and Environmental Hygiene*, **6**(11), 927–37.

Stetson, D.S., Albers, J.W., Silverstein, B.A. and Wolfe, R.A., 1992, Effects of age, sex and anthropometric factors on nerve conduction measures, *Muscle & Nerve*, **15**(10), 1095–1104.

Stetson, D.S., Silverstein, B.A., Keyserling, W.M., Wolfe, R.A. and Albers, J.W., 1993, Median sensory distal amplitude and latency: comparisons between nonexposed managerial/professional employees and industrial workers, *Am J Ind Med*, **24**(2), 175–89.

Stevens, J.C., Sun, S., Beard, C.M., O'Fallon, W.M. and Kuland, L.T., 1988, Carpal tunnel syndrome in Rochester, Minnesota, 1961 to 1980, *NEURA*, **38**, 134–8.

Stewart, A.F., Adler, M., Byers, C.M., Segre, G.V. and Broadus, A.E., 1982, Calcium homeostasis in immobilization: an example of resorptive hypercalciuria, *New Eng J Med*, **306**(19), 1136–40.

Stobbe, T.G., 1982, *The Development of a Practical Strength Testing Program for Industry*, Ph.D. thesis, Ann Arbor: Industrial and Operation Engineering, Industrial Health Science, University of Michigan.

Stock, S., 1991, Workplace ergonomic factors and the development of musculoskeletal disorders of the neck and upper limbs: a meta-analysis, *Am J Ind Med*, **19**, 87–107.

Strasser, A.L., 1988, Preplacement exams assess workers medical ability to perform a task, *Occ Health & Sfty*, **57**(3), 16.

Stroud, S.D. and Thompson, C.E., 1985, Hypothenar hammer syndrome: a commonly undetected occupational hazard, *Occup Health Nursing*, **33**(1), 31–2.

Stubbs, D.A., Buckle, P.W., Hudson, M.P. and Rivers, P.M., 1983, Back pain in the nursing profession II: the effectiveness of training, *Ergonomics*, **26**(8), 767–79.

Sundelin, G. and Hagberg, M., 1989, The effects of different pause types on neck and shoulder EMG activity during VDU work, *Ergonomics*, **32**(5), 527–37.

Sundelin, G. and Hagberg, M., 1992a, Effects of exposure to excessive draughts on myoelectric activity in shoulder muscles, *J Electromyography Kinesiology*, **2**(1), 36–41.

Sundelin, G. and Hagberg, M., 1992b, Electromyographic signs of shoulder muscle fatigue in repetitive arm work paced by the Methods Time Measurement System, *Scand J Work Environ Health*, **18**, 262–8.

Susser, M., 1991, What is a cause and how do we know one?: a grammar for pragmatic Lepidemiology, *Am J Epidemiol*, **133**(7), 635–48.

Szabo, R.M. and Gelberman, R.H., 1987, The pathophysiology of nerve entrapment syndromes, *J Hand Surg*, **12A**, 880–4.

Szabo, R.M. and Chidgey, L.K., 1989, Stress carpal tunnel pressures in patients with carpal tunnel syndrome and normal patients, *J Hand Surg*, **14A**(4), 624–7.

Szilagyi, A.D. and Wallace, M.J., Jr, 1990, *Organizational Behavior and Performance*, 5th Edn, Glenview, IL: Scott, Foresman.

Takala, E.P. and Viikari-Juntura, E., 1991, Loading of shoulder muscles in a simulated work cycle: comparison between sedentary workers with and without neck-shoulder symptoms, *Clin Biomech*, **6**, 145–52.

Takala, E.-P. and Viikari-Juntura, E., 1991, Muscle force, endurance and neck-shoulder symptoms of sedentary workers: an experimental study on bank cashiers with and without symptoms, *International Journal of Industrial Ergonomics*, **7**, 123–32.

Takala, E.-P., Viikari-Juntura, E. and Häkkänen, M., 1992, Pressure pain threshold on forearm: an indicator of work load?. In: Hagberg, M. and Kilbom, Å. eds. *International Scientific Conference on Prevention of Work related Disorders, PREMUS, Stockholm, Sweden, May 12–14*, Stockholm: National Institute of Occupational Health, pp. 211–3. (Arbete och hälsa, 17).

Tanaka, S. and McGlothlin, D.J., 1993, A conceptual quantitative model for prevention of work related carpal tunnel syndrome (CTS), *International Journal of Industrial Ergonomics*, **11**, 181–93.

Tanaka, S., Seligman, P., Halperlin, W., Thun, M., Timbrook, C.L. and Wasil, J.J., 1988, Use of workers' compensation claims data for surveillance of cumulative trauma disorders, *JOM*, **30**(6), 488–92.

Task Force on the Health Surveillance of Workers, 1986, Health surveillance of workers:

the report of the Task Force on Health Surveillance of Workers, *Can J Public Health*, **77**(Suppl 2), 91–9.

Taylor, F.W., 1911, *The Principles of Scientific Management*, New York: W.W. Norton & Company.

Theorell, T., Harms-Ringdahl, K., Ahlberg-Hulten, G. and Westin, B., 1991, Psychosocial job factors and symptoms from the locomotor system: a multicausal analysis, *Scand J Rehab Med*, **23**, 165–73.

Thompson, D., Lowerson, A. and Zalewski, M., 1987, The use of Hettinger test in preemployment screening, in Buckle, P. (Ed.) *Musculoskeletal Disorders at Work*, pp. 177–82, London: Taylor & Francis.

Thun, M., Tanaka, S., Smith, A.B., Halperin, W.E., Lee, S.T., Luggen, M.E. and Hess, E.V., 1987, Morbidity from repetitive knee trauma in carpet and floor layers, *Brit J Ind Med*, **44**, 611–20.

Thurman, J.E., Louzine, A.E. and Kogi, K., 1988, *Higher Productivity and a Better Place to Work: Practical Ideas for Owners and Managers of Small and Medium-Sized Industrial Enterprises: Action Manual*, Geneva: International Labour Office.

Tichauer, E.R., 1966, Some aspects of stress on forearm and hand in industry, *J Occup Med*, **8**(2), 63–71.

Tola, S., Riihimäki, H., Videman, T., Viikari-Juntura, E. and Hänninen, K., 1988, Neck and shoulder symptoms among men in machine operating, dynamic physical work and sedentary work, *Scand J Work Environ Health*, **14**, 299–305.

Toohey, A.K., LaSalle, T.L., Martinez, S. and Polisson, R.P., 1990, Iliopsoas bursitis: clinical features, radiographic findings, and disease associations, *Sem Arth Rh*, **20**(1), 41–7.

Toomingas, A., Hagberg, M., Jorulf, L., Nilsson, T., Burström, L. and Kihlberg, S., 1991, Outcome of the abduction external rotation test among manual and office workers, *Am J Ind Med*, **19**, 215–27.

Torell, G., Sanden, A. and Järvholm, B., 1988, Musculoskeletal disorders in shipyard workers, *J Soc Occup Med*, **38**(4), 109–12.

Trist, E.L. and Bamforth, K.W., 1951, Some social and psychological consequences of the longwall method of coal-getting, *Human Relat*, **4**(1), 3–38.

Uhthoff, H.K. and Sarkar, K., 1990, An algorithm for shoulder pain caused by soft-tissue disorders, *Clin Orthop*, **254**, 121–7.

Ulin, S.S., Ways, C.M., Armstrong, T.J. and Snook, S.H., 1990, Perceived exertion and discomfort versus work height with a pistol shaped screw driver, *Am Ind Hyg Assoc J*, **51**(11), 588–94.

Ursin, H., Endresen, I.M. and Ursin, G., 1988, Psychological factors and self-reports of muscle pain, *Eur J Appl Physiol*, **57**, 282–90.

van Boxtel, A. and van der Ven, J.R., 1978, Differential EMG activity in subjects with muscle contraction headaches related to mental effort, *Headache*, **17**, 233–7.

van der Beek, A.J., van Gaalen, L.C. and Frings-Dresen, M.H.W., 1992, Working postures and activities of lorry drivers: a reliability study of on-site observation and recording on a pocket computer, *Appl Ergon*, **23**(5), 331–6.

Vanggaard, L., 1975, Physiological reactions to wet-cold, *Aviat Space Environ Med*, **46**(January), 33–6.

Vanswearingen, J.M., 1983, Measuring wrist muscle strength, *J Orthop Sports Phys Ther*, **4**(4), 217–28.

Van Velzer, C.T., 1992, Economic ergonomic interventions: the benefits of job rotation. In: *The Economics of Ergonomics: Proceedings of the 25th Annual Conference of the Human Factors Association of Canada, October 25–28, Hamilton, Ontario*, Missisauga, Ont.: Human Factors Association of Canada, pp. 207–12.

Van Wely, P., 1970, Design and disease, *Appl Ergon*, **1**, 262–9.

Vartiainen, M., 1987, *The Hierarchical Development of Mental Regulation, and Training Methods*, Otaniemi: Helsinki University of Technology, Industrial Economics and Industrial Psychology, TKK Offset. (Report no. 100).

Vartiainen, M., Teikari, V. and Pulkkis, A., 1989, *Psykologinen Työnopetus*, Helsinki: Otakustantamo. (Document no. 516).

Veiersted, K.B., Westgaard, R.H. and Andersen, P., 1990, Pattern of muscle activity during stereotyped work and its relation to muscle pain, *Int Arch Occup Environ Health*, **62**(1), 31–41.

Venning, P.J., Walter, S.D. and Stitt, L.W., 1987, Personal and job-related factors as determinants of incidence of back injuries among nursing personnel, *JOM*, **29**(10), 820–5.

Verbeek, J., 1991, The use of adjustable furniture: evaluation of an instruction programme for office workers, *Appl Ergon*, **22**(3), 179–84.

Vicente, K.J., 1990, A few implications of an ecological approach to human factors, *Human Factors Society Bulletin*, **33**, 1–4.

Videman, T., 1982a, Experimental osteo-arthritis in the rabbit: comparison of different periods of repeated immobilization, *Acta Orthop Scand*, **53**(3), 339–47.

Videman, T., 1982b, The effect of running on the osteoarthritic joint: an experimental matched-pair study with rabbits, *Rheumatol Rehabil*, **21**(1), 1–8.

Vihma, T., Nurminen, M. and Mutanen, P., 1982, Sewing machine operators' work and musculo-skeletal complaints, *Ergonomics*, **25**(4), 295–8.

Viikari-Juntura, E., 1983, Neck and upper limb disorders among slaughterhouse workers, *Scand J Work Environ Health*, **9**, 283–90.

Viikari-Juntura, E., 1988, *Examination of the Neck: Validity of some Clinical, Radiological and Epidemiological Methods*, PhD thesis, Helsinki: NIOH.

Viikari-Juntura, E., Vuori, J., Silverstein, B.A., Kuosma, E. and Videman, T., 1991, A life-long prospective study on the role of psychosocial factors in neck-shoulder and low-back pain, *Spine*, **16**(9), 1056–61.

Viikari-Juntura, E., Kurppa, K., Kuosma, E., Huuskonen, M., Kuorinka, I., Ketola, R. and Könni, U., 1991, Prevalence of epicondylitis and elbow pain in the meat-processing industry, *Scand J Work Environ Health*, **17**, 38–45.

Vincent, M.J. and Tipton, M.J., 1988, The effects of cold immersion and hand protection on grip strength, *Aviat Space Environ Med* **59**, 738–41.

Vingård, E., 1991, Overweight predisposes to coxarthrosis, *Acta Orthop Scand*, **62**, 106–9.

Vingård, E., Alfredsson, L., Goldie, I. and Hogstedt, C., 1991a, Sports and osteoarthrosis of the hip, in Vingård, E. (Ed.) *Work, Sports, Overweight and Osteoarthrosis of the Hip: Epidemiological Studies*, Stockholm: National Institute of Occupational Health. (Arbete och Halsa, 25).

Vingård, E., Hogstedt, C., Alfredsson, L., Fellenuis, E., Goldie, I. and Koster, M., 1991b, Coxarthrosis and physical work load, *Scand J Work Environ Health*, **17**, 104–9.

Wachs, J.E. and Parker, J.E., 1987, Registered nurses' lifting behavior in the hospital setting. In: Asfour, S.S. ed. *Trends in Ergonomics/Human Factors IV: Proceedings of the Annual International Industrial Ergonomics and safety Conference held in Miami, Florida, 9–12 June*, part B, New York: North-Holland, pp. 883–90.

Waersted, M. and Westgaard, R.H., 1991, Working hours as a risk factor in the development of musculoskeletal complaints, *Ergonomics*, **34**(3), 265–76.

Waersted, M., Bjørklund, R. and Westgaard, R.H., 1987, Generation of muscle tension related to a demand of continuing attention. In: Knave, B., Widebäck P.-G. eds. *Working with Display Units 86: International Scientific Conference on Work with Display Units, Stockholm, Sweden*, New York: North-Holland, pp. 288–93.

Waersted, M., Bjørklund, R.A. and Westgaard, R.H., 1991, Shoulder muscle tension induced by two VDU-based tasks of different complexity, *Ergonomics*, **34**(2), 137–50.

Waikar, A., Lee, K., Sanyal, S., Parks, C. and Aghazadeh, F., 1990, Evaluation of workplaces for the risk of carpal tunnel syndrome. In: Das, B. ed. *Advances in*

Industrial Ergonomics and Safety II: Proceedings of the Annual International Ergonomics and Safety Conference, Montreal, 10–13 June, Bristol, Pa.: Taylor & Francis, pp. 207–14.

Wall, E., Massie, J., Kwan, M., Rydevik, B., Myers, R. and Garfin, S., 1992, Experimental stretch neuropathy, *J Bone Joint Surg*, **74B**(1), 126–9.

Walter, S.D., Hart, L.E., McIntosh, J.M. and Sutton, J.R., 1989, The Ontarian cohort study of running-related injuries, *Arch Intern Med*, **149**, 2561–4.

Wands, S.E. and Yassi, A., 1992, 'Let's talk back': a program to empower laundry workers, *Am J Indust Med*, **22**, 703–9.

Wangenheim, M., Samuelson, B. and Wos, H., 1986, ARBAN: a force ergonomic analysis method. In: Corlett, N., Wilson, J. and Manenica, I. eds. *The Ergonomics of Working Postures: Models, Methods and Cases: Proceedings of the First International Occupational Ergonomics Symposium, Zadar, Yugoslavia, 18–20 April 1985*, London: Taylor & Francis, pp. 243–55.

Waris, P., 1979, Occupational cervicobrachial syndromes, *Scand J Work Environ Health*, **5**(suppl 3), 3–14.

Waris, P., Kuorinka, I. and Kurppa, K., 1979, Epidemiologic screening of occupational neck and upper limb disorders: methods and criteria, *Scand J Work Environ Health*, **5**(suppl 3), 25–38.

Wasserman, D.E., 1987, *Human Aspects of Occupational Vibration*, New York: Elsevier Science Publishers. (Advances in Human Factors/Ergonomics, 8).

Waters, T.R., Putz-Anderson, V., Garg, A. and Fine L.J., 1993, Revised NIOSH equation for the design and evaluation of manual lifting tasks, *Ergonomics*, **36**(7), 749–76.

Weber, A., Fussler, C., O'Hanlon, J.F., Gierer, R. and Grandjean, E., 1980, Psycho-physiological effects of repetitive tasks, *Ergonomics*, **23**(11), 1033–46.

Weed, D.L., 1986, On the logic of causal inference, *AJE*, **123**(6), 965–79.

Wells, R., Moore, A. and Cholewicki, J., 1990a, Evaluation of upper limb stresses using musculoskeletal loads during a rotating light assembly task. In: *Advances in Industrial Ergonomics and Safety II: Proceedings of the Annual International Industrial Ergonomics and Safety Conference, Montreal, 10–13 June*, Philadelphy, Pa.: Taylor & Francis, pp. 183–90.

Wells, R., Ranney, D. and Moore, A., 1992, Relationship between forearm muscle pain/tenderness and work exposures: results from repetitive manual tasks. In: *Occupational Disorders of The Upper Extremities, Proceedings of a Seminar held at the University of Michigan, Sept 30–Oct 1*.

Wells, J.A., Zipp, J.F., Schuette, P.T. and McEleney, J., 1983, Musculoskeletal disorders among letter carriers: a comparison of weight carrying, walking and sedentary occupations, *J Occup Med*, **25**(11), 814–20.

Werner, C.O., Elmquist, D. and Ohlin, P., 1983, Pressure and nerve lesion in the carpal tunnel, *Acta Orthop Scand*, **54**, 312–6.

Westgaard, R.H. and Aarås, A., 1984, Postural muscle strain as a causal factor in the development of musculo-skeletal illness, *Appl Ergon*, **15**(3), 162–74.

Westgaard, R.H. and Aarås, A., 1985, The effect of improved workplace design on the development of work related musculoskeletal illnesses, *Appl Ergon*, **16**(2), 91–7.

Westgaard, R.H. and Björklund, R., 1987, Generation of muscle tension additional to postural load, *Ergonomics*, **30**, 911–23.

Westling, G. and Johansson, R.S., 1984, Factors Influencing the force control during precision grip, *Exp Brain Res*, **53**, 277–84.

Wheat, J.R., Graney, M.J., Shachtman, R.H., Ginn, G.L., Patrick, D.L. and Hulka, B.S., 1992, Does workplace health promotion decrease medical claims?, *Am J Prev Med*, **8**(2), 110–4.

WHO, 1985, *Identification and Control of Work related Diseases*, Geneva: World Health Organization. (Technical Report Series no. 714).

Wieslander, G., Norbäck, D., Göthe, C.-J. and Juhlin, L., 1989, Carpal tunnel syndrome (CTS) and exposure to vibration, repetitive wrist movements, and heavy manual work: a case-referent study, *Br J Ind Med*, **46**, 43–7.

Wiker, S., 1992, Posturally mediated perceptions of strain encountered when lifting: a preliminary analysis of the basis and value in predicting worker posture. In: Mattila, M and Karwowski, W. eds. *Computer Applications in Ergonomics, Occupational Safety & Health: Proceedings of the International Conference on Computer-aided Ergonomics and Safety, Tampere, Finland, 18–20 May*, Amsterdam: North-Holland, pp. 497–509.

Wiker, S.F., Chaffin, D.B. and Langolf, G.D., 1989, Shoulder posture and localized muscle fatigue and discomfort, *Ergonomics*, **32**, 211–37.

Wiker, S.F., Chaffin, D.B. and Langolf, G.D., 1990, Shoulder postural fatigue and discomfort: a preliminary finding of no relationship with isometric strength capability in a light-weight manual assembly task, *International Journal of Industrial Ergonomics*, **5**(2), 133–46.

Wikström, B.-O., 1993, Effects from twisted postures and whole-body vibration during driving, *International Journal of Industrial Ergonomics*, **12**, 61–75.

Wiktorin, C., Karlqvist, L., Winkel, J. and the Stockholm Music I Study Group, 1993, Validity of self-reported exposure to work postures and manual materials handling, *Scand J Work Environ Health*, **19**, 208–14.

Williams, V.W., Cope, R., Gaunt, W.D., Adelstein, E.H., Hoyt, T.S., Singh, A., Pressley, T.A., English, R., Schmacher, H.R. and Walker, S.E., 1987, Metacarpophalangeal arthropathy associated with manual labour (Missouri metacarpal syndrome), *Arth Rheum*, **30**(12), 1362–71.

Wilson, J.R., 1991, Participation: a framework and a foundation for ergonomics?, *J Occup Psychol*, **64**, 67–80.

Winkel, J., 1987, On the significance of physical activity in sedentary work. In: Knave, B. and Widebäck, P.-G. eds. *Work with Display Units 86: Selected Papers from the International Scientific Conference on Work with Display Units, Stockholm, May 12–15*, Amsterdam: North-Holland, pp. 229–36.

Winkel, J. and Bendix, T., 1984, A method for electromyographic analyses of muscular contraction frequencies, *Eur J Appl Physiol*, **53**, 211–37.

Winkel, J. and Westgaard, R., 1992a, Occupational and individual risk factors for shoulder-neck complaints, part I: guidelines for the practitioner, *International Journal of Industrial Ergonomics*, **10**, 79–83.

Winkel, J. and Westgaard, R., 1992b, Occupational and individual risk factors for shoulder-neck complaints, part II: the scientific basis (literature review) for the guide, *International Journal of Industrial Ergonomics*, **10**, 85–104.

Winkel, J., Dallner, M., Ericson, M., Fransson, C., Karlqvist, L., Nygård, C.-H., Selin, K., Hjelm, E.W., Wiktorin, C. and the Stockholm-Music I Study Group, 1991, Evaluation of a questionnaire for the estimation of physical load in epidemiologic studies: study design. In: Queinnec, Y. and Daniellou, F. eds. *Designing for Everyone: Proceedings of the Eleventh Congress of the International Ergonomics Association, Paris*, vol. 1, New York: Taylor & Francis, pp. 227–53.

Winn, F.J. and Habes, D.J., 1990, Carpal tunnel area as risk factor for carpal tunnel syndrome, *Muscle & Nerve*, **13**, 254–8.

Wisseman, C.L. and Badger, D., 1977, *Hazard Evaluation and Technical Assistance, Report no. TA 76–93, Eastman Kodak Co, Windsor, Colorado*, Cincinnati, OH: NIOSH, U.S. Dept. of Health, Education and Welfare, Center for Disease Control.

Wolfe, F., Smythe, H.A., Yunus, M.B., Bennet, R.M., Bombardier, C., Goldenberg, D.L., Tugwell, P., Campbell, S.M., Abeles, M., Clark, P., Fam, A.G., Farber, S.J., Fiechtner, J.J., Franklin, C.M., Gatter, R.A., Hamaty, D., Lessard, J., Lichtbroun, A.S., Masi, A.T., McCain, G.A., Reynolds, W.S., Romano, T.J., Russell, I.J. and

Sheon, R.P., 1990, The American College of Rheumatology 1990: criteria for the classification of fibromyalgia, *Arthritis Rheum*, **33**(2), 160–72.

Workplace Health Fund, 1992, *Ergonomics Education for Union Health Professionals*, New York: Workplace Health Fund.

Yankelovich, D., 1979, *Work, Values and the New Breed*, New York: Van Norstrand Reinholt.

Zink, K.J., 1991, Participatory ergonomics: some developments and examples from West Germany, in Noro, K. and Imada, A. (Eds.) *Participatory Ergonomics*, pp. 165–80, London: Taylor & Francis.

Zino, K., 1988, We've got to make it, *Parade Magazine*, (Sept. 4), 22–5.

Zipp, P., Haider, E., Halpern, N. and Rohmert, W., 1983, Keyboard design through physiological strain measurements, *Appl Ergon*, **14**(2), 117–22.

Appendix I

List of symptoms, disorders and diseases under WMSDs

Examples of the terminology used for WMSDs in the literature

Various terms have been used in the literature to designate this group of disorders. In French, a bibliographic search brought out approximately 40 different terms. A sample of some of the terms which can be found in the English and French literature is provided below.

English terminology

- Cumulative trauma disorder (CTD): mostly used in the USA
- Occupational cervicobrachial disorder (OCD): mostly used in Japan
- Occupational overuse syndrome (OOS): mostly used in Australia
- Repetitive strain injury (RSI): mostly used in Australia and Canada
- Work related musculoskeletal disorder (WMSD): more and more in use worldwide

French terminology

- Lésions musculo-tendineuses liées aux tâches répétitives
- Lésions par efforts répétés
- Lésions ostéo-articulaires
- Lésions musculaires et ostéo-articulaires
- Lésions péri-articulaires
- Affections du membre supérieur liées à des traumatismes répétés

- Lésions attribuables au travail répétitif
- Troubles cervicobrachiaux d'origine professionnele
- Troubles ostéo-musculaires
- Syndrome par suremploi
- Affections ostéo-articulaires

Examples of some of the disorders and symptoms called, or grouped under, CTD, RSI, OOS, WMSD (and so on) in the literature

In the literature, many disorders and symptoms have been grouped under, or called, CTD, RSI, and so on. Often one disorder can have more than one name: for example, nerve entrapment at carpal canal and carpal tunnel syndrome, or lateral epicondylitis and tennis elbow. A sample of what can be found in the literature follows.

Some disorders which could be considered as CTD, RSI, etc.

Tendon-related disorders

- Tendinitis/peritendinitis/tenosynovitis/insertion tendinitis (enthesopathy)/ synovitis of most joints, in particular shoulder, elbow, hand-wrist
- Epicondylitis (medial and lateral)
- De Quervain's disease (stenosing tenosynovitis)
- Dupuytren's contracture
- Trigger finger
- Ganglion cyst

Nerve-related disorders

- Carpal tunnel syndrome (CTS) (median nerve entrapment at the wrist)
- Cubital tunnel syndrome (ulnar nerve entrapment at the elbow)
- Guyon canal syndrome (ulnar nerve entrapment at the Guyon canal)
- Pronator teres syndrome (median nerve entrapment at the elbow)
- Radial tunnel syndrome (radial nerve entrapment at the elbow)
- Thoracic outlet syndrome (TOS) (neurogenic TOS = entrapment of the brachial plexus at different locations)
- Cervical syndrome (radiculopathy) (compression of nerve roots)
- Digital neuritis

Joint-related disorders

● Osteoarthritis of most joints/degenerative joint disease

Muscle-related disorders

● Tension neck syndrome
● Muscle sprain and strain
● Myalgia and myositis

Circulatory/vascular type disorders

● Hypothenar hammer syndrome
● Raynaud's syndrome (white fingers syndrome)

Bursa-related disorders

● Bursitis of most joints

Some disorders, whose names are popularised, are also considered CTDs, WMSDs, etc., such as: bowler's thumb, bricklayer's shoulder, carpenter's elbow, carpet layer's knee, cherry pitter's thumb, cotton twister's hand, cymbal player's shoulder, frozen shoulder, game keeper's thumb, golfer's elbow, jailor's elbow, jeweler's thumb, manure shoveler's hip, stitcher's wrist, telegraphist's cramp, tennis elbow, writer's cramp.

Appendix II

Acronyms, terms and definitions used in this book

Initials and acronyms used in this book

ACOM	American College of Occupational Medicine [now known as American College of Occupational and Environmental Medicine (ACOEM)]
ANSI	American National Standards Institute
APDF	Amplitude probability distribution function
BLS	Bureau of Labor Statistics
BMI	Body mass index
CAT-scan	Computerized axial tomography scanner
CEO	Chief executive officer
CI	Confidence intervals
CTD	Cumulative trauma disorder
CTS	Carpal tunnel syndrome
DIP	Distal interphalangeal joint
EEC	European Economic Community (now known as European Union)
EMG	Electromyography or Electromyogram
ETF	Ergonomic task force
EVA	Exposure variation analysis
GM	General Motors Corporation
HFS	Human Factors Society (now known as the Human Factors and Ergonomics Society)
HLA(-B27)	Histocompatibility locus antigens
IAP	Intra-abdominal pressure
ICD (and ICD-9)	International classification of disease, WHO
IDP	Intra-discal pressure

ILO	International Labor Organisation/International Labor Office
IR	Incidence rate
ISO	International Standardization Organization
JSR	Job strength rating
LBP	Low back pain
MCP	Metacarpophalangeal joint
MMPI	Minnesota multiphasic personality inventory
MPF	Mean power frequency
MPGS	Maximal power grip strength
MRI	Magnetic resonance imaging
MTM	Methods-time measurement
MTS	Motion and time study
MVC	Maximal voluntary contraction
n	Number of subjects in a population study
N	Newton
N/A	Not available
N.m	Newton acting at a distance of 1 metre from a joint
NCV	Nerve conduction velocity
NIOSH	National Institute for Occupational Safety and Health, USA
OCD	Occupational cervicobrachial disorder
OD	Organizational development
ODAM	Organisational design and management
OHS	Occupational health and safety
OOS	Occupational overuse syndrome
OR	Odds ratio
OSHA	Occupational Safety and Health Administration, USA
PE	Physical examination
PR	Prevalence rate
QWL	Quality of work life
RF	Risk factors
RMS	Root mean square values
ROM	Range of motion (joints)
RR	Relative risk
RSI	Repetitive strain injury
SD	Standard deviation
SMRs	Standardized morbidity ratios
SR	Severity rate
SSS	Stress symptom scores
TLV	Threshold limit value
TOS	Thoracic outlet syndrome
TQM	Total quality management
UAW	United Auto Workers
VDT	Visual (or video) display terminal

VDU	Visual (or video) display unit
VWF	Vibration-induced white fingers
WHO	World Health Organisation
WMSD	Work related musculoskeletal disorder

Terms and definitions used in this book

Some of the definitions used here were taken, *verbatim*, from *A Dictionary of Epidemiology* (Last, 1988) and *Research Methods in Occupational Epidemiology* (Checkoway *et al.*, 1989). We wish to thank the authors/editors for granting us permission to do so. Terms and definitions pertinent to chapter 3 can be found in Box 3.1.

95 per cent Confidence intervals: See Box 3.1.

Anthropometry: A science concerned with comparative measurements of the human body.

Asymptomatic: A disease or disorder showing or producing no symptoms.

Attributable fraction: See Appendix III.

Behavioural factors at work: Behavioural factors at work define the way in which workers go about their jobs. There is no value judgement as to the 'goodness' or 'badness' of how they are carrying out their work. Worker behaviour can be influenced by several considerations including knowledge, experience, training, supervision instruction or personal emotions.

Bias: Any trend in the collection, analysis, interpretation, publication, or review of data that can lead to conclusions that are systematically different from the truth. Many varieties of bias have been described. The term 'bias' does not necessarily carry an imputation of prejudice or other subjective factor, such as the experimenter's desire for a particular outcome. This differs from conventional usage in which bias refers to a partisan point of view (adapted from Last, 1988).

Biomechanics: A science which studies the application of mechanical laws on the living body.

Burden: The amount of a substance in the body or in some particular target (e.g. organ) at one point in time (Checkoway *et al.*, 1989).

Case definitions vs Diagnostic criteria: See subsection 5.2.1.

Case-control study: (Syn: case comparison study, case compeer study, case history study, case referent study, retrospective study). (Last, 1988). See Box 3.1.

Cohort study: (Syn: concurrent, follow-up, incidence, longitudinal, prospective study). (Last, 1988). See Box 3.1.

Confidence interval: See Box 3.1.

Confounder(s): A variable that, if not controlled, produces distortion in the estimated effect of the study exposure; in the absence of misclassification, such a variable will be associated with the study exposure and predictive of risk among the non-exposed (i.e. to be a confounder, a factor must be associated with both exposure and disease, even in the absence of study exposure). A confounder must not be an intermediate step in the causal pathway from exposure to disease (Checkoway *et al.*, 1989).

Controls or referents: See Box 3.1.

Cross-sectional study: (Syn: disease frequency survey, prevalence study). See Box 3.1.

Disability: Disability is any restriction or lack (resulting from impairment) of ability to perform an activity in a manner or within the range considered normal for a human being.

Discomfort: See subsection 2.3.2.

Disease: See subsection 2.3.2.

Disorder: See subsection 2.3.2.

Dose: The amount of a substance that is delivered to a target during some specified time interval (Checkoway *et al.*, 1989).

Dose-response relationship: A relationship in which a change in amount, intensity or duration of exposure is associated with a change – either an increase or a decrease – in risk of a specified outcome (Last, 1988).

Duration: Length of time a physical activity (such as a specific motion, exertion, activity, task or job) is performed (ANSI Z-365).

Eccentric contractions: A muscle lengthens whilst it is active (as in lowering a weight).

Effect modifier(s): (Syn: conditional variable, moderator variable). A factor that modifies the effect of a putative causal factor under study. For example, age is an effect modifier for many conditions, and immunization status is an effect modifier for the consequences of exposure to pathogenic organisms. Effect modification is detected by varying the selected effect measure for the factor under study across levels of another factor (Last, 1988).

Effect modifier bias: See subsection 2.3.3.

Epidemiologic study: See Box 3.1.

Epidemiology: The study of the distribution and determinants of health-related states or events in specified populations, and the application of this study to control of health problems (Last, 1988).

Ergonomics: A science concerned with developing and applying knowledge about human performance capabilities, limitation and other characteristics as they related to the design of the interfaces between people and work.
The goal of an ergonomics programme is to identify and prevent or reduce OHS problems including work related musculoskeletal disorders.

Microergonomics: See section 6.2.

Macroergonomics: See section 6.2.

Experimental study: See Box 4.1.

Exposure(s): The amount of a factor to which a group or individual was exposed; sometimes contrasted with dose, the amount that enters or interacts with the organism. Exposures may of course be beneficial rather than harmful, e.g. exposure to immunizing agents (Last, 1988).

Exposure-response: See subsection 3.1.3.

Fatigue: See subsection 2.3.2.

Handicap: Handicap is a disadvantage for a given individual resulting from an impairment or a disability that limits or prevents the fulfilment of a role that is normal (depending on age, gender, social and cultural factors) for that individual.

Hazard: A factor or exposure that may adversely affect health (Last, 1988).

Hawthorne effect: The effect (usually positive or beneficial) of being under study upon the persons being studied; their knowledge of the study often influences their behaviour. The name derives from work studies by Whitehead, Dickson, Roethlisberger, and others, in the Western Electric Plant, Hawthorne, Illinois, reported by Elton Mayo in *The Social Problems of an Industrial Civilization* (London: Routledge, 1949) (Last, 1988).

Healthy worker effect: primary/secondary: A phenomenon observed initially in studies of occupational diseases: Workers usually exhibit lower overall death rates than the general population, due to the fact that the severely ill and disabled are ordinarily excluded from employment, because either the fittest members of the population are selected for employment (primary healthy worker effect) or because those workers that become ill leave the workforce (secondary healthy worker effect) (adapted from Last, 1988).

Illness: See subsection 2.3.2.

Impairment: Impairment is any loss or abnormality of psychological, physiological or anatomical structure or function.

Incidence: The number of new events, e.g. new cases of a disease in a defined population, within a specified period of time (adapted from Last, 1988). See Box 3.1 and Equation 5.1 for more details.

Individual susceptibility: Individual susceptibility is defined as an increased vulnerability to musculoskeletal disorders in a person caused by disease, genetic code or lack of fitness.

Information/measurement bias: See subsection 2.3.3

Laboratory study: See Box 3.1.

Observational study: See Box 3.1.

Odds ratio: See Box 3.1.

Organizational factor(s) or variable(s): Work organization deals with the way in which work is structured, supervised and processed. It deals with the institutional features of work such as the nature of the organizational chart, who is the boss, power, authority, responsibilities, how work gets done, the nature of tasks including such features as workload and content. It is the objective nature of the work process.

Work organization factors are seen as the objective aspects of how work is organized, supervised and carried out (see also psychosocial factors).

Outcome(s): In OHS, an outcome is a worker's manifestation (reaction) to a work task, a work environment, etc., for example an improved or deteriorated health status is an outcome, or absenteeism is another outcome, etc. All the possible results that may stem from exposure to a causal factor, or from preventive or therapeutic interventions (adapted from Last, 1988).

Pain: See subsection 2.3.2.

Pathology: Medical science concerned with all aspects of disease, but with special reference to the underlying causal structural and functional changes in tissues and organs of the body (adapted from Hensyl, 1990).

Pathophysiology: The physiology of disordered function (Dorland's *Illustrated Medical Dictionary*, 1974).

Posture: The position of the limbs or the carriage of the body as a whole (Hensyl, 1990).

Power (statistical or study power): A characteristic of a statistical hypothesis test, denoting the probability that the null hypothesis will be rejected if it is indeed false (Last, 1988).

Prevalence: The number of instances of a given disease or other condition in a given population at a designated time. See Box 3.1 and Equation 5.2 for more details (adapted from Last, 1988).

Prevention (primary, secondary, tertiary): Authorities on preventive medicine do not agree on the precise boundaries between these levels, nor on how many levels can be distinguished, but the differences of opinion are semantic rather than substantive.

An epidemiologic interpretation of the distinction is: primary prevention is aimed at reducing incidence (the number of new cases arising) of disease and other departures from good health; secondary prevention aims to reduce prevalence (numbers of existing cases) by shortening the duration; and tertiary prevention is aimed at reducing complications (adapted from Last, 1988).

In medicine, primary prevention may mean controlling/reducing the problem at the source, secondary prevention may mean treatment of existing problems and tertiary prevention may imply rehabilitation.

Prospective cohort study: See cohort study

Psychosocial factor(s) or variable(s): Psychosocial factors at work are the subjective aspects as perceived by the workers and the managers. They often have the same names as the work organization factors, but are different in that they carry 'emotional' value for the worker. Thus, the nature of supervision can have positive or negative psychosocial effects (emotional stress), while the work organizational aspects are just descriptive of how the supervision is accomplished and do not carry emotional value (see also organisational factors). Psychosocial factors are the individual subjective perceptions of the work organization factors.

Referents: See 'Controls'.

Rehabilitation: Restoration, following disease, illness or injury, of the ability to function in a normal or near normal manner (including gainful employment) (Hensyl, 1990).

Relative risk: See Box 3.1.

Reliability (measurement): The degree of stability exhibited when a measurement is repeated under identical conditions. Reliability refers to the degree to which the results obtained by a measurement procedure can be replicated. Lack of reliability may arise from divergences between observers or instruments of measurement or instability of the attribute being measured (Last, 1988).

Risk factor(s): An aspect of personal behaviour or lifestyle, an environmental exposure (including work) or an inborn or inherited characteristic, which on the basis of epidemiologic evidence is known to be associated with health-related conditions (WMSDs in our case) considered important to prevent. The term 'risk factor' is rather loosely used, with any of the following meanings:

1. An attribute or exposure that is associated with an increased probability of a specified outcome, such as the occurrence of a disease. Not necessarily a causal factor. A risk marker.
2. An attribute or exposure that increases the probability of occurrence of disease or other specified outcome. A determinant.
3. A determinant that can be modified by intervention, thereby reducing the probability of occurrence of disease or other specified outcomes. To avoid confusion, it may be referred to as a modifiable risk factor.

(Adapted from Last, 1988)

In the context of this book, risk factor denotes a factor present at work which may have a relation to the development of WMSDs.

Screening: Screening is the application of at least one test (or examination) to individuals in order to sort out apparently well persons who are probably developing the disorder from those who are not. Screening tests are not diagnostic; persons with suspicious screening findings are referred to their physicians for diagnosis (Last, 1988).

Pre-employment screening: Screening before any offer of employment is made.

Preplacement screening: Screening before job placement, after an offer of employment has been made.

Selection bias: See subsection 2.3.3.

Signs: Any objective evidence of a disease, i.e. such evidence as is perceptible to the examining physician, as opposed to the subjective sensations (symptoms) of the patient (Dorland's *Illustrated Medical Dictionary*, 1974).

Study factor: See Box 3.1.

Study design: See Box 3.1.

Surrogate variables (outcomes): See subsection 2.3.3.

Surveillance: The ongoing systematic collection, analysis and interpretation of health and exposure data in the process of describing and monitoring a health event. Surveillance data are used to determine the need for occupational safety and health action and to plan, implement and evaluate ergonomic interventions and programmes (adapted from Klaucke *et al.*, 1988).

Symptoms: Any morbid phenomenon or departure from the normal in function, appearance or sensation, experienced by the patient and indicative of disease (Hensyl, 1990).

Syndrome: A set of symptoms which occur together; the sum of signs of any morbid state; a symptom complex (Dorland's *Illustrated Medical Dictionary*, 1974).

Target population: The collection of individuals about which we want to make inferences. The term is sometimes used to indicate the population from which a sample is drawn and sometimes to denote any 'reference' population about which inferences are required (adapted from Last, 1988).

Validity: An expression of the degree to which a measurement measures what it purports to measure (Last, 1988).

Work organization: See organizational factors or variables.

Work hardening: See section 9.5.

Appendix III

Attributable fractions for WMSDs

The attributable fraction, sometimes also called the etiologic fraction, is a value calculated to estimate the amount of increased risk that can be 'attributed' to the exposure examined in a particular study (in the context of the study's given outcome, the exposure factor and the population at hand).

The attributable fraction can be calculated for the exposed subjects in the study, in which case it represents the proportion by which the risk of the outcome at hand would be reduced for the exposed group if the exposure were eliminated in that study. This is called the 'attributable fraction (exposed)' (AF_e). Alternatively, the 'attributable fraction (population)' is the proportion by which the risk of the outcome for the entire population would be reduced if the exposure were eliminated in the particular context at hand (Last, 1988). We will concern ourselves here only with the attributable fraction (exposed).

Theoretically, this measure is based on the following calculations and assumes that causes other than the exposure under study have had equal effect on the exposed and unexposed subjects (Last, 1988).

$$AF_e = \frac{I_e - I_u}{I_e} \text{ or } AF_e = \frac{RR - 1}{RR}$$

where I_e is the incidence amongst exposed subjects and I_u the incidence amongst unexposed subjects or where RR is the ratio of I_e / I_u.

It must be remembered that in order for the attributable fraction to be correctly interpreted one must keep it in the context of the particular study in which it was calculated, the given outcome, the exposure factor and the population at hand. Further, one must also consider the width of the confidence intervals for the risk ratio to understand that the true value for the attributable fraction (exposed) would also have a similar width around the value calculated.

Author index

Subject index